Zeolites

Zeolites

Editors

Magdalena Król
Paulina Florek

MDPI • Basel • Beijing • Wuhan • Barcelona • Belgrade • Manchester • Tokyo • Cluj • Tianjin

Editors

Magdalena Król	Paulina Florek
Faculty of Materials Science and Ceramic	Faculty of Materials Science and Ceramic
AGH University of Science and Technology	AGH University of Science and Technology
Kraków	Kraków
Poland	Poland

Editorial Office
MDPI
St. Alban-Anlage 66
4052 Basel, Switzerland

This is a reprint of articles from the Special Issue published online in the open access journal *Crystals* (ISSN 2073-4352) (available at: www.mdpi.com/journal/crystals/special_issues/clay-based_zeolites).

For citation purposes, cite each article independently as indicated on the article page online and as indicated below:

LastName, A.A.; LastName, B.B.; LastName, C.C. Article Title. *Journal Name* **Year**, *Volume Number*, Page Range.

ISBN 978-3-0365-3412-1 (Hbk)
ISBN 978-3-0365-3411-4 (PDF)

Contents

About the Editors

Magdalena Król

Magdalena Król is a professor at the Faculty of Materials Science and Ceramic at the AGH University of Science and Technology (AGH-UST) in Poland. Her scientific activities are focused on the structural study of silicate and aluminosilicate materials using vibrational spectroscopy, application of natural sorbents for immobilization of dangerous ions, and obtainment of new composite materials. She pays attention to materials for the protection of the environment, with particular emphasis on minerals from the zeolite group. Magdalena Król has co-authored more than 70 publications in scientific journals, several book chapters, and many conference presentations.

Paulina Florek

Paulina Florek completed her M. Eng. in 2019 and is currently a PhD student of chemical sciences at the University of Science and Technology in Krakow. She conducts research with professor Włodzimierz Mozgawa and professor Magdalena Król at the Department of Silicates Chemistry and Macromolecular Compounds. Together with the team, she studies the properties of geopolymers and the immobilization of heavy metal ions in ceramic materials. Her research is mainly based on structural studies using research techniques such as FT-IR spectroscopy, Raman spectroscopy, and X-ray diffraction.

Preface to "Zeolites"

Zeolites and zeolite-based materials are characterized by unique physical and chemical properties that can be used for numerous and important applications. These properties arise directly from their specific structure, i.e., the system of channels and chambers characterized by well-defined molecular dimensions. If thermal stability, chemical resistance, the possibility of generating active sites, and sorption and ion-exchange properties are taken into account, it is not surprising that zeolites play an extremely important role in various applications of chemical technology, and their importance continues to increase.

This Special Issue aims to attract original contributions in topics related to zeolites, covering aspects such as the origin of such materials, their characterization, and their applications in different areas. It presents reports on the unique methods of synthesis of zeolites and improvement of existing ones, as well as structural characterization and their potential use in various fields. I believe that this collection outlines the current state of the art in the field of zeolitic materials and will become a source of new ideas resulting in the development of this group of minerals.

I would like to thank the contributing authors and colleagues acting as peer reviewers for their efforts in preparing high-quality manuscripts.

Magdalena Król, Paulina Florek
Editors

Editorial

Natural vs. Synthetic Zeolites

Magdalena Król

Faculty of Materials Science and Ceramic, AGH University of Science and Technology, 30 Mickiewicza Av., 30-059 Krakow, Poland; mkrol@agh.edu.pl

Received: 14 July 2020; Accepted: 15 July 2020; Published: 17 July 2020

Abstract: This brief review article describes the structure, properties and applications of natural and synthetic zeolites, with particular emphasis on zeolites obtained from natural or waste materials. Certainly, such short work does not exhaust the complexity of the problem, but it sheds light on some outstanding issues on this subject.

Keywords: natural zeolite; zeolite synthesis; zeolite characterization; zeolite application

---

1. What Are Zeolites?

Zeolites can be defined in two ways. They are hydrated tectoaluminosilicates with the general formula [1,2]:

$$M_xM'_yN_z\,[T_mT'_n \ldots O_{2(m+n)-\varepsilon}\,(OH)_{2z}]\,(OH)_{br}\,(aq)_p \cdot qQ, \quad (1)$$

where M, M′ are exchangeable and non-exchangeable cations, respectively; N are non-metallic cations (generally removable on heating); (aq) represents chemically bonded water (or other strongly held ligands of T-atoms); Q are sorbate molecules; T, T′ are Si and Al, but also Be, B, Ga, Ge and P, among others. This formula is particularly useful when describing natural zeolites [2], but also those synthesized from natural or waste materials, due to their complex chemical composition.

On the other hand, zeolites can be more graphically, as shown in Figure 1, defined as crystalline inorganic polymers consisting of $[SiO_4]$ and $[AlO_4]$ tetrahedra, having the structure filled ions and water molecules, having great freedom of movement.

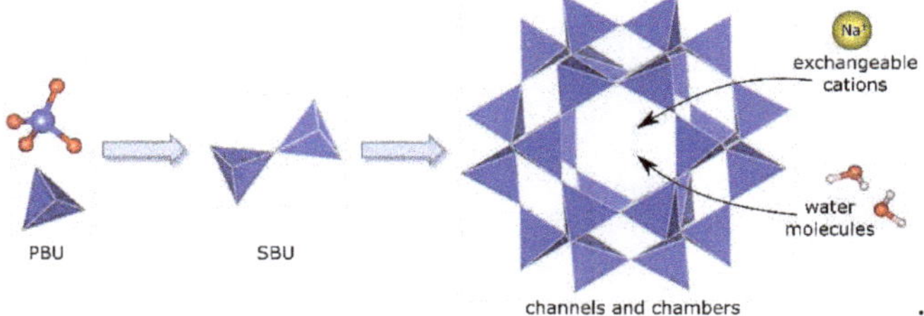

Figure 1. Scheme of zeolite structure.

The specific structure of the zeolites gives them a number of unique properties. The most important regarding potential uses include [3,4]:

- Low density and large volume of free spaces;
- The existence of channels and chambers of strictly defined dimensions (shape-selectivity);
- High degree of hydration and the presence of so-called "zeolite water";
- High degree of crystallinity;

- Possibility of sorption of molecules and ions;
- Ion exchange capacity;
- Catalytic properties.

These properties, caused by zeolites, arouse great interest among chemists, technologists and mineralogists. Although it may seem that the years of the most intensive research on this group of minerals have already passed, interest in them is not decreasing. Constantly conducted research on the specific properties of zeolites show comprehensive possibilities of using this type of mineral, as shown in Figure 2. Of course, all applications of zeolites cannot be mentioned (only examples will be presented in this paper), but even those presented, for example, give a picture of the great importance of zeolites in various applications. In addition, other properties and possibilities of practical use are being discovered.

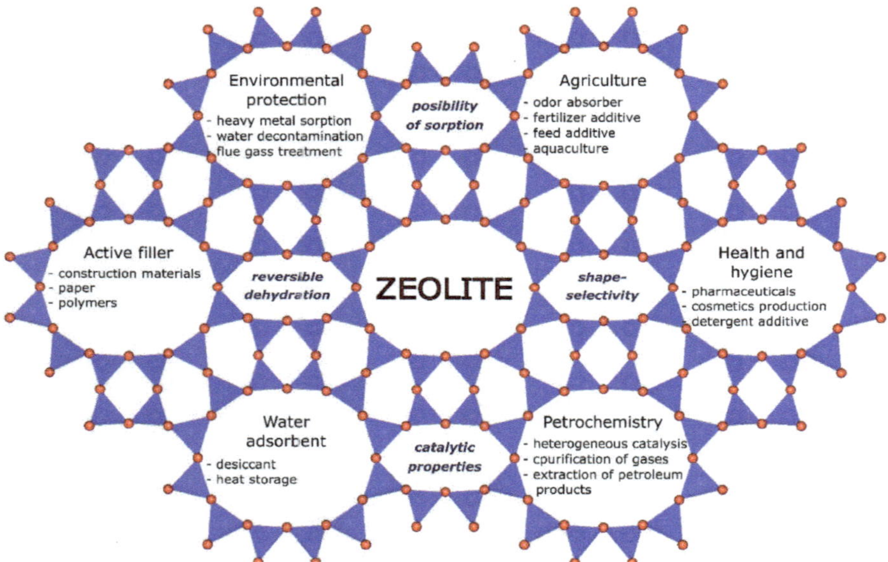

Figure 2. Zeolite applications.

2. Natural Zeolites

Natural zeolites are hydrothermal and of mainly volcanic origin. They can occur both in crystallized forms found in igneous and metamorphic rocks, as well as in grains of smaller diameters accumulated in sedimentary rocks [5]. Ocean bottom sediments are relatively huge and rich in zeolites, but these deposits are so far inaccessible to humans. However, these minerals may also constitute important components of tuffs or clay. Such surface retention of zeolite sediments, and therefore relatively simple mining using the opencast method, creates perfect conditions for their wider use. It should be mentioned here that the zeolites naturally occurring in nature, possessing operational significance, are: clinoptilolite; mordenite; chabazite.

Application Examples

Natural zeolites have a high selectivity for heavy metal ions [6–8] and ammonium ions [9]. Thus, zeolites have found important uses in environmental protection and agriculture. Wastewater treatment from heavy metal ions [10,11] or radioactive isotopes [12] can take place in sorption columns filled with zeolite. Ammonium ions contained in municipal, industrial and agricultural wastes can also be removed in a similar way [13]. In agriculture, zeolites can be used as carriers of agrochemical compounds, in the treatment of soil and fish ponds and as a feed additive [2,14]. Attempts to modify the structure to give them catalytic [15] or antibacterial [16] properties are also made. After all, they are also widely used in many households as pet litter.

On the other hand, natural zeolites have limited applications in industry because, as already mentioned, their properties are strictly dependent on their crystal structure. The main disadvantage is that the channel diameters are too small (in the case of clinoptilolite, which is the most common in nature, it is 0.30–4 nm [17]), which do not allow for the adsorption of larger gas molecules and organic compounds. In addition, zeolite deposits are a non-renewable resource. The need for the synthesis of molecular sieves and adsorbents with very specific parameters means that numerous attempts were made to obtain zeolites in the laboratory.

3. Synthetic Zeolites

Zeolites have been recognized as minerals of natural origin, but currently more than one hundred different types of zeolite structures are known which can be obtained synthetically [17]. Under natural conditions zeolites were formed as a result of the reaction of volcanic ash with the waters of the basic lakes. This process lasted several thousand years. In laboratory conditions, an attempt can be made to imitate hydrothermal processes using elevated temperature or pressure and using natural raw materials and/or synthetic silicates. The synthesis reaction requires appropriate equipment, clean substrates and energy. As a result, the price of the product may be much higher than the price of natural zeolite. Therefore, research often focuses on the search for cheaper and available substrates for the production of zeolites, while striving to reduce the cost of the reaction itself. The current trends in research on the synthesis of zeolites are shaped by environmental aspects, which implies the use of natural or waste raw materials for this purpose.

3.1. *Substrates and Products*

Various natural silica carriers, such as clay minerals (kaoline [18–20], haloisite [20,21]), volcanic glasses (perlite [22–24], pumice [25]) or diatomites [26], are used in zeolite synthesis. On the other hand, zeolites are widely obtained from aluminosilicate waste materials, such as fly ash [27–29], rice husk [30] or expanded perlite waste [31].

Of course, synthesis using raw materials with a complex chemical composition will not give the product 100% purity and zeolites obtained in this way are excluded from many important commercial applications; however, the use of natural raw materials for the production of zeolites has economic advantages when compared with the use of synthetic substrates.

3.2. *Synthesis Methods*

The oldest of the works on the synthesis of aluminosilicates under hydrothermal conditions dates back to the 1950s of the last century [32]. They show that, by heating the aluminosilicate raw materials in the presence of alkaline solutions within a few hours or days, depending on the type of raw materials and process conditions (temperature, pressure), a final product can be obtained. Today, many different methods for the synthesis of zeolites are known. The most important of them should be mentioned:

- Hydrothermal synthesis (at normal or elevated pressure);
- Molten salt method [33];
- Fusion method [34];
- Alkali activation [35–37];
- Microwave-assisted synthesis [38];
- Synthesis by dialysis [39].

The first method is relatively commonly used. This process generally reflects the natural conditions in which rocks containing zeolite minerals were formed. Hydrothermal (80–350 °C) synthesis of zeolites requires the supply of components that are a source of Si and Al, followed by treatment with an alkaline solution (pH > 8.5). The reactions, during which processes such as dissolution, condensation, gelatinization and crystallization take place [40] are carried out in autoclaves, are often at elevated pressure. The appropriate control of process parameters favors the formation of desired products.

The estimated cost of zeolite material obtained by the above methods is between natural and synthetic zeolite. However, taking into account the fact that fees for the storage and utilization of waste will probably increase, the implementation of one of these technologies may prove to be the most cost-effective solution.

3.3. Zolitization Products

Synthetic zeolites, as with natural ones, are diverse in structure and properties. This is influenced by the composition of the raw materials and the synthesis conditions. As mentioned above, the crystallization of individual types of zeolites is a function of such parameters as reaction time, temperature and pressure, as well as the chemical composition of the reaction mixture, including the concentration of reagents [41]. The type of phases formed in such systems is relatively well known. At high temperatures and at elevated pressure, mainly analcime, as shown in Figure 3a, zeolite Na-P1, as shown in Figure 3b, and hydroxysodalite, as shown in Figure 3c, in various quantitative proportions are obtained, depending on the synthesis parameters. At temperatures <100 °C, zeolite X, as shown in Figure 4a, Na-P1 zeolite, as shown in Figure 4b, and hydroxysodalite can be obtained. In addition, zeolite A can be obtained in reaction systems with a high proportion of aluminum (Si/Al < 1), as shown in Figure 4c. A less frequently used fusion method leads to the synthesis of sodalite and cancrite.

(**a**) (**b**) (**c**)

Figure 3. Microstructures of zeolites obtained in hydrothermal conditions at elevated pressure: (**a**) analcime; (**b**) zeolite Na-P1; (**c**) hydroxysodalite.

(**a**) (**b**) (**c**)

Figure 4. Microstructures of zeolites obtained in hydrothermal conditions at low temperatures (<100 °C): (**a**) zeolite X; (**b**) zeolite Na-P1; (**c**) zeolite A.

It is worth noting that, when comparing the microstructures of materials obtained under different conditions, it can be stated that, at low temperatures, crystallites of smaller sizes and less developed morphology are formed.

From a practical point of view, zeolites with a larger pore size are more useful. Hence, in order to obtain material with good sorption properties, it is important to choose process parameters that promote the formation of zeolite X or zeolite Na-P1. The possibility of obtaining other zeolite phases can be regulated by the addition of organic templates, which generate higher process costs.

An important ecological approach is planning the synthesis conditions in such a way as to limit the production of highly alkaline wastewater.

3.4. Benefits of Synthetic Zeolite

Numerous scientific studies confirm the benefits of synthetic zeolites compared to natural ones. The efficiency of natural zeolites in the removal of radioactive waste from the environment (Cs and Sr radionuclides) was found to be lower than that of synthetic zeolites [42,43]. Synthetic zeolites also show much higher adsorption capacity for heavy metal ions (eg. Cd^{2+}, Pb^{2+}, Zn^{2+}, Cu^{2+}, Ni^{2+}, Cr^{3+}) than the natural zeolites [44,45].

Another advantage of synthetic zeolites over natural ones is the significantly larger pore size. This allows for the sorption of larger molecules, which extends the range of potential applications. For example, it was found that synthetic zeolites have two-times higher oil sorption capacities than the natural clinoptilolite, so synthetic zeolites are a promising alternative for natural mineral sorbents for land-based petroleum spills cleanup [46]. Furthermore, zeolites with smaller pore sizes, used as catalysts, suffer from pore blockage and ultimately poisoning and deactivation, while zeolites with large interconnected channels remain stable much longer in reactions [47].

During the synthesis of zeolites, the Al content can be adjusted. Zeolites with low Si/Al ratio are much more polar and thus exhibit stronger sorption capacities. Zeolites with high silicon content are also characterized by greater power of active centers, which promotes them for catalytic applications [48]. On the other hand, high silica zeolites have a more homogeneous surface characteristic and exhibit hydrophobic properties. They have been used in reactions in which resulting water poisons the acid sites of the catalysts [49].

It should be noted that the synthetic zeolites are not without disadvantages. The worrisome problem is that synthetic zeolites are mostly in the form of finely grained crystalline and highly dispersive powder (single crystals have a size to a few microns), which certainly limits their use. For natural zeolites, after mining the deposit, isolation processes, such as crushing and pellet formation, are usually used to obtain zeolites in usable form. In contrast, the synthetic ones can be used in the form of hard, wear-resistant granules. There are many reports related to receiving zeolite granulates, but for the most part they are not yet used in practical applications.

4. Summary

In agreement with the scientific evidence presented in the related literature so far, it can be generally stated that both sedimentary rocks containing zeolites and zeolites synthesized from aluminosilicate raw materials may be regarded as useful for many industries, including agriculture and environmental protection. Synthetic zeolites are the major alternate materials to natural ones. Synthetic zeolites can tailor physical and chemical characteristics to serve many applications more closely and they are more uniform in quality than their natural equivalent. However, conclusive statements on the exact applications and benefits of clay- and waste-based zeolites should be carefully investigated and analyzed for each specific material as the mechanisms of action correlate with the specific material's physical and chemical properties.

Due to the different origins of substrates and different methods of synthesis, the products have various chemical and phase compositions. Constantly conducted research on the specific properties of zeolites show comprehensive possibilities of using this type of mineral. Further research is necessary to tackle yet undetermined topics, such as:

- Accurate analysis of used raw material composition;

- Selection of appropriate parameters of zeolite synthesis;
- Comprehensive structural characterization of zeolites;
- Potential modification of zeolites to improve their properties;
- Zeolite granulation;
- Assessment of ion exchange and sorption potential;
- Behavior and mechanisms in zeolite-catalyzed reactions.

Additional comprehension of the above problems will allow advances in the applicability of zeolites in providing new low cost, eco-friendly solutions to industrial, agricultural or everyday applications.

Funding: This work was partially supported by The National Science Centre Poland under grant no. 2016/21/D/ST8/01692.

Conflicts of Interest: The authors declare no conflict of interest.

References

1. Meier, W.M. Zeolites and Zeolite-Like Materials. In Proceedings of the Seventh International Zeolite Conference, Tokyo, Japan, 17–22 August 1986; Elsevier: Amsterdam, The Netherlands, 1986; pp. 13–22.
2. Reháková, M.; Čuvanová, S.; Dzivák, M.; Rimár, J.; Gaval'ova, Z. Agricultural and agrochemical uses of natural zeolite of the clinoptilolite type. *Curr. Opin. Solid State Mater. Sci.* **2004**, *8*, 397–404. [CrossRef]
3. Breck, D.W. *Zeolite Molecular Sieves: Structure, Chemistry, and Use*; Wiley: Hoboken, NJ, USA, 1973.
4. Ciciszwili, G.W.; Andronikaszwili, T.G.; Kirow, G.N.; Filizowa, Ł.D. *Zeolity Naturalne*; Wydawnictwo Naukowo-Techniczne: Warszawa, Poland, 1990.
5. Gottardi, G.; Galli, E. *Natural Zeolites, Mineral and Rocks 18*; Springer: Berlin/Heidelberg, Germany, 1985.
6. Dyer, A. Ion-exchange properties of zeolites. In *Zeolites and Ordered Mesoporous Materials: Progress and Prospects, Studies in Surface Science and Catalysis 157*; Čejka, J., Bekkum, H., Eds.; Elsevier: Amsterdam, The Netherlands, 2005; pp. 181–204.
7. Sprynskyy, M.; Buszewski, B.; Terzyk, A.P.; Namieśnik, J. Study of the selection mechanism of heavy metal (Pb^{2+}, Cu^{2+}, Ni^{2+}, and Cd^{2+}) adsorption on clinoptilolite. *J. Colloid Interface Sci.* **2006**, *304*, 21–28. [CrossRef] [PubMed]
8. Gorimbo, J.; Taenzana, B.; Muleja, A.A.; Kuvarega, A.T.; Jewell, L.L. Adsorption of cadmium, nickel and lead ions: Equilibrium, kinetic and selectivity studies on modified clinoptilolites from the USA and RSA. *Environ. Sci. Pollut. Res. Int.* **2018**, *25*, 30962–30978. [CrossRef]
9. Vassileva, P.; Voikova, D. Investigation on natural and pretreated Bulgarian clinoptilolite for ammonium ions removal from aqueous solutions. *J. Hazard. Mater.* **2009**, *170*, 948–953. [CrossRef] [PubMed]
10. Ciosek, A.L.; Luk, G.K. An innovative dual-column system for heavy metallic ion sorption by natural zeolite. *Appl. Sci.* **2017**, *7*, 795. [CrossRef]
11. Belova, T.P. Adsorption of heavy metal ions (Cu^{2+}, Ni^{2+}, Co^{2+}, and Fe^{2+}) from aqueous solutions by natural zeolite. *Heliyon* **2019**, *5*, e02320. [CrossRef] [PubMed]
12. Belousov, P.; Semenkova, A.; Egorova, T.; Romanchuk, A.; Zakusin, S.; Dorzhieva, O.; Tyupina, E.; Izosimova, Y.; Tolpeshta, I.; Chernov, M.; et al. Cesium sorption and desorption on glauconite, gentonite, zeolite, and diatomite. *Minerals* **2019**, *9*, 625. [CrossRef]
13. Ivanova, E.; Karsheva, M.; Koumanova, B. Adsorption of ammonium ions onto natural zeolite. *J. Chem. Technol. Metall.* **2010**, *45*, 295–302.
14. Sangeetha, C.; Baskar, P. Zeolite and its potential uses in agriculture: A critical review. *Agric. Rev.* **2016**, *37*, 101–108. [CrossRef]
15. Junaid, A.S.M.; Yin, H.; Koenig, A.; Swenson, P.; Chowdhury, J.; Burland, G.; McCaffrey, W.C.; Kuznicki, S.M. Natural zeolite catalyzed cracking-assisted light hydrocarbon extraction of bitumen from Athabasca oil sands. *Appl. Catal. A* **2009**, *354*, 44–49. [CrossRef]
16. Copcia, V.E.; Luchian, C.; Dunca, S.; Bilba, N.; Hristodor, C.M. Antibacterial activity of silver-modified natural clinoptilolite. *J. Mater. Sci.* **2011**, *46*, 7121–7128. [CrossRef]
17. Baerlocher, C.; McCusker, L.B. Database of Zeolite Structures. Available online: http://www.iza-structure.org/databases/ (accessed on 15 July 2020).

18. Prasetyoko, H.D.; Santoso, M.; Qoniah, I.; Leaw, W.L.; Firda, P.B.D. A review on synthesis of kaolin-based zeolite and the effect of impurities. *J. Chin. Chem. Soc-Taip.* **2020**, *67*, 911–936.
19. Hadi NurPereira, P.M.; Ferreira, B.F.; Oliveira, N.P.; Nassar, E.J.; Ciuffi, K.J.; Vicente, M.A.; Trujillano, R.; Rives, V.; Gil, A.; Korili, S.; et al. Synthesis of zeolite A from metakaolin and its application in the adsorption of cationic dyes. *Appl. Sci.* **2018**, *8*, 608.
20. Biel, O.; Rożek, P.; Florek, P.; Mozgawa, W.; Król, M. Alkaline activation of kaolin group minerals. *Crystals* **2020**, *10*, 268. [CrossRef]
21. Gualtieri, A.F. Synthesis of sodium zeolite from natural halloysite. *Phys. Chem. Miner.* **2001**, *28*, 719–728. [CrossRef]
22. Dyer, A.; Tangkawanit, S.; Rangsriwatananon, K. Exchange diffusion of Cu^{2+}, Ni^{2+}, Pb^{2+} and Zn^{2+} into analcime synthesized from perlite. *Microporous Mesoporous Mater.* **2004**, *75*, 273–279. [CrossRef]
23. Christidis, G.E.; Phapaliars, I.; Kontopoulos, A. Zeolitization of perlite fines: Mineralogical characteristics of the end product and mobilization of chemical elements. *Appl. Clay Sci.* **1999**, *15*, 305–324. [CrossRef]
24. Rujiwatra, A. A selective preparation of phillipsite and sodalite from perlite. *Mater. Lett.* **2004**, *58*, 2012–2015. [CrossRef]
25. Burriesci, N.; Crisafulli, M.L.; Saija, L.M. Hydrothermal synthesis of zeolites from rhyolitic pumice of different geological origins. *Mater. Lett.* **1983**, *2*, 74–78. [CrossRef]
26. Garcia, G.; Cardenasa, E.; Cabrera, S.; Hedlund, J.; Mouzon, J. Synthesis of zeolite Y from diatomite as silica source. *Microporous Mesoporous Mater.* **2016**, *219*, 29–37. [CrossRef]
27. Querol, X.; Moreno, N.; Umaña, J.C.; Alastuey, A.; Hernández, E.; López-Soler, A.; Plana, F. Synthesis of zeolites from coal fly ash: An overview. *Int. J. Coal Geol.* **2002**, *50*, 413–423. [CrossRef]
28. Kunecki, P.; Panek, R.; Wdowin, M.; Bień, T.; Franus, W. Influence of the fly ash fraction after grinding process on the hydrothermal synthesis efficiency of Na-A, Na-P1, Na-X and sodalite zeolite types. *Int. J. Coal Sci. Technol.* **2020**. [CrossRef]
29. Ojha, K.; Pradhan, N.C.; Samanta, A.N. Zeolite from fly ash: Synthesis and characterization. *Bull. Mater. Sci.* **2004**, *27*, 555–564. [CrossRef]
30. Prasetyoko, D.; Ramli, Z.; Endud, S.; Hamdan, H.; Sulikowski, B. Conversion of rice husk ash to zeolite beta. *Waste Manag.* **2006**, *26*, 1173–1179. [CrossRef] [PubMed]
31. Król, M.; Mozgawa, W.; Morawska, J.; Pichór, W. Spectroscopic investigation of hydrothermally synthesized zeolites from expanded perlite. *Microporous Mesoporous Mater.* **2014**, *196*, 216–222. [CrossRef]
32. Barrer, R.M.; White, E.A.D. The hydrothermal chemistry of silicates. Part II. Synthetic crystalline sodium aluminosilicate. *J. Chem. Soc. (Resumed)* **1952**, *286*, 1561–1571. [CrossRef]
33. Park, M.; Choi, C.L.; Lim, W.T.; Kim, M.C.; Choi, J.; Heo, N.H. Molten-salt method for the synthesis of zeolitic materials: I. Zeolite formation in alkaline molten-salt system. *Microporous Mesoporous Mater.* **2000**, *37*, 81–89. [CrossRef]
34. Zhang, M.; Zhang, H.; Xu, D.; Han, L.; Niu, D.; Tian, B.; Zhang, J.; Zhang, L.; Wu, W. Removal of ammonium from aqueous solutions using zeolite synthesized from fly ash by a fusion method. *Desalination* **2011**, *271*, 111–121. [CrossRef]
35. Takeda, H.; Hashimoto, S.; Yokoyama, H.; Iwamoto, Y.; Iwamoto, Y. Characterization of zeolite in zeolite-geopolymer hybrid bulk materials derived from kaolinitic clays. *Materials* **2013**, *6*, 1767–1778. [CrossRef]
36. Król, M.; Minkiewicz, J.; Mozgawa, W. IR spectroscopy studies of zeolites in geopolymeric materials derived from kaolinite. *J. Mol. Struct.* **2016**, *1126*, 200–206. [CrossRef]
37. Rożek, P.; Król, M.; Mozgawa, W. Geopolymer-zeolite composites: A review. *J. Clean. Prod.* **2019**, *230*, 557–579. [CrossRef]
38. Querol, X.; Alastuey, A.; Lopez-Soler, A.; Plana, P.; Andres, J.M.; Juan, R.; Perrer, P.; Ruiz, C. A fast method for recycling fly ash: Microwave-assisted zeolite synthesis. *Env. Sci. Technol.* **1997**, *31*, 2527–2533. [CrossRef]
39. Tanaka, H.; Eguchi, H.; Fujimoto, S.; Hino, R. Two-step process for synthesis of a single phase Na–A zeolite from coal fly ash by dialysis. *Fuel* **2006**, *85*, 1329–1334. [CrossRef]
40. Cundy, C.S.; Cox, P.A. The hydrothermal synthesis of zeolites: Precursors, intermediates and reaction mechanism. *Microporous Mesoporous Mater.* **2005**, *82*, 1–78. [CrossRef]
41. Gonthier, S.; Gora, L.; Güray, I.; Thompson, R.W. Further comments on the role of autocatalytic nucleation in hydrothermal zeolite syntheses. *Zeolites* **1993**, *13*, 414–418. [CrossRef]

42. Lonin, A.Y.; Levenets, V.V.; Neklyudov, I.M.; Shchur, A.O. The usage of zeolites for dynamic sorption of cesium from wastewaters of nuclear power plants. *J. Radioanal. Nucl. Chem.* **2015**, *303*, 831–836. [CrossRef]
43. Abdel Moamen, O.A.; Ismail, I.M.; Abdelmonem, N.; Abdel Rahmana, R.O. Factorial design analysis for optimizing the removal of cesium and strontium ions on synthetic nano-sized zeolite. *J. Taiwan Inst. Chem. Eng.* **2015**, *55*, 133–144. [CrossRef]
44. He, K.; Chen, Y.; Tang, Z.; Hu, Y. Removal of heavy metal ions from aqueous solution by zeolite synthesized from fly ash. *Environ. Sci. Pollut. Res.* **2016**, *3*, 2778–2788. [CrossRef]
45. Kozera-Sucharda, B.; Gworek, B.; Kondzielski, I. The simultaneous removal of zinc and cadmium from multicomponent aqueous solutions by their sorption onto selected natural and synthetic zeolites. *Minerals* **2020**, *10*, 343. [CrossRef]
46. Bandura, L.; Franus, M.; Józefaciuk, G.; Franus, W. Synthetic zeolites from fly ash as effective mineral sorbents for land-based petroleum spills cleanup. *Fuel* **2015**, *147*, 100–107. [CrossRef]
47. Nizami, A.S.; Ouda, O.K.M.; Rehan, M.; El-Maghraby, A.M.O.; Gardy, J.; Hassanpour, A.; Kumar, S.; Ismail, I.M.I. The potential of Saudi Arabian natural zeolites in energy recovery technologies. *Energy* **2016**, *108*, 162–171. [CrossRef]
48. Ward, J.W. The nature of active sites on zeolites: I. The decationated Y zeolite. *J. Catal.* **1967**, *9*, 225–236. [CrossRef]
49. Corma, A. State of the art and future challenges of zeolites as catalysts. *J. Catal.* **2003**, *216*, 298–312. [CrossRef]

MDPI

Article

First Occurrence of Willhendersonite in the Lessini Mounts, Northern Italy

Michele Mattioli * and Marco Cenni

Department of Pure and Applied Sciences, Campus Scientifico Enrico Mattei, University of Urbino Carlo Bo, Via Cà le Suore 2/4, 61029 Urbino, Italy; marco.cenni@tiscali.it
* Correspondence: michele.mattioli@uniurb.it

Abstract: Willhendersonite is a rare zeolite, with very few occurrences reported globally (Terni Province, Italy; the Eifel Region, Germany; Styria, Austria). Moreover, the data available from these sites are very limited and do not allow a detailed picture of this zeolite's mineralogical and chemical characteristics. In this work, a new willhendersonite occurrence is reported from the Tertiary volcanic rocks of the Lessini Mounts, northern Italy. Morphology, mineralogy and chemical composition of selected crystals were studied by scanning electron microscopy (SEM), energy-dispersive X-ray (EDX), X-ray Diffraction (XRD), and electron probe microanalyser (EPMA). Willhendersonite occurs within basanitic rocks as isolated, colorless, transparent crystals with prismatic to flattened morphologies. Individual crystals often grow together to form small elongated clusters and trellis-like aggregates. The diffraction pattern exhibits 33 well-resolved diffraction peaks, all of which can be indexed to a triclinic cell with unit cell parameters a = 9.239(2) Å; b = 9.221(2) Å; and c = 9.496(2) Å, α = 92.324(2)°, β = 92.677(2)°, γ = 89.992° (Space Group $P\bar{1}$). The chemical data point to significant variability from Ca-rich willhendersonite $(K_{0.23}Na_{0.03})_{\Sigma=0,26}Ca_{1.24}$ $(Si_{3.06}Al_{3,00}Fe^{3+}{}_{0.01})_{\Sigma=6,07}$ $O_{12}\cdot 5H_2O$) to Ca-K terms $(K_{0.94}Na_{0.01})_{\Sigma=0,95}Ca_{0.99}$ $(Si_{3.07}Al_{2.93}Fe^{3+}{}_{0.00})_{\Sigma=6,00}O_{12}\cdot 5H_2O$). Willhendersonite from the Lessini Mounts highlights the existence of an isomorphous series between the Ca-pure crystals and Ca-K compositions, possibly extended up to a potassic end-member.

Keywords: willhendersonite; chabazite; zeolites; Lessini Mounts

Citation: Mattioli, M.; Cenni, M. First Occurrence of Willhendersonite in the Lessini Mounts, Northern Italy. *Crystals* **2021**, *11*, 109. https://doi.org/10.3390/cryst11020109

Academic Editor: Magdalena Król

Received: 31 December 2020
Accepted: 24 January 2021
Published: 26 January 2021

1. Introduction

Willhendersonite was first described by Peacor et al. [1], who studied specimens from mafic potassic lava at San Venanzo (Terni, Italy) and from a limestone xenolith within basalt at Mayen (Eifel area, Germany). Both these samples correspond to the schematic chemical formula $CaKAl_3Si_3O_{12}\cdot 5H_2O$, with space group $P\bar{1}$ and unit cell parameters a = 9.206; b = 9.216; and c = 9.500 Å, α = 92.34, β = 92.70, γ = 90.12° [1,2]. Structural investigations [2,3] on single crystals from the German locality revealed the close structural relationship to chabazite, $CaAl_2Si_4O_{12}\cdot 6H_2O$, which has the same framework topology with framework type code CHA (chabazite, willhendersonite [4–6]). However, the Si/Al ordering leads to a reduction of the framework symmetry from topological $R\bar{3}m$ to topochemical $R\bar{3}$. Moreover, the presence of different degrees of ordering (Si,Al) and the position of extra-framework cations and water molecules further reduce the real symmetry to $P\bar{1}$ [2,7,8]. A further occurrence was reported by Walter et al. [9] from Wilhelmsdorf (Styria, Austria), and a chemical and crystallographic description of a Ca-rich willhendersonite from a melilitite plug at Colle Fabbri, a second locality within the volcanic rocks of the Terni province, was given by Vezzalini et al. [10]. Recently, single crystals of willhendersonite from Bellerberg (eastern Eifel District, Germany) were studied by X-ray diffraction methods between 100 and 500 °K, showing a phase transition from triclinic to rhombohedral symmetry [11]. These changes in the framework are accompanied by migration of cations, partly assuming unfavorably low coordinations in the high-temperature structure due to the loss of H_2O molecules. Re-hydration at room temperature yields the triclinic structure of willhendersonite, although the single crystals become polysynthetically twinned [11].

Willhendersonite from Terni was chemically analyzed by Peacor et al. [1]. The data yield the empirical formula, based on 12 oxygens of $K_{0.90}Ca_{1.01}Al_{2.93}Si_{3.08}O_{12}{\cdot}5.43H_2O$, which compares favorably with the ideal formula $KCaAl_3Si_3O_{12}{\cdot}5H_2O$ (Z = 2). A large number of crystals from Colle Fabbri were chemically analyzed by Vezzalini et al. [10]. The obtained data showed chemical variability ranging from the holotype sample composition ($CaKAl_3Si_3O_{12}{\cdot}5H_2O$), through intermediate compositions, to a Ca-pure term. The most common Ca/Ca + K ratio among the different varieties ranges from 1.0 to 0.9 and represents Ca-pure terms. The ratio 0.6 to 0.5, which is similar to that reported in the literature for willhendersonite from San Venanzo, is also quite common. The rarest compositions are the intermediate ones. The value of R is always close to 0.50, with calcium as the dominant extra-framework cation (0.98–1.54 apfu) [10]. The potassium content is relevant in samples from San Venanzo and Mayen (0.9 apfu and 0.74 apfu, respectively) but is very low (0.03 to 0.30 apfu) in the sample from Colle Fabbri [1,10]. In any case, all the data available in the literature for this zeolite come from only three areas (Terni, Styria, and Eifel). For this reason, it is of great importance to increase the amount of data, finding crystals of willhendersonite also in other locations.

The Tertiary basalts of the Lessini Mounts (Veneto Volcanic Province, Northern Italy) are known as suitable host rocks for the growth of secondary mineral associations [12] Zeolites such as gmelinite, analcime, chabazite, and phillipsite have already been observed in the basaltic rock of Monte Calvarina [13]. Chabazite, phillipsite, harmotome, and analcime were discovered in the vugs of volcanite outcropping near Fittà [14]. Fibrous erionite and offretite with potential toxicological implications have been recently discovered in several localities of the Lessini Mounts [12,15–18]. Among these common zeolites, rare species such as willhendersonite and yugawaralite were recently found in northern Italy [12]. This study aims to present a mineralogical and chemical characterization of willhendersonite, found for the first time within vesicles of the basanitic rocks in the Lessini Mounts.

2. Geological Background

The Veneto Volcanic Province (Northern Italy) covers an area of about 2000 km^2 (Figure 1) and is the result of extensive volcanic activity that occurred from the Tertiary [19–21]. Several magmatic pulses occurred between the Late Paleocene and the Miocene, with the most significant part of the eruptions taking place in submarine environments [19–22]. The Veneto Volcanic Province can be subdivided into four main volcanic districts based on different tectono-magmatic features (Figure 1). They are the Lessini Mounts, the Marostica Hills, the Berici Hills, and the Euganean Hills. The main products are volcaniclastic rocks, hyaloclastites, pillow lavas, and lava flows of a mafic to ultramafic composition. Rocks of more acidic compositions are rare and only occur in the Euganean Hills.

The Lessini Mounts are located within an NNW-trending extensional structure, namely the Alpone-Agno graben [23,24], bounded to the west by the NNW-SSE Castelvero normal fault (Figure 1). The volcanic sequence of the Lessini Mounts has a thickness of up to 400 m and is mainly represented by tuffs and lava flows, with several column-jointing eruptive necks and subordinate hyaloclastites and pillow lavas. The most abundant rock-types are basanites and alkali olivine basalts, while transitional basalts, tholeiites, nephelinite, hawaiites, trachy-basalts, and basaltic andesites are less abundant [20,21,25]. Petrological and geochemical data [20,21,25,26] indicate a within-continental-plate character for the magmatism in agreement with regional geodynamics, which places this magmatism in a context dominated by a tensional system.

Most of the Lessini volcanic rocks are often deeply weathered and show cavities and vugs of variable sizes that are almost always lined by a thin, microcrystalline crust, which is the substratum of well-shaped, secondary minerals [12]. The secondary phases are mainly zeolites and clay minerals, which represent ~90 vol.% of the total secondary minerals; other silicates (apophyllite, gyrolite, prehnite, pectolite) are very rare, as are oxides (quartz) and carbonates (calcite, aragonite). Clay minerals are generally the first minerals that precipitated along the walls, whereas the core of the vesicles commonly contains well-shaped zeolites. The coating thickness is usually less than 0.5 mm, while the zeolite crystals range in size from <1 mm to 1 cm. Typical vesicle infillings consist of chabazite, phillipsite-harmotome, analcime, natrolite, gmelinite, and offretite, and these are the most represented types. Heulandite, stilbite, and erionite are less common, whereas willhendersonite and yugawaralite are very rare zeolite species.

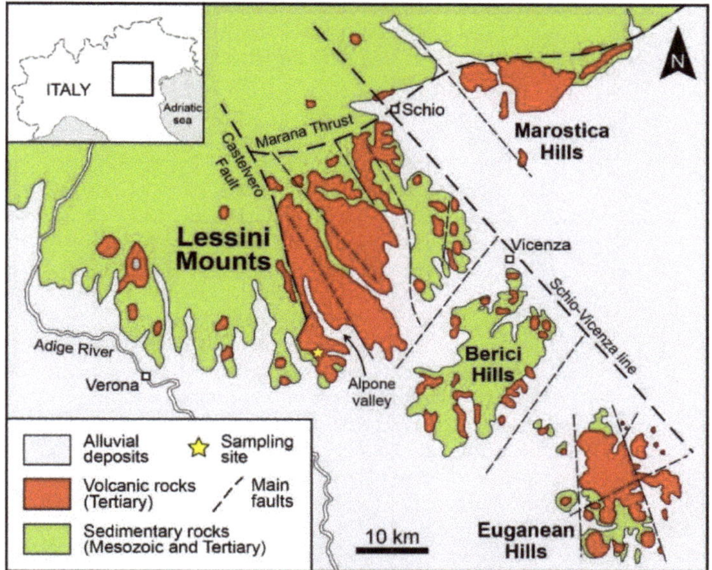

Figure 1. Simplified geological map of the Veneto Volcanic Province (modified from de Vecchi et al. [20]) showing the Lessini Mounts and the locations of the sampling site.

3. Materials and Methods

The investigated crystals were obtained from a suite of more than 300 samples of volcanic rocks from Lessini Mounts, Italy. The zeolite willhendersonite can be found in small cavities of basanitic rocks predominantly filled by chabazite and clay minerals.

Willhendersonite crystals were initially recognized from their physical properties using a binocular microscope, (Nikon TK-1270E, Tokyo, Japan) while powder X-ray diffraction (XRD) was used to confirm their mineralogical composition. Pure crystals were selected from each sample and extracted from the matrix under a binocular microscope. The separated crystals were carefully cleaned and repeatedly treated in an ultrasonic bath to remove any impurities and the microspheres of clay minerals already observed during the microscopic investigations. After cleaning, an aliquot of the separated crystals was carefully pulverized in an agate mortar. All of the powder samples were prepared by side-loading an aluminum holders to obtain a quasi-random orientation. The XRD patterns were recorded using a Philips X'Change PW 1830 X-ray diffractometer (Philips X'PERT, Malvern Panalytical, Almelo, The Netherlands); Cu Kα radiation), with a monochromator on secondary optics. The samples were run between 2° and 65° 2θ. The analytical conditions were a 35 kV accelerating potential, a 30 mA filament current, a 0.02° step, and a counting time of 1 s/step.

Morphological observations and semi-quantitative chemical compositions were performed by Scanning Electron Microscopy (SEM) and Energy Dispersion Spectroscopy

(EDS) using a Philips 515 equipped with EDAX 9900 (Eindhoven, The Netherlands) and a Jeol 6400 (Jeol, Japan) with an Oxford Link Isis. The operating conditions were a 15 kV accelerating potential and a 2 to 15 nA beam current. A defocused electron beam and a shortened accumulation time (from 100 s down to 50 s) were used to minimize the alkaline metals' migration. The standards used were natural minerals and synthetic phases.

Quantitative chemical compositions were determined by an Electron Probe Microanalyser (EPMA), a CAMECA Camebax 799 (Cameca sas, cedex, France), on the other aliquot of the previously separated and cleaned crystals, fixed with epoxy resin on polished thin sections. Operating conditions were 15 kV and 15 nA using the wavelength-dispersive method; errors were ±2–5% for major and ±5–10% for minor components. Standards comprised a series of pure elements, simple oxides, or simple silicate compositions. The analyses were selected on the basis of their low E% value (E% is the balance error of the electrical charges: $[Al - (K + 2Ca)]/(K + 2Ca)$ [27]). Zeolites with an E% > 10 were rejected.

4. Results

In the Lessini basanitic rocks, willhendersonite generally occurs as euhedral to subhedral small crystals (about 0.1–0.2 mm) with a prismatic, tabular, and flattened morphology on {001}, often with rectangular sections. Individual crystals often grow together to form small elongated clusters of characteristic twinned combinations (Figure 2). In these twinned aggregates, the crystals grow in three diverse orientations, each with faces nearly perpendicular to those of the others, which typically leads to the so-called "trellis-like" aggregates [1]. The crystals are colorless and perfectly transparent, with vitreous luster on crystal faces. The cleavage is perfect, parallel to {100}, {010}, and {001}. These three cleavage planes are equivalent to the rhombohedral cleavage of isostructural chabazite. Willhendersonite from the Lessini Mounts grows on a substrate mainly consisting of clay minerals. These latter typically form layers along the walls of pore spaces with botryoidal habits and appear in a wide range of colors varying from white, pink, yellow, brown, and green to black. Clay minerals are also present as micrometric spherules of a characteristic green-blue to red color, corresponding to saponite composition with a tri-octahedral structure [12].

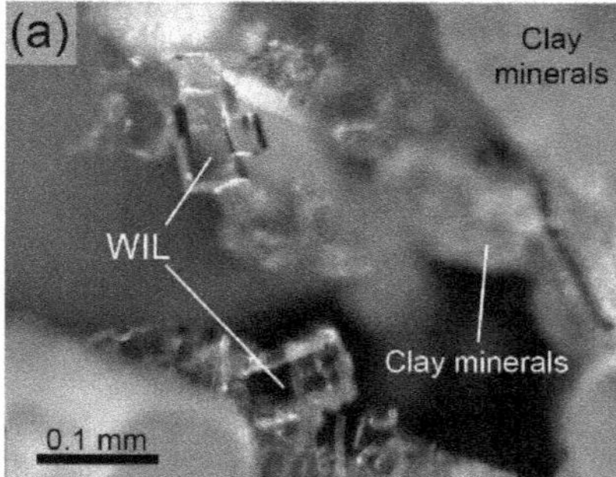

Figure 2. Willhendersonite from the Lessini Mounts: (**a**) stereomicroscopic image of colorless, transparent individual crystals of willhendersonite (WIL) associated to form small elongated clusters and grown on clay minerals; (**b**) SEM photomicrograph of willhendersonite twinned crystal aggregates (WIL) on a substrate of clay minerals with botryoidal shape and micrometric spherules of tri-octahedral saponite.

Powder X-ray diffraction data of separated willhendersonite crystals are listed in Table 1, while their diffraction pattern is shown in Figure 3. The unit-cell parameters of willhendersonite were determined by least-squares refinement of powder-diffractometer data, which was carried out in the space group $P\overline{1}$ starting from the positional parameters of the framework atoms by Tillmanns et al. [2]. High-purity quartz (a = 4.9137, c = 5.4053 Å) has been used for the refinement as an internal standard. The diffraction pattern exhibits

33 well-resolved diffraction peaks, all of which can be indexed to a triclinic cell with unit cell parameters a = 9.239(2) Å; b = 9.221(2) Å; and c = 9.496(2) Å, α = 92.324(2)°, β = 92.677(2)°, γ = 89.992° (Space Group P$\bar{1}$). The diffractogram perfectly fits that of the holotype willhendersonite and does not show other peaks relating to impurities, testifying to the purity of the analyzed crystals.

Table 1. X-ray powder diffraction data for willhendersonite from the Lessini Mounts.

2 Theta	I/I_o	d (Å)	h	k	l	2 Theta	I/I_o	d (Å)	h	k	l
9.32	34.77	9.4789	0	0	1	27.36	6.15	3.2601	−2	2	0
9.61	97.78	9.1915	0	1	0	28.89	6.17	3.0871	0	0	3
9.64	100	9.1591	1	0	0	29.47	20.59	3.0275	1	2	2
13.63	13.46	6.4914	1	1	0	30.04	22.37	2.9664	1	0	3
17.11	23.01	5.1774	1	1	1	30.75	56.75	2.9054	−1	−1	3
18.83	9.99	4.7407	0	0	2	31.42	8.74	2.8458	0	3	1
19.41	10.24	4.5979	2	0	0	31.77	15.93	2.8171	3	0	1
20.72	15.93	4.2871	0	−1	2	31.91	35.21	2.8028	−3	1	1
21.72	27.29	4.0873	−1	2	0	34.86	9.71	2.5741	−2	−2	2
21.68	25.71	4.1202	−2	1	0	35.47	15.13	2.5201	1	−3	2
22.63	16.21	3.9247	−1	−1	2	36.86	7.21	2.4384	1	2	3
23.16	23.16	3.8546	−2	−1	1	39.46	7.05	2.2814	−1	−1	4
23.46	25.29	3.8262	1	−1	2	41.55	8.75	2.1717	−3	3	0
23.86	17.57	3.7104	−1	2	1	44.22	6.95	2.0464	−4	1	2
23.96	15.76	3.7461	2	−1	1	49.47	6.97	1.8408	0	1	5
26.36	7.09	3.3813	−2	0	2	54.53	7.26	1.6863	0	−4	4

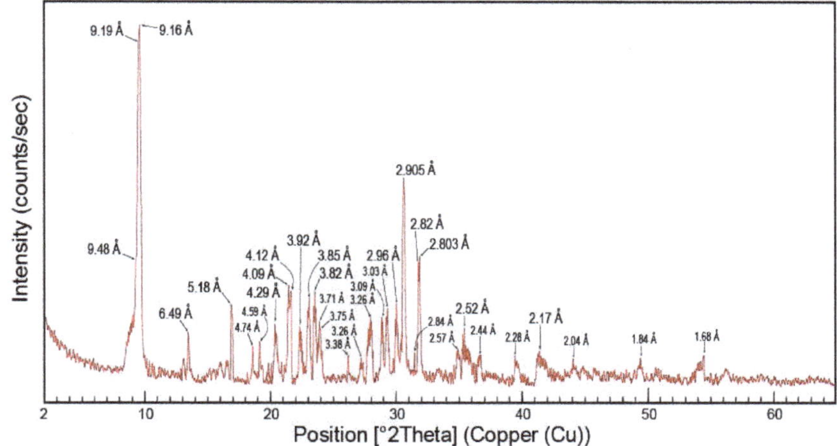

Figure 3. X-ray powder diffraction pattern for willhendersonite from the Lessini Mounts. All the diffraction peaks refer to willhendersonite.

A representative number of crystals from different specimens were extracted, carefully cleaned, and analyzed for chemical composition. The resultant analyses are presented in Tables 2 and 3. The collected chemical data show significant variability, mainly in the extra-framework cation content. Based on the Ca/(Ca + K) ratio, two different groups of crystals can be distinguished.

The first group of crystals correspond to a Ca-rich willhendersonite (Table 2), with an average chemical composition of $(K_{0.29}Na_{0.03})_{\Sigma=0,32}Ca_{1.24}(Si_{3.06}Al_{3,00}Fe^{3+}{}_{0.01})_{\Sigma=6,07}O_{12}\cdot 5H_2O)$. The Ca/(Ca + K) ratio varies from 0.78 to 0.84 (average 0.81) and calcium is the dominant extra-framework cation (1.21–1.29 apfu), while potassium is very low (0.24–0.34 apfu) and the sodium is always <0.1 apfu.

Crystals of the second group are Ca-K willhendersonite (Table 3) and are characterized by an average chemical composition of $(K_{0.94}Na_{0.01})_{\Sigma=0,95}Ca_{0.99}(Si_{3.07}Al_{2.93}Fe^{3+}{}_{0.00})_{\Sigma=6,00}O_{12}\cdot5H_2O)$. They have a Ca/(Ca + K) ratio in the range of 0.49–0.54 (average 0.51), with significant calcium (1.04–0.96 apfu) and potassium (0.9–0.98 apfu) contents, while the sodium is always <0.1 apfu. For all samples, the tetrahedral content R (R = Si/(Si + Al)) is near to 0.50 (average 0.51).

Table 2. Representative chemical compositions of Ca-rich willhendersonite from Lessini Mounts. * The H_2O content is calculated by difference to 100; Mg, Sr, and Ba were checked but were always below the detection limit; E% is the balance error of the electrical charges: [Al − (K + 2Ca)]/(K + 2Ca); R is the tetrahedral content (Si/(Si + Al)) and St-d is standard deviation.

	WIL1	WIL2	WIL5	WIL8	WIL9	WIL10	WIL14	WIL15	WIL19	WIL20	Average	St-d
					Ca-rich willhendersonite							
SiO_2	35.34	35.31	34.68	35.12	35.74	34.99	35.51	35.73	34.85	34.88	35.22	0.37
Al_2O_3	29.41	29.31	29.55	29.16	28.89	29.13	29.45	28.87	29.46	29.44	29.27	0.24
K_2O	2.75	2.65	2.88	2.31	2.55	3.01	2.13	2.65	2.45	2.56	2.59	0.26
CaO	13.24	13.36	13.75	13.84	13.01	12.96	13.22	13.24	13.45	13.35	13.34	0.28
Na_2O	0.15	0.27	0.13	0.37	0.25	0.25	0.12	0.18	0.12	0.15	0.20	0.08
Fe_2O_3	0.12	0.08	0.11	0.2	0.08	0.05	0.12	0.11	0.05	0.04	0.10	0.05
H_2O *	18.99	19.02	18.90	19.00	19.48	19.61	19.45	19.22	19.62	19.58	19.29	0.29
Total	81.01	80.98	81.10	81.00	80.52	80.39	80.55	80.78	80.38	80.42	80.79	0.27
					Cation on the basis of 12 framework oxygens						Average	St-d
Si	3.05	3.05	3.05	3.04	3.13	3.05	3.07	3.09	3.03	3.04	3.06	0.03
Al	3.00	2.99	3.06	2.97	2.99	2.99	3.00	2.94	3.02	3.09	3.00	0.04
Fe^{3+}	0.01	0.00	0.01	0.01	0.01	0.00	0.01	0.01	0.00	0.00	0.01	0.00
ΣT	6.06	6.04	6.11	6.03	6.12	6.05	6.08	6.04	6.06	6.13	6.07	0.04
Ca	1.23	1.24	1.29	1.28	1.22	1.21	1.22	1.23	1.25	1.24	1.24	0.03
K	0.30	0.29	0.31	0.26	0.29	0.34	0.24	0.29	0.27	0.28	0.29	0.03
Na	0.03	0.05	0.02	0.06	0.04	0.04	0.02	0.03	0.02	0.03	0.03	0.01
ΣC	1.55	1.58	1.63	1.60	1.55	1.59	1.48	1.55	1.55	1.55	1.56	0.04
Total	7.61	7.62	7.74	7.63	7.68	7.64	7.56	7.59	7.60	7.68	7.63	0.06
E%	0.09	0.08	0.05	0.05	0.09	0.09	0.12	0.07	0.09	0.11		
Ca/(Ca + K)	0.80	0.81	0.81	0.83	0.81	0.78	0.84	0.81	0.82	0.81	0.81	0.02
R	0.50	0.51	0.50	0.51	0.51	0.50	0.51	0.51	0.50	0.50	0.51	0.00

Table 3. Representative chemical compositions of Ca-K willhendersonite from Lessini Mounts. * The H_2O content is calculated by difference to 100; Mg, Sr, and Ba were checked but were always below the detection limit; E% is the balance error of the electrical charges: [Al − (K + 2Ca)]/(K + 2Ca); R is the tetrahedral content (Si/(Si + Al)) and St-d is standard deviation.

	WIL4	WIL7	WIL12	WIL13	WIL22	WIL23	WIL24	WIL27	WIL29	WIL31	Average	St-d
					Ca-K willhendersonite							
SiO_2	34.81	34.45	34.86	34.91	34.23	34.57	35.02	34.96	35.12	34.82	34.78	0.28
Al_2O_3	28.12	28.42	28.09	28.02	28.66	28.38	27.86	27.98	27.87	28.46	28.19	0.28
K_2O	8.11	8.45	8.66	8.12	7.99	8.13	8.55	8.71	8.05	8.44	8.32	0.27
CaO	10.75	10.42	10.25	10.77	11.02	10.79	10.35	10.16	10.22	10.46	10.52	0.29
Na_2O	0.04	0.05	0.02	0.04	0.01	0.06	0.08	0.04	0.04	0.01	0.04	0.02
Fe_2O_3	0.05	0.01	0.02	0.04	0.05	0.01	0.03	0.06	0.01	0.01	0.03	0.02
H_2O *	18.12	18.20	18.10	18.10	18.04	18.06	18.11	18.09	18.69	17.80	18.13	0.22
Total	81.88	81.80	81.90	81.90	81.96	81.94	81.89	81.91	81.31	82.20	81.87	0.22

Table 3. *Cont.*

	WIL4	WIL7	WIL12	WIL13	WIL22	WIL23	WIL24	WIL27	WIL29	WIL31	Average	St-d
			Cation on the basis of 12 framework oxygens								Average	St-d
Si	3.07	3.05	3.08	3.08	3.02	3.05	3.09	3.09	3.11	3.06	3.07	0.03
Al	2.92	2.96	2.92	2.91	2.98	2.95	2.90	2.91	2.91	2.95	2.93	0.03
Fe^{3+}	0.00	0.00	0.00	0.00	0.00	0.00	0.00	0.00	0.00	0.00	0.00	0.00
ΣT	5.99	6.01	6.00	5.99	6.00	6.00	5.99	6.00	6.01	6.01	6.00	0.01
Ca	1.02	0.99	0.97	1.02	1.04	1.02	0.98	0.96	0.97	0.99	0.99	0.03
K	0.91	0.95	0.98	0.91	0.90	0.91	0.96	0.98	0.91	0.95	0.94	0.03
Na	0.01	0.01	0.00	0.01	0.00	0.01	0.01	0.01	0.01	0.00	0.01	0.00
ΣC	1.93	1.95	1.95	1.94	1.94	1.94	1.96	1.95	1.89	1.93	1.94	0.02
Total	7.93	7.96	7.95	7.93	7.94	7.94	7.95	7.95	7.90	7.94	7.94	0.02
E%	−0.01	0.01	0.00	−0.01	0.00	0.00	−0.01	0.00	0.02	0.01		
Ca/(Ca + K)	0.53	0.51	0.50	0.53	0.54	0.53	0.50	0.49	0.52	0.51	0.51	0.01
R	0.51	0.51	0.51	0.51	0.50	0.51	0.52	0.51	0.52	0.51	0.51	0.00

5. Discussion and Conclusions

Alteration phenomena in the basaltic rocks from the Lessini Mounts resulted in distinctive secondary mineral assemblages representing a multi-stage hydrothermal alteration process [12]. The occurrence of secondary minerals, their frequency, and their associations may be significantly different on both the outcrop- and sample-scale, with variability that can be in the order of a few centimeters. In the earliest stages, clay and silica minerals precipitate along the vesicles' inner walls, followed by the fine-grained zeolites erionite, offretite, analcime, natrolite, heulandite, and stilbite. The final stage is marked by the large, well-shaped zeolites (phillipsite-harmotomo, gmelinite, chabazite) associated with rare zeolites yugawaralite and willhendersonite. In particular, as regards willhendersonite, it is important to remember that, in the world, only three finds of this zeolite are reported (Terni, Eifel, Styria), and the data available in the literature are very few (practically absent for that of Styria). The morphological and chemical data presented here aim to implement this rare zeolite knowledge, found for the first time in northern Italy.

Willhendersonite from the Lessini Mounts is morphologically very similar to that of the other three known localities. Crystals are colorless, transparent, and mainly tabular and flattened, and they are often twinned to form trellis-like intergrowths. The cleavages are perfect, well developed, and parallel to {100}, {010}, and {001}, as indexed on the triclinic cell.

The calculated cell parameters' values were found to be relatively consistent with literature data (Table 4). In particular, unit-cell values of willhendersonite from Lessini are very similar to the Mayen and Bellberg samples, while those from Terni and Styria show slightly lower unit-cell parameters. No significant differences were found in the unit-cell parameters with respect to the holotype willhendersonite studied by Tillmanns et al. [2].

Table 4. Unit-cell parameters of willhendersonite from the Lessin Mounts, compared with literature data. Values of estimated standard deviations are given in parentheses.

	a (Å)	*b* (Å)	*c* (Å)	α (°)	β (°)	γ (°)	V (Å³)
Lessini (this work)	**9.239(2)**	**9.221(2)**	**9.496(2)**	**92.324(2)**	**92.677(2)**	**89.992(2)**	**808.9(5)**
Literature data							
Terni [1]	9.138	9.178	9.477	92.31	92.50	90.05	793.4
Terni [10]	9.180(3)	9.197(3)	9.440(3)	91.42(2)	91.72(2)	90.05(2)	796.4(5)
Styria [9]	9.16	9.17	9.49	92.3	92.9	90.4	795.4
Mayen [1]	9.21(2)	9.23(2)	9.52(2)	92.4(1)	92.7(1)	90.1(1)	807.7(30)
Mayen [2]	9.206(2)	9.216(2)	9.500(4)	92.34(3)	92.70(3)	90.12(3)	804.4(4)
Bellberg [11]	9.248(5)	9.259(5)	9.533(5)	92.313(5)	92.761(5)	89.981(5)	814.7(8)

Regarding the chemical composition, two different groups of crystals were distinguished based on the extra-framework cation content (Figure 4). The Ca-rich group (red circles in Figure 4) has a composition that straddles those observed in the Colle Fabbri samples [10]. In particular, willhendersonite from the Lessini Mounts has a composition that includes the intermediate terms from Colle Fabbri (Will-int, [10]) but extends towards the Ca-richest terms, expanding the compositional range so far known in the literature.

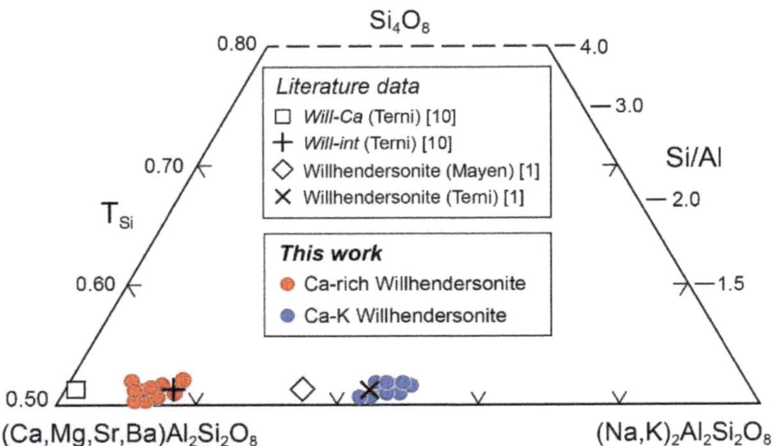

Figure 4. Compositional ternary diagram illustrating major variations of willhendersonite compositions. Red and blue circles are crystals from the Lessini Mounts. Components are: S_4O_8 (molecular proportions [Si − 2(Ca + Mg + Sr + Ba) − 2(Na + K)]/4, $(Ca,Mg,Sr,Ba)Al_2Si_2O_8$ (molecular proportions Ca + Mg + Sr + Ba), and $(Na,K)_2Al_2Si_2O_8$ (molecular proportions (Na + K)/2).

The Ca-K crystals (blue circles in Figure 4), on the other hand, are characterized by comparable quantities of K and Ca and, consequently, are distributed in the central sector of the ternary diagram of Figure 4. Their composition comprises the San Venanzo samples [1] but extends towards richer terms in K, defining greater chemical variability in extra-framework cation content.

In the willhendersonite from the Lessini Mounts, the K content is low in the Ca-rich group (average of 0.29 apfu) and it is notably high in the Ca-K group (average 0.94 apfu). In comparison, the K content is significant in the samples from San Venanzo and Mayen (0.90 and 0.74 apfu, respectively) and very low in the sample from Colle Fabbri (0.03–0.30 apfu). In the Lessini samples, the Ca content has an average of 1.24 apfu in the Ca-rich group and 0.99 apfu in the Ca-K group, whereas the Ca content is notably high in the Colle Fabbri sample (1.43–1.28 apfu) and it is low in the samples from San Venanzo and Mayen (1.01 and 1.06 apfu, respectively).

The new chemical data of willhendersonite from the Lessini Mounts, together with the literature data on Mayen [1,2] and the occurrence of a Ca-pure willhendersonite reported in Vezzalini et al. [10], highlight the presence of a considerable compositional variability (probably not yet fully defined) of this zeolitic species. The existence of a continuous isomorphous series between Ca and K end-members, suggested by Vezzalini et al. [10], has been reinforced by the description of Lessini samples with intermediate compositions. This suggests a redefinition of willhendersonite as a series extending from a Ca end-member to compositions with equal proportions of Ca and K. According to [10], the absence of the K-pure willhendersonite end-member means that, at present, the series is incomplete. Further analyses are planned to investigate the distribution of this rare zeolite in the Lessini secondary minerals and to better understand the chemical variability of willhendersonite to define the existence of a Ca-K isomorphous series.

According to the literature data [12], the sampling area of willhendersonite (south-western sector of the Alpone valley, Figure 1) is characterized by the occurrence of other zeolites, all of which are of Ca-rich type (e.g., Ca-chabasite, Ca-phillipsite, Ca-gmelinite,

Ca-erionite, heulandite), followed by extensive crystallization of calcite. The presence of Ca-rich species in this area of the Lessini Mounts might suggest that the zeolite-forming fluids could have interacted with the underlying calcareous marine sedimentary rocks through which the basalts erupted. Hydrothermal fluids may have permeated the sedimentary and volcanic pile, leaching Ca out of the calcareous sedimentary rocks. Furthermore, as suggested by experimental works [28], with a likely high concentration of carbonate ions in solution, the zeolites' growth rate would increase in response to the increased dissolution of silica from the surrounding rocks. In this context, the composition of the underlying rocks through which the volcanics erupted seems to play an important role in zeolite formation.

Author Contributions: Conceptualization, methodology, and analysis, M.M. and M.C.; writing—original draft preparation, M.M.; writing—review and editing, M.M. and M.C.; supervision, project administration, and funding acquisition, M.M. All authors have read and agreed to the published version of the manuscript.

Funding: This work was funded in the framework of the 2018 research programs of the Department of Pure and Applied Sciences of the University of Urbino Carlo Bo (project "New asbestiform fibers: mineralogical and physical-chemical characterization", responsible M. Mattioli).

Data Availability Statement: Data is contained within the article. The data presented in this study can be seen in the content above.

Acknowledgments: We would like to thank Amleto Longhi for giving us the idea of studying secondary minerals in the Lessini basalt. We are also grateful to Franco Bressan, Dino Righetti, and Guglielmino Salvatore for their help in the field and for providing us with high-quality mineral specimens. Field work was supported by the Mineralogical and Geological Association of Verona AGMV. Special thanks go to E. Salvioli Mariani and L. Valentini for their kind help and advices on the SEM observations and EDS analysis, and R. Carampin for his great expertise with the EMP.

Conflicts of Interest: The authors declare no conflict of interest. The funders had no role in the design of the study; in the collection, analyses, or interpretation of data; in the writing of the manuscript, or in the decision to publish the results.

References

1. Peacor, D.R.; Dunn, P.J.; Simmons, W.B.; Tillmanns, E.; Fischer, R.X. Willhendersonite, a new zeolite isostructural with chabazite. *Am. Mineral.* **1984**, *69*, 186–189.
2. Tillmanns, E.; Fischer, R.X.; Baur, W.H. Chabazite-type framework in the new zeolite willhendersonite, $KCaAl_3Si_3O_{12}·5H_2O$. *Neues Jahrb. Mineral. Mon.* **1984**, 547–558.
3. Tillmanns, E.; Fischer, R.X. Über ein neues Zeolithmineral mit geordnetem Chabasitgerüst. *Z. Krist.* **1982**, *159*, 125–126.
4. Gottardi, G.; Galli, E. *Natural Zeolites*; Springer: Berlin/Heidelberg, Germany, 1985.
5. Passaglia, E.; Sheppard, R.A. The Crystal Chemistry of Zeolites. In *Reviews in Mineralogy and Geochemistry*; Bish, D.L., Ming, D.W., Eds.; Mineralogical Society of America: Washington, DC, USA, 2001; Volume 45, pp. 69–116.
6. Baerlocher, C.; McCusker, L.B. Database of Zeolite Structures, Published by the International Zeolite Association, Hosted at Lab. Crystallography, ETH Zürich, Switzerland. 2007. Available online: http://www.iza-structure.org/databases/ (accessed on 10 February 2017).
7. Gottardi, G. Mineralogy and crystal chemistry of zeolites. In *Natural Zeolites*; Sand, L.B., Mumpton, F.A., Eds.; Pergamon Press: Oxford, UK, 1978; pp. 31–44.
8. Bish, D.L.; Ming, D.W. Natural Zeolites: Occurrence, Properties, Applications. In *Reviews in Mineralogy and Geochemistry*; Bish, D.L., Ming, D.W., Eds.; Mineralogical Society of America: Washington, DC, USA, 2001; Volume 45, 654p.
9. Walter, F.; Postl, W. Willhendersonit vom Stradner Kogel, südlich Gleichenberg, Steiermark. *Mitt. Abt. Miner. Landesmus. Joanneum* **1984**, *52*, 39–43.
10. Vezzalini, G.; Quartieri, S.; Galli, E. Occurrence and crystal structure of a Ca-pure willhendersonite. *Zeolites* **1997**, *19*, 75–79. [CrossRef]
11. Fischer, R.X.; Kahlenberg, V.; Lengauer, C.L.; Tillmanns, E. Thermal behavior and structural transformation in the chabazite-type zeolite willhendersonite, $KCaAl_3Si_3O_{12}·5H_2O$. *Am. Mineral.* **2008**, *93*, 1317–1325. [CrossRef]
12. Mattioli, M.; Cenni, M.; Passaglia, E. Secondary mineral assemblages as indicators of multi stage alteration processes in basaltic lava flows: Evidence from the Lessini Mountains, Veneto Volcanic Province, Northern Italy. *Per. Mineral.* **2016**, *85*, 1–24.
13. Galli, E. La phillipsite barifera ("wellsite") di M. Calvarina (Verona). *Per. Mineral.* **1972**, *41*, 23–33.
14. Passaglia, E.; Tagliavini, A.; Gutoni, R. Offretite and other zeolites from Fittà (Verona, Italy). *Neues Jahrb. Mineral. Mon.* **1996**, 418–428.

15. Giordani, M.; Mattioli, M.; Dogan, M.; Dogan, A.U. Potential carcinogenic erionite from Lessini Mounts, NE Italy: Morphological, mineralogical and chemical characterization. *J. Toxicol. Environ. Health A* **2016**, *79*, 808–824. [CrossRef]
16. Cangiotti, M.; Battistelli, M.; Salucci, S.; Falcieri, E.; Mattioli, M.; Giordani, M.; Ottaviani, M.F. Electron paramagnetic resonance and transmission electron microscopy study of the interactions between asbestiform zeolite fibers and model membranes. *J. Toxicol. Environ. Health A* **2017**, *80*, 171–187. [CrossRef] [PubMed]
17. Giordani, M.; Mattioli, M.; Ballirano, P.; Pacella, P.; Cenni, M.; Boscardin, M.; Valentini, L. Geological occurrence, mineralogical characterization and risk assessment of potentially carcinogenic erionite in Italy. *J. Toxicol. Environ. Health B* **2017**, *20*, 81–103. [CrossRef] [PubMed]
18. Mattioli, M.; Giordani, M.; Arcangeli, P.; Valentini, L.; Boscardin, M.; Pacella, A.; Ballirano, P. Prismatic to Asbestiform Offretite from Northern Italy: Occurrence, Morphology and Crystal-Chemistry of a New Potentially Hazardous Zeolite. *Minerals* **2018**, *8*, 69. [CrossRef]
19. Barbieri, G.; De Zanche, V.; Sedea, R. Vulcanismo Paleogenico ed evoluzione del semigraben Alpone-Agno (Monti Lessini). *Rend. Soc. Geol. It.* **1991**, *14*, 5–12.
20. De Vecchi, G.P.; Sedea, R. The Paleogene basalt of the Veneto region (NE Italy). *Mem. Inst. Geol. Min. Univ. Padua* **1995**, *47*, 253–274.
21. Bonadiman, C.; Coltorti, M.; Milani, L.; Salvini, L.; Siena, F.; Tassinari, R. Metasomatism in the lithospheric mantle and its relationships to magmatism in the Veneto Volcanic Province, Italy. *Per. Mineral.* **2001**, *70*, 333–357.
22. Barbieri, G.; De Zanche, V.; Medizza, F.; Sedea, R. Considerazioni sul vulcanismo terziario del Veneto occidentale e del Trentino meridionale. *Rend. Soc. Geol. It.* **1982**, *4*, 267–270.
23. Zampieri, D. Tertiary extension in the southern Trento Platform, Southern Alps, Italy. *Tectonics* **1995**, *14*, 645–657. [CrossRef]
24. Zampieri, D. Segmentation and linkage of the Lessini Mountains normal Faults, Southern Alps, Italy. *Tectonophysics* **2000**, *319*, 19–31. [CrossRef]
25. Milani, L.; Beccaluva, L.; Coltorti, M. Petrogenesis and evolution of the Euganean magmatic complex, Veneto Region, North-East Italy. *Eur. J. Mineral.* **1999**, *11*, 379–399. [CrossRef]
26. Beccaluva, L.; Bonadiman, C.; Coltorti, M.; Salvini, L.; Siena, F. Depletion events, nature of metasomatizing agent and timing of enrichment processes in lithospheric mantle xenoliths from the Veneto Volcanic Province. *J. Petrol.* **2001**, *42*, 173–187. [CrossRef]
27. Passaglia, E. The crystal chemistry of chabazites. *Am. Mineral.* **1970**, *55*, 1278–1301.
28. Hawkins, D.B. Kinetics of glass dissolution and zeolite formation under hydrothermal conditions. *Clays Clay Miner.* **1981**, *29*, 331–340. [CrossRef]

Article

First Report on the Geologic Occurrence of Natural Na–A Zeolite and Associated Minerals in Cretaceous Mudstones of the Paja Formation of Vélez (Santander), Colombia

Carlos Alberto Ríos-Reyes [1,*], German Alfonso Reyes-Mendoza [2], José Antonio Henao-Martínez [3], Craig Williams [4] and Alan Dyer [5]

1 Escuela de Geología, Universidad Industrial de Santander, Bucaramanga 680001, Colombia
2 Facultat de Ciències de la Terra, Universitat de Barcelona, 680001 Barcelona, Spain; georeycol@yahoo.es
3 Escuela de Química, Universidad Industrial de Santander, Bucaramanga 680001, Colombia; jahenao@uis.edu.co
4 School of Applied Sciences, University of Wolverhampton, Wolverhampton WV1 1LY, UK; c.williams@wlv.ac.uk
5 Retired Prof Founder Member of the British Zeolite Association, Wolverhampton WV1 1LY, UK; aldilp@aol.com
* Correspondence: carios@uis.edu.co

Abstract: This study reports for the first time the geologic occurrence of natural zeolite A and associated minerals in mudstones from the Cretaceous Paja Formation in the urban area of the municipality of Vélez (Santander), Colombia. These rocks are mainly composed of quartz, muscovite, pyrophyllite, kaolinite and chlorite group minerals, framboidal and cubic pyrite, as well as marcasite, with minor feldspar, sulphates, and phosphates. Total organic carbon (TOC), total sulfur (TS), and millimeter fragments of algae are high, whereas few centimeters and not biodiverse small ammonite fossils, and other allochemical components are subordinated. Na–A zeolite and associated mineral phases as sodalite occur just beside the interparticle micropores (honeycomb from framboidal, cube molds, and amorphous cavities). It is facilitated by petrophysical properties alterations, due to processes of high diagenesis, temperatures up to 80–100 °C, with weathering contributions, which increase the porosity and permeability, as well as the transmissivity (fluid flow), allowing the geochemistry remobilization and/or recrystallization of pre-existing silica, muscovite, kaolinite minerals group, salts, carbonates, oxides and peroxides. X-ray diffraction analyses reveal the mineral composition of the mudstones and scanning electron micrographs show the typical cubic morphology of Na–A zeolite of approximately 0.45 mμ in particle size. Our data show that the sequence of the transformation of phases is: Poorly crystalline aluminosilicate → sodalite → Na–A zeolite. A literature review shows that this is an unusual example of the occurrence of natural zeolites in sedimentary marine rocks recognized around the world.

Keywords: natural zeolite A; mineralogy; mudstones; crystal; sedimentary environment

Citation: Ríos-Reyes, C.A.; Reyes-Mendoza, G.A.; Henao-Martínez, J.A.; Williams, C.; Dyer, A. First Report on the Geologic Occurrence of Natural Na–A Zeolite and Associated Minerals in Cretaceous Mudstones of the Paja Formation of Vélez (Santander), Colombia. *Crystals* **2021**, *11*, 218. https://doi.org/10.3390/cryst11020218

Academic Editor: Magdalena Król

Received: 7 February 2021
Accepted: 19 February 2021
Published: 22 February 2021

1. Introduction

The Swedish mineralogist Axel Fredrick Cronstedt discovered in 1756 the stilbite in allusion to their visibly lost water when heated and named the group of zeolites (from the Greek words "zeo" meaning "to boil" and "lithos" meaning "stone"). Nearly 250 years have passed since this discovery, and geologists traditionally have considered that natural zeolites are formed during aqueous fluids reaction with rocks in a variety of geological environments [1–6]. Several occurrences of natural zeolites have been reported worldwide as accessory minerals in the vugs and cavities of basalts and other basic igneous rocks, as major constituents of many bedded pyroclastic deposits and are thought to be among the most widespread and abundant authigenic silicate minerals in sedimentary rocks [1,4]. Most natural zeolites form during diagenetic processes in sedimentary rocks [7–9]. Zeolites occurring in volcanic lava flow cavities are formed either during lava pile burial metamorphism [10–12], continental basalts' hydrothermal alteration [11,13] or diagenesis in areas of high heat flow caused by active geothermal systems [12,14,15]. Zeolites, as products of hydrothermal crystallization, are generally known from active volcanic rock-associated geothermal systems. Very little work has been published on zeolite occurrences related to late stage pegmatite crystallization [16] in hydrothermal ore veins [17] as alteration products along fault planes [18] and in hydrothermal fractures and veins in granites and gneisses [19]. Zeolites are crystalline, microporous, hydrated aluminosilicates of alkaline or alkaline earth metals. The framework consists of $[SiO_4]^{4-}$ and $[AlO_4]^{5-}$ tetrahedra, which corner-share to form open structures; such tetrahedra are linked to each other by sharing all of the oxygen to form interconnected cages and channels containing mobile water molecules and alkali and/or alkaline earth cations [2,3,20,21]. Zeolites have been widely used as catalysts, adsorbents, and ion exchangers in many technical applications due to their exceptional properties such as extremely high adsorption capacity, catalyzing action, thermal stability, and resistance in different chemical environments [2,22–31]. Zeolites have been extensively used in various technological applications, which include oil refining processes such as catalytic cracking [32,33], as molecular sieves for separating and sorting molecules [34], as adsorbents for water, soil, and air purification [27,35–39] for removing radioactive contaminants [40], for harvesting waste heat and solar heat energy [41], as detergents [42], as antibacterial [43], as drug delivery for oral and topical administration [44,45] or feeding additives for farm animals [46]. They have become worthy of being called the mineral of the future of several countries around the world that have made significant progress in exploring and exploiting this mineral. However, only a few of the natural zeolites in the world are found in sufficient quantities and have the purity required by the industry. The US Geological Survey has reported natural zeolites' worldwide occurrence in the USA, Japan, Korea, Bulgaria, Czechoslovakia, Romania, Hungary, Russia, Croatia, Serbia, South Africa, Italy, Germany, Turkey, and China, the latter having the greatest worldwide production clinoptilolite, mordenite, heulandite, chabazite, phillipsite, and laumontite. At least 60 species of natural zeolites are known to exist, occurring naturally in soils, sediments, and rocks [47], predominantly concentrated in those rocks and soils of volcanic origin. The aim of this study is to report for the first time the geologic occurrence of natural zeolite A and associated minerals in the urban area of Vélez (Santander), Eastern Cordillera, Colombia (Figure 1).

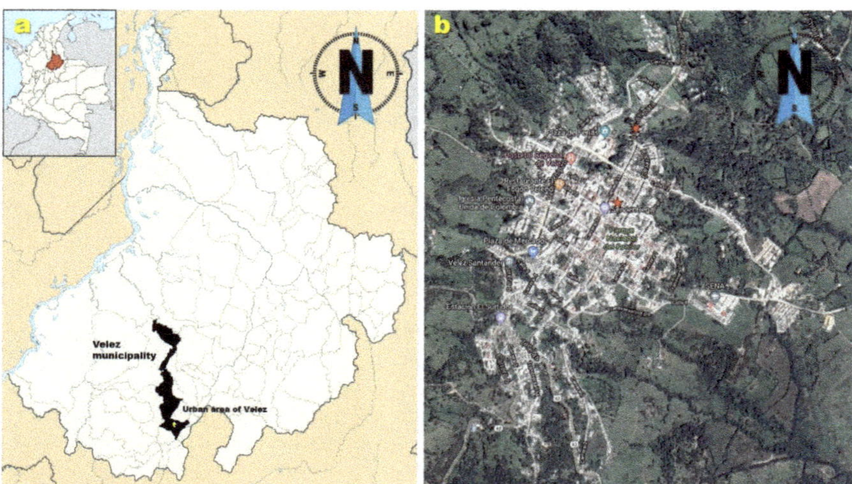

Figure 1. (**a**) Location of the municipality of Velez (Santander), Colombia, and its urban area. (**b**) Satellite image of the urban area of Vélez, showing the sampling localities indicated by red stars (adapted and modified from Google Maps [48]).

2. Materials and Methods

The investigated samples correspond to mudstones and regoliths, of road cuts and excavations for houses, from the La Paja Formation, in the Eastern Cordillera, over the Northern Andes. A preliminary visual inspection of the samples was performed by conventional petrography (results not shown). More detailed analyses were performed by means of X-ray powder diffraction (XRPD), field emission gun-environmental scanning electron microscopy/energy dispersive X-ray spectroscopy (FEG-ESEM/EDS), and Fourier transform infrared with attenuated total reflection (FTIR-ATR) spectroscopy. The bulk mineralogical composition was determined via XRPD using a BRUKER D8 ADVANCE X-ray diffractometer equipped with operating in Da Vinci geometry and equipped with an X-ray tube (Cu-Kα1 radiation: λ = 1.5406 Å), a one-dimensional LynxEye detector (with aperture angle of 2.93°), a divergent slit of 0.6 mm, two soller axials (primary and secondary) of 2.5°, and a nickel filter. All samples were milled in an agate mortar to a particle size of less than 50 μm and then mounted on a sample holder of polymethylmethacrylate (PMMA) using the filling front technique prior to the XRPD analysis. Data collection was carried out at 40 kV and 30 mA in the 2θ range of 3.5–70°, with a step size of 0.01526° (2θ) and counting time of 1 s/step. Phase identification was performed using the crystallographic database Powder Diffraction File (PDF-2) from the International Centre for Diffraction Data (ICDD) and the Crystallographica Search-Match program. The microstructure and chemical composition were examined using the back-scattered electron (BSE) imaging and EDS analysis on a FEI QUANTA 650 FEG-ESEM, under the following analytical conditions: Magnification = 100–20000×, WD = 9.0–11.0 mm, HV = 20 kV, signal = BSE in ZCONT mode, detector = BSED, EDS Detector EDAX APOLO X with a resolution of 126.1 eV (in. Mn Kα). Structural characterization from the functional groups was performed by FTIR-ATR, using a computer model THERMO SCIENTIFIC IS50, with diamond crystal in the spectral range 400–4000 cm^{-1}.

3. Results

3.1. Field Occurrence

The urban area of Vélez (Santander), Colombia, is sitting on mudstones (Figure 2a,b) of the Paja Formation of the Barremian to Lower Aptian age, which is petrographic and biostratigraphically divided into four segments (Kip1, Kip2, Kip3, and Kip4) and has a large detrital contribution (<40%, primary quartz and muscovite and few microcline), with aluminum-silicates of Na, K, Fe, Ti, Ca, Mg, and P (in crystalline and other amorphous phases = 30%, and biogenic particles (approximately 25%), in an open sea (not of platform), and of low energy level [49]. According to Lazar et al. [50], they can also be classified as ar-fMs-Cs and ca-cMs-Zs shales. Macquaker and Adams [51] considered that they range from sand and silt-bearing clay-rich mudstone to clay-dominated mudstone. Reyes-Mendoza [49] indicated dark gray and carbonaceous mudstones, rather than shales for the absence of fissility, with important microfabric features, such as massive facies dominant, thinly microbial laminae, wavy micro-lamination, not parallel and discontinuous, curled to silty lenticular lamination, massive fossiliferous, locally micro-tempestites, as well as nodules and/or concretions, towards the top grain-growing slightly, beds (sandy facies), and thin layers of fine sandstone. Meanwhile, in the bottom near muddy sequence no evaporitic facies (gypsum) happens. Other abundant minerals are kaolinite and pyrophyllite, iron sulfides (primary framboidal or euxinic pyrite, <5 mμ; autigenic cubic pyrite, and amorphous or anhedral marcasite), organic matter (laminated, amorphous, discrete or in organo-mineralogical aggregates, OMA) with high TOC (>4.64%), sulfur (4.80%), sulphates and inorganic phosphates, amines and aliphatic hydrocarbons. They are not considered calcareous (Ca < 4.30%), with dolomite-ankerite, scarce calcite and some irregular lenses or hollow nodules of several oxides-oxyhydroxides and sodalite, hydrosodalite, as well as crystoballite [49]. Other phyllosilicates are clinocloro, berlinite, chamosite, nacrite (assembly that could indicate a low-grade regional metamorphism or high diagenesis with pyrophyllite, up to 21%), dickite, illite, gibbsite, halloysite, kaolinite, phlogopite, vermiculite, phengite and zeolite as products of weathering, the last of them described in detail in this study. Abundant millimetric fragments of phaeophyta and punctually gyrogonites of karophites occur, which indicate a bloom of algae in the superior column of the water. Other allochemicals are ammonites (in molds <5 cm), consolidated intraclasts (<500 μm), few foraminiferous, agglutinated and bilobulates, pyritized radiolarians, calcareous filaments as well as bone remains in the lower and middle part. Primary intragranular, intergranular intercrystalline, and low porosity (<5%) was identified compared with secondary porosity associated with structural discontinuities: Stratification, lamination, jointing, irregular fractures, and probably regional foliation [52]. Reyes-Mendoza [49] evaluated and mapped eight colluviums units (Figure 2c,e) based on their sedimentology, morphological position and paleosoils of beige-orange colors, as well as thicknesses of 1.5–2.2 m under the old houses of the municipality of Vélez (historical center itself), protected by the hard surfaces of the urban development.

Figure 2. Occurrence of (**a**,**b**) a mudstone outcrop of the Kip3 segment of the Paja Formation along the cutting slope in the San Luis neighborhood, and (**c**–**e**) a regolith of the Kip2 segment of the Paja Formation and excavation in the Las Nieves neighborhood.

3.2. Framework of Zeolite LTA

As shown in Figure 3, zeolite Lynde type A (LTA) displays a three-dimensional framework obtained when periodic building units (PerBUs) (which are built using the sodalite cage (or β-cage) consisting of 24 T atoms (six 4-rings, four 6-rings, three 6-2 units or four 1-4-1 units)), related by pure translations along the cube axes, are linked through double 4-rings (D4Rs) for one cube face. It is made of secondary building units 8 or 4-4 or 6-2 or 6 or 1-4-1 or 4 and composite building units D4R, SOD (or β-cage), and LTA (or α-cage) [53]. The pore diameter is defined by an eight-member oxygen ring and is small at 4.2 Å, which leads into a larger cavity of minimum free diameter 0.41 Å [54]. It contains cages and channels in their negatively charged frameworks due to the substitution of Si^{4+} by Al^{3+} and cations that can enter its porous structure to balance the charge of its structural framework. The general chemical formula of zeolite LTA is $[Na_{12}(Al_{12}Si_{12}O_{48}]27H_2O]$, which shows that it has 12 tetrahedra in every cell unit, occupied by 12 Na atoms and 27 H_2O molecules [55], with a cubic unit cell (a = 11.9Å), space group Pm-3m [53], and Si/Al ratio of 1 [56].

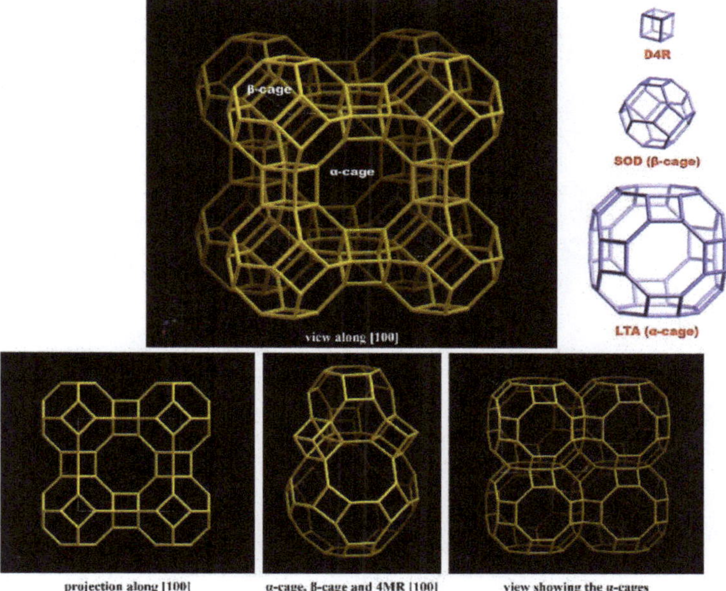

Figure 3. The structural framework of zeolite Lynde type A (LTA), showing its characteristic cages and channels. The vertices mark the positions of the T atoms, the lines symbolize the oxygen bridges between them. Diamond Crystal Structure Software Version 2.1 (structural data published by the International Zeolite Association).

3.3. X-Ray Diffraction

Figure 4 illustrates the characteristic X-ray diffraction pattern of the analyzed mudstone, which reveals high intensity peaks associated with the presence of dickite, illite, and pyrophyllite. Na–A zeolite and sodalite show low intensity peaks. Low intensity peaks correspond to quartz, goethite, and lepidocrocite. Experimental work by Heller-Kallai and Lapides [57] showed that the transformation of kaolinite and metakaolinite with aqueous sodium hydroxide under hydrothermal conditions, with sodalite as the primary product was obtained from kaolinite, whereas the use of metakaolinite promoted the formation of mainly Na–A zeolite, in some cases associated with minor amounts of faujasite. The XRD patterns show the occurrence of a mixture of different mineral phases, including Na–A zeolite and sodalite: Quartz (12.1–19.1 wt%); muscovite (1.5 wt%); pyrophyllite (4.0–19.6 wt%); lepidocrocite (2.8 wt%); kaolinite (1.0 wt%); dickite (13.2 wt%); illite (7.0–11.1 wt%); goethite (10.1 wt%); titanomagnetite (1.0 wt%); zeolite A (2.5 wt%); zeolite X (0.1–4.1 wt%); zeolite SSZ-35 (8.6 wt%). Crystalline phases (55.4–62.9 wt%); amorphous, etc. (37.1–44.6 wt%).

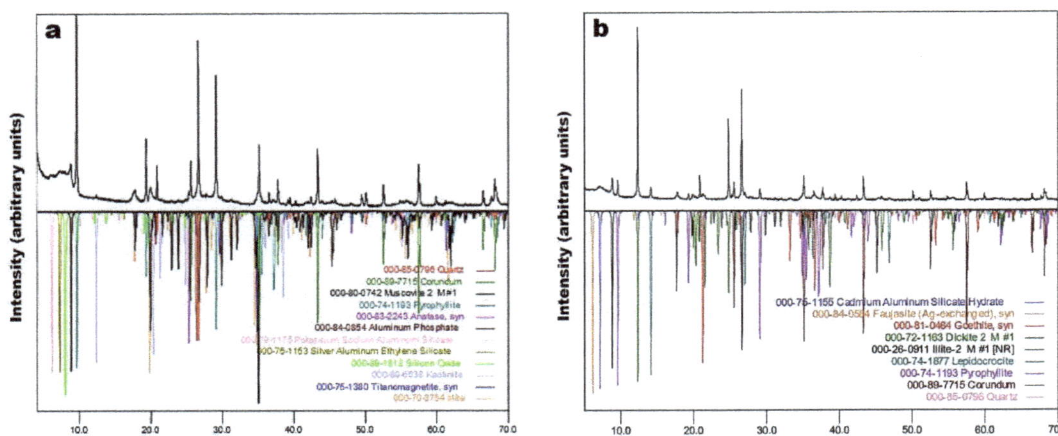

Figure 4. XRD pattern of the analyzed (**a**) mudstone and (**b**) regolith. Qtz: Quartz; Ms: Muscovite; Pyr: Pyrophyllite; Lpd: Lepidocrocite; Kao: Kaolinite; Dck: Dickite; Ill: Illite; Goe: Goethite; TiMag: Titanomagnetite; LTA: Na–A zeolite; FAU: Faujasite; SSZ: Zeolite SSZ-35.

3.4. Scanning Electron Microscopy

SEM micrographs (Figure 5) illustrate the occurrence of the mudstone of the Kip3 segment of the Paja Formation. The morphological features of this rock show randomly interstratified clay minerals with developed sheet structures, bedding planes, and crystal morphology, as well as migration pathways and microporosity (Figure 5a). The bedding planes are broken by rounded and cubic micropores associated to framboidal and cubic pyrite, respectively (Figure 5b). Figure 5c shows the occurrence of aggregates of Na–A zeolite and sodalite. Na–A zeolite displays a typical cubic morphology, whereas sodalite develops aggregates of randomly oriented and intersecting blade-shaped crystals. Ríos et al. [58] described spherical agglomerates of sodalite growing out onto the surface of well-developed cubes of Na–A zeolite displaying fluorite-type interpenetration twinning, which indicates that it is a thermodynamically metastable phase which was successively replaced by a more stable phase as sodalite. Figure 5d,f illustrates the typical appearance of kaolinite occurring as aggregates composed of pseudohexagonal plates, which also constitutes the rock matrix.

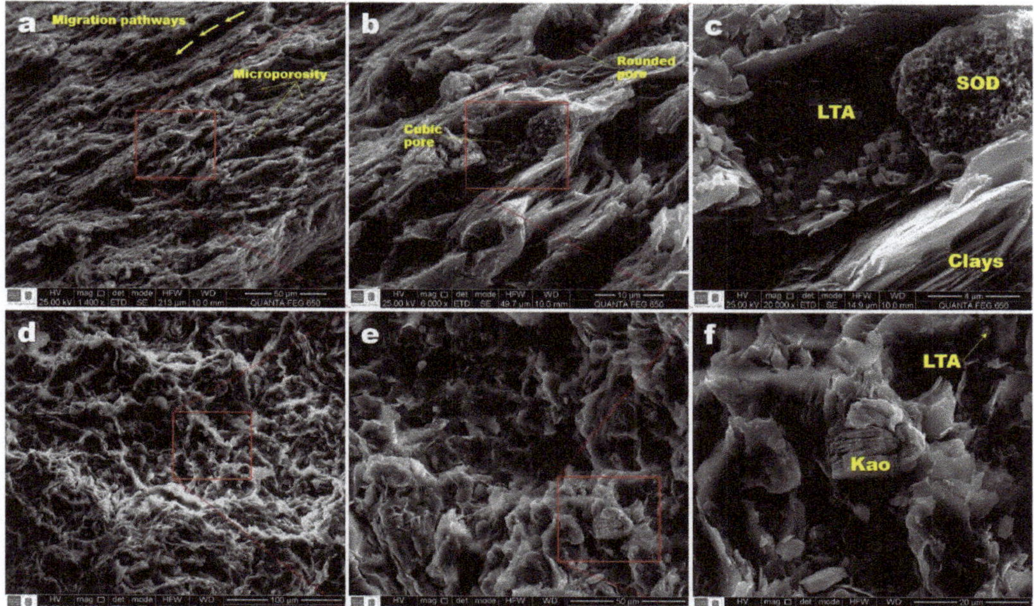

Figure 5. Secondary electron (SE) images of the (**a**) morphological features of the wavy micro-lamination, not parallel and discontinuous, fabric of the analyzed mudstone, (**b**) microporosity, (**c**) aggregates of Na–A zeolite and sodalite and clays, and (**d**–**f**) aggregates of clays of kaolinite-type in the matrix. LTA: Na–A zeolite; SOD: Sodalite; Kao: Kaolinite.

Figure 6 illustrates SE images of the occurrence of aggregates of cubic Na–A zeolite and honeycomb sodalite in the pore space left by the oxidation of cubic and framboidal pyrite (upper part), and the typical EDS spectra of Na–A zeolite and associated mineral phases (lower part). Clay minerals occurring in the matrix of the mudstone reveal the presence of high concentrations of Al (16.51 wt%), Si (29.01 wt%), and O (24.78 wt%). Na–A zeolite shows high concentrations of Al (17.97 wt%), Si (27.14 wt%), and O (9.63 wt%). Sodalite shows high concentrations of Al (17.65 wt%), Si (35.30 wt%), and O (16.38 wt%). The peaks of Fe, Mg, Na, K, S, and Ti (results not shown) represent the contribution of associated mineral phases such as pyrite, Ti-oxides, and feldspars. The peak of Au in the spectra is attributed to the gold coating.

The SE image and EDS spectrum of the analyzed Na–A zeolite are shown in Figure 7. Na–A zeolite displays a typical cubic morphology, which is broken by the presence of aggregates of sodalite, which develops from randomly oriented and intersecting blade-shaped crystals growing at expenses of Na–A zeolite. It shows high concentrations of Al (17.97 wt%), Si (27.14 wt%), and O (9.63 wt%). The peaks of Fe, Mg, Na, K, S, and Ti (results not shown) represent the contribution of associated mineral phases such as pyrite, Ti-oxides, and feldspars. The peak of Au in the spectra is attributed to the gold coating.

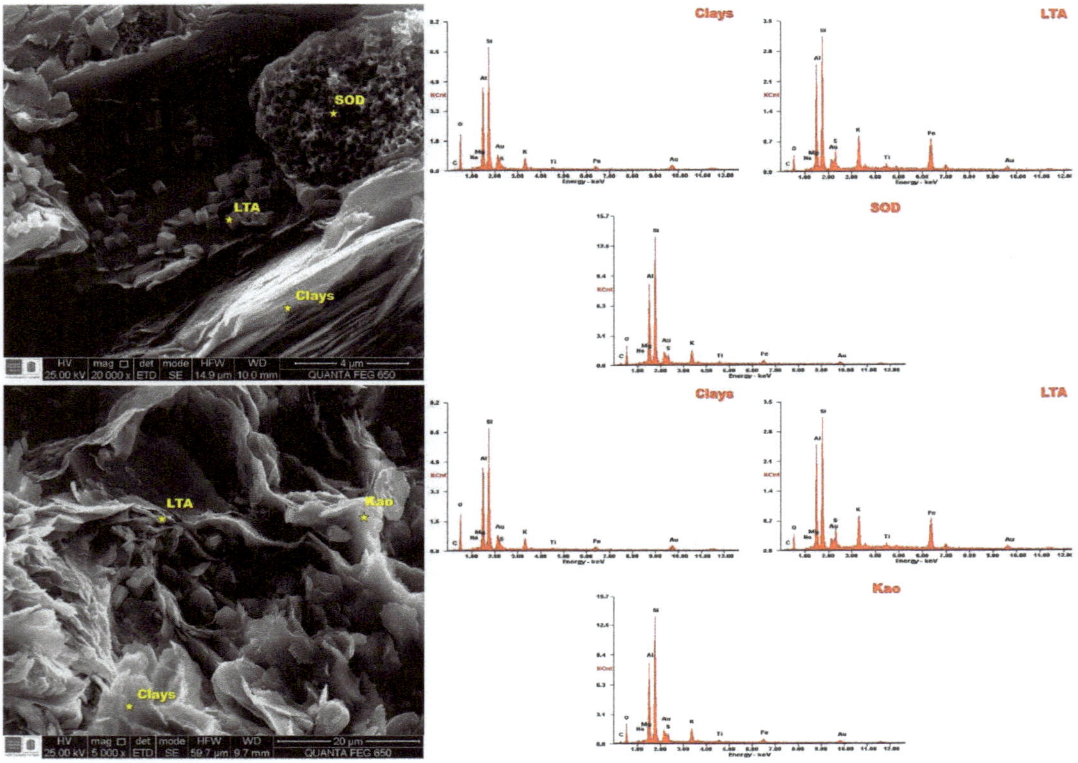

Figure 6. SE images of the analyzed aggregates of Na–A zeolite, sodalite, and clays. LTA: Na–A zeolite; SOD: Sodalite; Kao: Kaolinite.

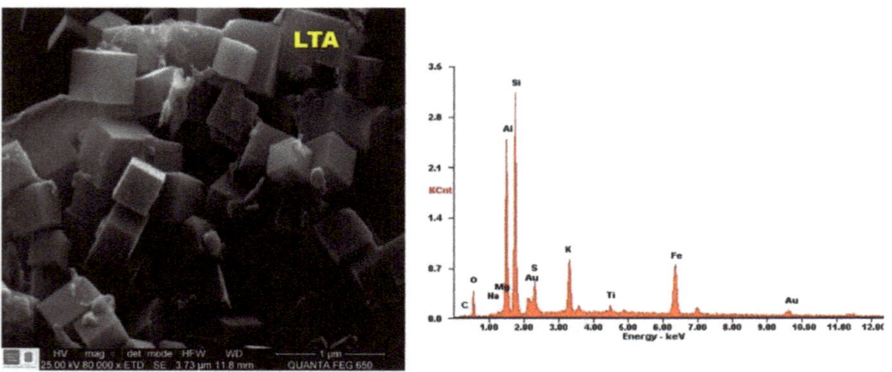

Figure 7. SE image (**left**) and EDS spectrum (**right**) of Na–A zeolite (LTA).

3.5. Fourier Transform Infrared with Attenuated Total Reflection Spectroscopy

FTIR spectra in Figure 8 show the typical vibration bands of the analyzed mudstone and regolith. We recognize the characteristic OH-stretching vibrations of kaolinite at 3698 cm^{-1} (surface OH stretching) and 3620–3629 cm^{-1} (inner OH stretching). However, the vibration band at 3620 cm^{-1} can also be attributed to gibbsite. The vibration bands at 1621 cm^{-1} (kaolinite), 1630 cm^{-1} (goethite or calcite), 1420–1423 cm^{-1} (calcite or siderite), 979 cm^{-1} (gibbsite), 987 cm^{-1} (chlorite), 929 and 907 cm^{-1} (margarite or kaolinite), 820–822 cm^{-1} (phlogopite or margarite), 792–796 cm^{-1} (quartz or goethite), 746 cm^{-1} (muscovite or dickite), 693 cm^{-1} (quartz, kaolinite, phlogopite or margarite), 687 cm^{-1} (kaolinite), 518 cm^{-1} (phlogopite), 512 cm^{-1} (margarite), 441–443 cm^{-1} (potassium feldspar, chlorite or margarite), 415 cm^{-1} (margarite), 411 cm^{-1} (muscovite), and 401 cm^{-1} (margarite) were also observed. The major structural groups present in zeolites can be detected from their FTIR patterns. The band at 1621–1630 cm^{-1} can be attributed to zeolitic water. However, the region of 1500–400 cm^{-1} is a fingerprint indicating structural features of zeolite frameworks. On the other hand, characteristic bands of zeolitic materials appeared, including the asymmetric Al–O stretch located in the region of 1250–950 cm^{-1} (with the bands at 979 and 987 cm^{-1} assigned to sodalite) and their symmetric Al–O stretch located in the region of 660–770 cm^{-1}. Similar results have been reported by several studies [37,59–63]. There is a well-defined peak at 693 cm^{-1} assigned to symmetric T–O–T vibrations of the sodalite framework in good agreement with the results reported by Flaningen et al. [59]. The peak at 687 cm^{-1} can be attributed to the Na–A zeolite. The bands in the region of 500–650cm^{-1} are related to the presence of the double rings (D4R and D6R) in the framework structure of sodalite. The bands in the region of 420–500 cm^{-1} are related to internal tetrahedron vibrations of Si–O and Al–O of sodalite. The bands at 441–443 cm^{-1} are related to internal tetrahedron vibrations of Si–O and Al–O of sodalite (T–O–T) bending modes of the sodalite framework. The detailed FTIR assignments for sodalite have been summarized by Barnes et al. [60] and later revised by Zhao et al. [64]. The bands in the region of 400–420 cm^{-1} are related to the pore opening or motion of the tetrahedral rings, which form the pore opening of zeolites [2].

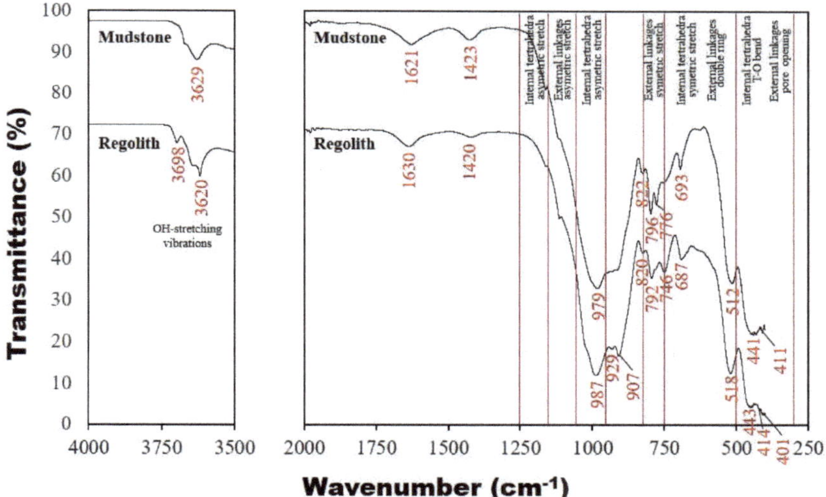

Figure 8. FTIR spectrum of the analyzed mudstone and regolith.

3.6. Discussion on the Mechanism of Formation of the Na–A Zeolite in Mudstone and Regolith

One of the main challenges facing geosciences is to reconstruct the environments in which minerals are formed. Several research groups have performed experimental work to successfully bridge the gap between the general formation mechanisms of crystalline phases and the crystallization of complex structures, such as those of zeolites [21]. We evaluate the geological characteristics of mudstone and regolith systems, taking in consideration not only their mineralogy but also the environmental conditions, which can be strongly affected by several factors, revealing a complex history of chemical reactions in the origin of zeolitic materials. The Cretaceous Paja Formation formed in closed basin environments with low water circulation and small oscillations in the relative sea-level [65,66]. For dissolution of mudstone grains and precipitation of authigenic minerals to occur on the weathered horizon of the La Paja Formation are circulating fluids for the transport of nutrients (ions), geochemical processes, and biogenic activity, which modifies porosity and permeability that will allow fluid migration. According to Passaglia et al. [67], the main factors that control the species and precipitation of diagenetic zeolites are the texture and composition of the host rock, composition of the fluids, and temperature. Na-enriched zeolites, such as the Na–A zeolite, tend to form through the rock-strongly alkaline (pH > 10) fluid interaction which are in a hydrologically closed system [22,67]. The SiO_2 and Al_2O_3 sources for the formation of Na–A zeolite is speculative, taking into account that volcanic glass, the source for many diagenetic zeolites, is not present in La Paja Formation rocks. However, the Na–A zeolite may form where the supply of mobile ions nearly equals the supply of Si and Al [68]. The chemical weathering of mudstones can be characterized by a sequence of chemical reactions between percolating groundwater and rock-forming minerals, with pyrite probably playing a very important role. Pyrite and other sulphides were affected by oxidation (by oxygen coming from the ground surface) to form sulfuric acid (H_2SO_4), sulfate minerals (e.g., gypsum), and several iron oxides such as goethite and hematite, developing an oxidation zone. Framboidal (raspberry-like) pyrite is found representing the most problematic form in terms of the effects of pyrite oxidation. Taking into account the fact that the oxidation of pyrite is primarily surface controlled, the reactivity of pyrite increases as the grain size decreases and the relative surface area increases [69]. The H_2SO_4, along with atmospheric agents, salts, inorganic and biogenic acids, dissolves rock-forming minerals, developing a dissolved zone. The few carbonates (mainly dolomite), which are buffering minerals (the acidity of the medium was neutralized) were dissolved in rhombohedral molds. However, it is probably that this did not occur in the analyzed samples where framboidal (raspberry-like) pyrite suffered oxidation. Taking into account the fact that alkaline pH tends to inhibit pyrite dissolution by blocking the access of the oxidizing agent (O_2) to pyrite surfaces [70,71]. Nanobacteria were observed, which possibly degrade amino acids and produce sulphites that nourish plants. Therefore, the weathering is strongly conditioned by physical, chemical, and biological processes. Consequently, the acidity of the system is neutralized and other elements silicon, aluminum, sodium, potassium, calcium, iron, magnesium, titanium, and phosphorus are released, which enter into the solution or precipitate within the mudstone or regolith (originally composed of clay minerals (about 50–60 wt%) and other minerals, especially quartz and carbonates), originating secondary minerals, such as new clays, zeolites, iron oxides, and hydroxides, etc. The clay minerals in mudstone are largely kaolinite and illite, which are excellent indicators of the environment due to their sensitivity to slight changes in the composition, temperature, and pH of their surroundings [72]. Different clay minerals have been used as starting materials with an appropriate $SiO_2/A1_2O_3$ ratio for studying the formation of zeolites at the laboratory scale [2–64,73–90]. However, kaolinite has been of interest in several studies. It varies in terms of the structural make up from one deposit to another, consists of dioctahedral 1:1 layer structures with a composition of $AI_2Si_2O_5[OH]_4$. This variation affects the ordering of the kaolinite structure and their subsequent chemical reactivity. Previous studies reveal that the improvement of the properties of the kaolinite by chemical methods is difficult due to its low reactivity, taking into account the fact that it is not significantly affected under acid or alkaline conditions [63,91–93]. Therefore, kaolinite is usually used after calcination at temperatures between 550–950 °C [37,91,92,94–96] to obtain a more reactive phase (metakaolinite). However, this clay mineral is not stable under highly alkaline conditions and different zeolitic materials can form. Several authors have reported the synthesis of kaolinite-based zeolites, including LTA [96–98], FAU [85,91,96,98,99], GIS [96,100], CAN [64,101], SOD [64,101,102], and JBW [37,103,104].

According to Zhao et al. [64], there are two major chemical processes involved in the reaction between kaolinite and alkaline solutions: Dissolution of kaolinite, releasing Si and Al, followed by the formation of zeolitic materials. In this study, we suggest that the Na–A zeolite clearly formed by precipitation from relatively low temperature fluids in pores of the analyzed samples, which is supported by the experimental work performed by Ríos [37], who obtained the Na–A zeolite along with sodalite and cancrinite under hydrothermal alkaline conditions at 100 °C during a period of reaction time of 6–120 h. It is seen that a higher alkali concentration during the synthesis process leads to a faster dissolution of the original kaolinite, accompanied by more crystalline zeolitic materials. The Na and Mg-enriched geochemistry of the analyzed mudstone and regolith provides additional evidence for alteration, reflecting the low temperature hydrothermal growth of zeolitic materials, which, therefore, are not related to the original sedimentary mineralogy. The genesis of sodic zeolites, such as the Na–A zeolite along with sodalite and cancrinite, was affected by parameters as the reaction temperature and time between the starting materials (mudstone or regolith) and low temperature fluids, the alkalinity (alkaline conditions or high Na content) of the fluids, and the very high Na/K ratio and the necessary SiO_2/Al_2O_3 ratio of the starting materials. Ríos and co-workers [37,58,63,105–109] performed the experimental work in order to simulate the geological conditions and chemical reactions expected during the formation of zeolites. The results of these studies reveal the hydrothermal transformation of kaolinite, with the production of several zeotypes, including sodalite, cancrinite, and Na–A zeolites in agreement with the previous studies. However, Ríos et al. [58] considered that sodalite and cancrinite represent a metastable phase via the Na–A zeolite forming reaction. In the case of a mudstone or regolith system as that reported in the present study, we consider that the chemistry of evolution of the Na–A zeolite was affected by several factors. The reaction history of the formation of the Na–A zeolite reported in this study can be summarized by the following stages: (1) Dissolution of aluminosilicate minerals (clays) in the mudstone or regolith releasing SiO_2 and Al_2O_3, (2) formation of intermediate metastable phases (sodalite), (3) occurrence of simultaneous reactions of precipitation and dissolution of a gel phase, nucleation, and growth of Na–A zeolite that reached chemical equilibrium. The Na–A zeolite is typically crystallized from amorphous aluminosilicate precursors in aqueous in the presence of alkali metals. A simple scheme of the crystallization of an amorphous aluminosilicate hydrogel to the Na–A zeolite is given in Figure 9. At the beginning, a dissolution of the aluminosilicate minerals (clays) in the mudstone or regolith occurred, releasing SiO_2 and Al_2O_3, with the production of an amorphous gel characterized by the presence of small olygomers. A dissolution process promotes the formation of the nutrients (ionic species), which then are transported to the nucleation sites, indicating that the ionic species are not static, since they necessarily need to move (transportation) to the nucleation sites. A nucleation process produced an equilibrated gel. During nucleation, the hydrogel composition and structure are significantly affected by thermodynamic and kinetic parameters. A polymerization of SiO_4 tetrahedra proceeds, which is represented by TO_4 primary tetrahedral building units that have been joined, revealing how they link together to form larger structures. A polymerization is the process that forms the Na–A zeolite precursors, which contains tetrahedral of Si and Al randomly distributed along the polymeric chains which are cross-linked to provide cavities large enough to accommodate the charge balancing alkali ions. During crystal growth, the TO_4 units were linked, with the formation of 4-ring and 6-ring, composed by 4 and 6 tetrahedral atoms, respectively, to create a large structure (secondary building unit) such as the SOD β-cage. The crystallization of Na–A zeolite occurred by linking the same secondary building units together. A phase transformation occurred as represented by the sequence of reaction: Poorly crystalline aluminosilicate → sodalite → Na–A zeolite.

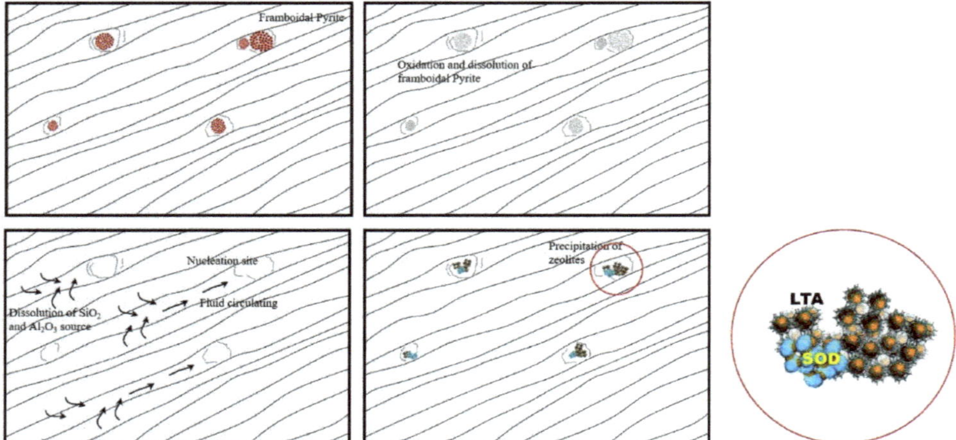

Figure 9. Sketch showing the reconstructed sequence of mineral reactions taking place in the studied mudstone and regolith.

4. Conclusions

In this study, an unusual example of the occurrence of natural zeolites in sedimentary marine rocks was recognized. Na–A zeolite and sodalite occur in mudstones and regoliths of the weathering zone of the Cretaceous La Paja Formation, Vélez (Santander), Colombia.

These rocks consist of quartz, muscovite, abundant pyrophyllite, kaolinite, and chlorite group minerals, pyrite, marcasite, minor feldspar, sulphates, and phosphates, with a high content of TOC, TS, and algae, as well as subordinated ammonite fossils and other allochemical components. High diagenesis (temperatures up to 80–100 °C), weathering and fluid flow allowed the geochemistry remobilization and/or recrystallization of pre-existing mineral phases.

Our model suggests that the genesis of sodic zeolites, such as Na–A zeolite and sodalite, was affected by parameters as the reaction temperature and time between the starting materials (mudstone or regolith) and low temperature fluids, the alkalinity (alkaline conditions or high Na content) of the fluids, as well as the very high Na/K ratio and the necessary SiO_2/Al_2O_3 ratio of the starting materials. With time, the nucleation and growth process of zeolitic phases involved the dissolution of early metastable phases (Na–A zeolite) and the crystallization of sodalite separated in the same media. However, additional work is necessary to determine their environmental conditions of formation.

Author Contributions: Conceptualization, validation, writing—review and editing, visualization, supervision, funding acquisition, C.A.R.-R.; conceptualization, formal analysis, investigation, methodology, writing—original draft, visualization, project administration, funding acquisition, G.A.R.-M. conceptualization, validation, investigation, writing—review and editing, supervision, funding acquisition, J.A.H.-M.; conceptualization, validation, investigation, writing—review and editing supervision, C.W.; conceptualization, validation, investigation, writing—review and editing, supervision, A.D. All authors have read and agreed to the published version of the manuscript.

Funding: This research was funded by the Universidad Industrial de Santander through their research facilities at the Microscopy, Spectroscopy, and X-Rays laboratories of the Guatiguará Technology Park, and their staff for the analytical service provided for data acquisition.

Institutional Review Board Statement: Not applicable.

Informed Consent Statement: Not applicable.

Data Availability Statement: All data are included in the manuscript.

Acknowledgments: The present study was initially supported as a kind contribution, by the field-work and preliminary analysis up scaling for the comprehensive characterization of such muddy rocks and its role in susceptibility to mass movements, making part of the doctoral thesis in Earth Sciences by G.A. Reyes-Mendoza, at the Universitat de Barcelona. The presence of these so particular crystalline phases caught his attention and the collected-preserved fine grain samples were delivered to zeolite experts, for their proper characterization and mineral genesis. We gratefully acknowledge the Vicerrectoría de Investigación y Extensión of the Universidad Industrial de Santander for the use of their research facilities. Authors thank the Microscopy, Spectroscopy, and X-Rays laboratories of the Guatiguará Technology Park, and their staff for the analytical service provided for data acquisition. The authors would also like to acknowledge the anonymous referees for their critical and insightful reading of the manuscript and are most grateful to the above-named people and institutions for support.

Conflicts of Interest: The authors declare no conflict of interest.

References

1. Mumpton, F.A. First reported Occurrence of Zeolites in Sedimentary Rocks of Mexico. *Am. Mineral.* **1973**, *58*, 287–290.
2. Breck, D.W. *Zeolite Molecular Sieves: Structure, Chemistry and Use*; John Wiley: New York, NY, USA, 1974; 313p.
3. Gottardi, G.; Galli, E. *Natural Zeolites*; Springer: Berlin/Heidelberg, Germany, 1985; 409p.
4. Ming, D.W.; Mumpton, F.A. Zeolites in soils. In *Minerals in Soil Environments*; Dixon, J.B., Weed, S.B., Eds.; Soil Science Society of America: Madison, WI, USA, 1989; pp. 873–911.
5. Tsitsishvili, G.; Skhirtladze, N.; Andronikashvili, T.; Tsitsishvili, V.; Dolidze, A. Natural zeolites of Georgia: Occurrences, properties, and application. *Prep. Catal. V Sci. Bases Prep. Heterog. Catal. Proc. Fifth Int. Symp.* **1999**, *125*, 715–722. [CrossRef]
6. Weisenberger, T. Zeolites in fissures of crystalline basement rocks. Ph.D. Thesis, Universitát Freiburg, Freiburg, Germany, 2009. Unpublished.
7. Boles, J.R.; Coombs, D.S. Zeolite facies alteration of sandstones in the Southland Syncline, New Zealand. *Am. J. Sci.* **1977**, *277*, 982–1012. [CrossRef]
8. Hay, R.L.; Sheppard, R.A. Occurrence of Zeolites in Sedimentary Rocks: An Overview. *Rev. Miner. Geochem.* **2001**, *45*, 217–234. [CrossRef]
9. Langella, A.; Cappelletti, P.; Gennaro, R.D. Zeolites in Closed Hydrologic Systems. *Rev. Miner. Geochem.* **2001**, *45*, 235–260. [CrossRef]
10. Neuhoff, P.S.; Fridriksson, T.; Arnórsson, S. Porosity evolution and mineral paragenesis during low-grade metamorphism of basaltic lavas at Teigarhorn, Eastern Iceland. *Am. J. Sci.* **1999**, *299*, 467–501. [CrossRef]
11. Ballance, P.F.; Waiters, W.A. Hydrothermal alteration, contact metamorphism, and authigenesis in Ferrar Supergroup and Beacon Supergroup rocks, Carapace Nunatak, Allan Hills, and Coombs Hills, Victoria Land, Antarctica. *New Zealand J. Geol. Geophys.* **2002**, *45*, 71–84. [CrossRef]
12. Weisenberger, T.; Selbekk, R.S. Multi-stage zeolite facies mineralization in the Hvalfjördur area, Iceland. *Acta Diabetol.* **2008**, *98*, 985–999. [CrossRef]
13. Walker, G.P.L. Zeolite Zones and Dike Distribution in Relation to the Structure of the Basalts of Eastern Iceland. *J. Geol.* **1960**, *68*, 515–528. [CrossRef]
14. Lagat, J. Hydrothermal alteration mineraology in geotermal fields with case examples from Olkaria Domes Getothermal Field, Kenya. In Proceedings of the Short Course IV on Exploration for Geothermal Resources, organized by UNU-GTP, KenGen and GDC, Lake Naivasha, Kenya, 1–22 November 2009.
15. Kralj, P.; Rychagov, S.; Kralj, P. Zeolites in volcanic-igneous hydrothermal systems: A case study of Pauzhetka geothermal field (Kamchatka) and Oligocene Smrekovec volcanic complex (Slovenia). *Environ. Earth Sci.* **2010**, *59*, 951–956. [CrossRef]
16. Orlandi, P.; Scortecci, P.B. Minerals of the Elba pegmatites. *Mineral. Rec.* **1985**, *16*, 353–364.
17. Deer, W.A.; Howie, R.A.; Wise, W.S.; Zussman, J. *An Introduction to Rock-Forming Minerals*; The Geological Society: London, UK, 2004; 696p.
18. Vincent, M.W.; Ehlig, P.L. Laumontite mineralization in rocks exposed north of San Andreas Fault at Cajon Pass, southern California. *Geophys. Res. Lett.* **1988**, *15*, 977–980. [CrossRef]
19. Weisenberger, T.; Bucher, K. Zeolites in fissures of granites and gneisses of the Central Alps. *J. Metamorph. Geol.* **2010**, *28*, 825–847. [CrossRef]
20. Barrer, R.M. *Hydrothermal Chemistry of Zeolites*; Academic Press: New York, NY, USA, 1982; 360p.
21. Szostak, R. *Molecular Sieves*; Springer: Berlin/Heidelberg, Germany, 1989; p. 359.
22. Booker, N.A.; Cooney, E.L.; Priestley, A.J. Ammonia removal from sewage using natural Australian zeolite. *Water Sci. Technol.* **1996**, *34*, 17–24. [CrossRef]
23. Dixit, L.; Prasada, T.S.R. New approach to acid catalysis and hydrocarbon—Zeolite interactions. *Stud. Surf. Sci. Catal.* **1998**, *113*, 313–319.

24. Loiola, A.; Andrade, J.; Sasaki, J.; da Silva, L. Structural analysis of zeolite NaA synthesized by a cost-effective hydrothermal method using kaolin and its use as water softener. *J. Colloid Interface Sci.* **2012**, *367*, 34–39. [CrossRef]
25. Nuić, I.; Trgo, M.; Medvidović, N.V. The application of the packed bed reactor theory to Pb and Zn uptake from the binary solution onto the fixed bed of natural zeolite. *Chem. Eng. J.* **2016**, *295*, 347–357. [CrossRef]
26. Liu, G.-H.; Wang, Y.; Zhang, Y.; Xu, X.; Qi, L.; Wang, H. Modification of natural zeolite and its application to advanced recovery of organic matter from an ultra-short-SRT activated sludge process effluent. *Sci. Total. Environ.* **2019**, *652*, 1366–1374. [CrossRef]
27. Pinto, G.C.; Ríos, C.A.; Vargas, L.Y. Comparative study on the use of a natural and modified Ecuadorian zeolites of the Cayo Formation on the remediation of an oil-polluted soil. *Ctf–Cienc. Tecnol. Y Futuro* **2019**, *9*, 93–104. [CrossRef]
28. Tran, Y.T.; Lee, J.; Kumar, P.; Kim, K.-H.; Lee, S.S. Natural zeolite and its application in concrete composite production. *Compos. Part. B Eng.* **2019**, *165*, 354–364. [CrossRef]
29. Vargas, A.M.; Cipagauta, C.C.; Molina, D.R.; Ríos, C.A. A comparative study on diclofenac sodium release from surfactant-modified natural zeolites as a pharmaceutical excipient. *Mater. Chem. Phys.* **2020**, *256*, 123644. [CrossRef]
30. Alabbad, E.A. Efficacy assessment of natural zeolite containing wastewater on the adsorption behaviour of Direct Yellow 50 from; equilibrium, kinetics and thermodynamic studies. *Arab. J. Chem.* **2021**, *14*, 103041. [CrossRef]
31. Rey, V.; Ríos, C.; Vargas, L.; Valente, T. Use of natural zeolite-rich tuff and siliceous sand for mine water treatment from abandoned gold mine tailings. *J. Geochem. Explor.* **2021**, *220*, 106660. [CrossRef]
32. Vogt, E.T.C.; Weckhuysen, B.M. Fluid catalytic cracking: Recent developments on the grand old lady of zeolite catalysis. *Chem. Soc. Rev.* **2015**, *44*, 7342–7370. [CrossRef]
33. Suganuma, S.; Katada, N. Innovation of catalytic technology for upgrading of crude oil in petroleum refinery. *Fuel Process. Technol.* **2020**, *208*, 106518. [CrossRef]
34. Cheng, X.-W.; Wang, J.; Huang, Q.; Long, Y.-C. Modified natural STI zeolite–A potentially useful molecular sieve. *Pharmacogenetics* **2007**, *170*, 2080–2085. [CrossRef]
35. Pitcher, S.; Slade, R.; Ward, N. Heavy metal removal from motorway stormwater using zeolites. *Sci. Total. Environ.* **2004**, *334*, 161–166. [CrossRef] [PubMed]
36. Santos, S.; Machado, R.; Correia, M.J.N.; Carvalho, J.R. Treatment of acid mining waters. *Miner. Eng.* **2004**, *17*, 225–232. [CrossRef]
37. Ríos, C.A. Synthesis of Zeolites from Geological Materials and Industrial Wastes for Potential Application in Environmental Problems. Ph.D. Thesis, University of Wolverhampton, Wolverhampton, UK, 2008.
38. Wang, S.; Peng, Y. Natural zeolites as effective adsorbents in water and wastewater treatment. *Chem. Eng. J.* **2010**, *156*, 11–24. [CrossRef]
39. Yang, Y.; Xu, W.; Zhang, F.; Low, Z.-X.; Zhong, Z.; Xing, W. Preparation of highly stable porous SiC membrane supports with enhanced air purification performance by recycling NaA zeolite residue. *J. Membr. Sci.* **2017**, *541*, 500–509. [CrossRef]
40. Marinin, D.V.; Brown, G.N. Studies of sorbent/ion-exchange materials for the removal of radioactive strontium from liquid radioactive waste and high hardness groundwaters. *Waste Manag.* **2000**, *20*, 545–553. [CrossRef]
41. Fujii, S.; Horie, N.; Nakaibayashi, K.; Kanematsu, Y.; Kikuchi, Y.; Nakagaki, T. Design of zeolite boiler in thermochemical energy storage and transport system utilizing unused heat from sugar mill. *Appl. Energy* **2019**, *238*, 561–571. [CrossRef]
42. Cardoso, A.M.; Horn, M.B.; Ferret, L.S.; Azevedo, C.M.; Pires, M. Integrated synthesis of zeolites 4A and Na–P1 using coal fly ash for application in the formulation of detergents and swine wastewater treatment. *J. Hazard. Mater.* **2015**, *287*, 69–77. [CrossRef]
43. Hrenovic, J.; Milenkovic, J.; Ivankovic, T.; Rajic, N. Antibacterial activity of heavy metal-loaded natural zeolite. *J. Hazard. Mater.* **2012**, *201–202*, 260–264. [CrossRef]
44. Faras, T.; Ruiz-Salvador, A.R.; Rivera, A. Interaction studies between drugs and a purified natural clinoptilolite. *Microporous Mesoporous Mater.* **2003**, *61*, 117–125. [CrossRef]
45. Cerri, G.; Farina, M.; Brundu, A.; Daković, A.; Giunchedi, P.; Gavini, E.; Rassu, G. Natural zeolites for pharmaceutical formulations: Preparation and evaluation of a clinoptilolite-based material. *Microporous Mesoporous Mater.* **2016**, *223*, 58–67. [CrossRef]
46. Papaioannou, D.; Katsoulos, P.; Panousis, N.; Karatzias, H. The role of natural and synthetic zeolites as feed additives on the prevention and/or the treatment of certain farm animal diseases: A review. *Microporous Mesoporous Mater.* **2005**, *84*, 161–170. [CrossRef]
47. Rouquerol, J.; Avnir, D.; Fairbridge, C.W.; Everett, D.H.; Haynes, J.M.; Pernicone, N.; Ramsay, J.D.F.; Sing, K.S.W.; Unger, K.K. Recommendations for the characterization of porous solids (Technical Report). *Pure Appl. Chem.* **1994**, *66*, 1739–1758. [CrossRef]
48. Google Maps. 2018. Available online: https://www.google.com/maps/?hl=es (accessed on 31 October 2018).
49. Reyes-Mendoza, G.A. Evaluación detallada e impacto de rocas lodosas meteorizadas. El caso de Vélez (Santander, Colombia). In *Póster y Resumen, en Memorias de la XIII Semana Técnica de Geología, Ingeniería Geológica y Geociencias*; Universidad de Caldas: Manizales, Colombia, 2018; 3p.
50. Lazar, O.R.; Bohacs, K.M.; Macquaker, J.H.S.; Schieber, J.; Demko, T.M. Capturing Key Attributes of Fine-Grained Sedimentary Rocks in Outcrops, Cores, and Thin Sections: Nomenclature and Description Guidelines. *J. Sediment. Res.* **2015**, *85*, 230–246. [CrossRef]
51. Macquaker, J.H.; Adams, A. Maximizing Information from Fine-Grained Sedimentary Rocks: An Inclusive Nomenclature for Mudstones. *J. Sediment. Res.* **2003**, *73*, 735–744. [CrossRef]

52. Reyes-Mendoza, G.A. *Las Rocas Lodosas de Vélez: De la Meteorización a Los Deslizamientos*; Ponencia oral. Fest Station—UIS Inteligente. Seminario U18 Fest—Ideas para transformar El Mundo; Universidad Industrial de Santander: Bucaramanga, Colombia, 23 September 2019.
53. Baerlocher, C.; McCusker, L. Database of Zeolite Structures. 2020. Available online: http://www.iza-structure.org/databases/ (accessed on 15 April 2020).
54. Schüring, A.; Auerbach, S.M.; Fritzsche, S.; Haberlandt, R. On entropic barriers for diffusion in zeolites: A molecular dynamics study. *J. Chem. Phys.* **2002**, *116*, 10890–10894. [CrossRef]
55. García-Soto, A.R.; Rodríguez-Niño, G.; Trujillo, C.A. Zeolite LTA synthesis: Optimising synthesis conditions by using the modified sequential simplex method Síntesis de Zeolita LTA: Optimización de las condiciones de síntesis, usando el método simplex secuencial modificado. *Ing. E Investig.* **2013**, *33*, 22–27.
56. García, A.L.; López, C.M.; Garcia, L.V.; De Goldwasser, M.R.; Casanova, J.D.C. Improvements in the synthesis of zeolites with low Si/Al ratio from Venezuelan sodium silicate for an environmentally friendly process. *Ing. E Investig.* **2016**, *36*, 62–69. [CrossRef]
57. Heller-Kallai, L.; Lapides, I. Reactions of kaolinites and metakaolinites with NaOH—comparison of different samples (Part 1). *Appl. Clay Sci.* **2007**, *35*, 99–107. [CrossRef]
58. Ríos, C.A.; Williams, C.D.; Fullen, M.A. Nucleation and growth history of zeolite LTA as-synthesized from kaolinite by two different methods. *Appl. Clay Sci.* **2009**, *42*, 446–454. [CrossRef]
59. Flaningen, E.M.; Khatami, H.A.; Szymanski, H.A. Infrared Structural Studies of Zeolite Frameworks. In *Molecular Sieve Zeolites, Advances in Chemistry 101*; Flanigen, E.M., Sand, L.B., Eds.; American Chemical Society: Washington, DC, USA, 1971; pp. 201–229.
60. Barnes, M.C.; Addai-Mensah, J.; Gerson, A.R. A methodology for quantifying sodalite and cancrinite phase mixtures and the kinetics of the sodalite to cancrinite phase transformation. *Microporous Mesoporous Mater.* **1999**, *31*, 303–319. [CrossRef]
61. Markovic, S.; Dondur, V.; Dimitrijevic, R. FTIR spectroscopy of framework aluminosilicate structures: Carnegieite and pure sodium nepheline. *J. Mol. Struct.* **2003**, *654*, 223–234. [CrossRef]
62. Byrappa, K.; Suresh-Kumar, B.V. Characterization of Zeolites by Infrared Spectroscopy. *Asian J. Chem.* **2007**, *19*, 4933–4935.
63. Ríos, C.A.; Williams, C.D.; Castellanos, O.M. Nucleation and growth mechanism of sodalite and cancrinite from kaolinite-rich clay under low-temperature hydrothermal conditions. *Mater. Res.* **2013**, *16*, 424–438.
64. Zhao, H.; Deng, Y.; Harsh, J.B.; Flury, M.; Boyle, J.S. Alteration of kaolinite to cancrinite and sodalite by simulated hanford tank waste and its impact on cesium retention. *Clays Clay Miner.* **2004**, *52*, 1–13. [CrossRef]
65. Forero-Onofre, H.; Sarmiento-Rojas, L. *La Facies Evaporítica de la Formación Paja*; Etayo, F., Laverde, F., Eds.; Proyecto Cretácico; Ingeominas: Bogota, Colombia, 1985; Volume 16, pp. 1–16.
66. Rivera, H.A.; Le Roux, J.P.; Sánchez, L.K.; Mariño-Martínez, J.E.; Salazar, C.; Barragán, J.C. Palaeoredox conditions and sequence stratigraphy of the Cretaceous storm-dominated, mixed siliciclastic-carbonate ramp in the Eastern Cordillera Basin (Colombia): Evidence from sedimentary geochemical proxies and facies analysis. *Sediment. Geol.* **2018**, *372*, 1–24. [CrossRef]
67. Passaglia, E.; Vezzalini, G.; Carnevali, R. Diagenetic chabazites and phillipsites in Italy: Crystal chemistry and genesis. *Eur. J. Miner.* **1990**, *2*, 827–840. [CrossRef]
68. Dickinson, W.W.; Grapes, R.H. Authigenic chabazite and implications for weathering in Sirius Group Diamictite, Table Mountain, Dry Valleys, Antartica. *J. Seidmentary Res.* **1997**, *67*, 815–820.
69. Cripps, J.C.; Taylor, R.K. The engineering properties of mudrocks. *Q. J. Eng. Geol. Hydrogeol.* **1981**, *14*, 325–346. [CrossRef]
70. Nicholson, V.R.; Gillham, W.R.; Reardon, J.E. Pyrite oxidation in carbonate-buffered solution: Experimental kinetics. *Geochim. Et Cosmochim. Acta* **1988**, *52*, 1077–1085. [CrossRef]
71. Nicholson, R.V.; Gillham, R.W.; Reardon, E.J. Pyrite oxidation in carbonate-buffered solution: Rate control by oxide coatings. *Geochim. Et Cosmochim. Acta* **1990**, *54*, 395–402. [CrossRef]
72. Chaudhri, A.R.; Mahavir, S. Clay Minerals as Climate Change Indicators—A Case Study. *Am. J. Clim. Change* **2012**, *1*, 231–239. [CrossRef]
73. Ruiz, R.M.; Blanco, C.G.; Pesquera, C.; González, F.; Benito, I.; López, J. Zeolitization of a bentonite and its application to the removal of ammonium ion from waste water. *Appl. Clay Sci.* **1997**, *12*, 73–83. [CrossRef]
74. Baccouche, A.; Srasra, E.; El Maaoui, M. Preparation of Na–P1 and sodalite octahydrate zeolites from interstratified illite–smectite. *Appl. Clay Sci.* **1998**, *13*, 255–273. [CrossRef]
75. Cañizares, P.; Duran, A.; Dorado, F.; Carmona, M. The role of sodium montmorillonite on bounded zeolite-type catalysts. *Appl. Clay Sci.* **2000**, *16*, 273–287. [CrossRef]
76. Gualtieri, A.F. Synthesis of sodium zeolites from a natural halloysite. *Phys. Chem. Miner.* **2001**, *28*, 719–728. [CrossRef]
77. Boukadir, D.; Bettahar, N.; Derriche, Z. Synthesis of zeolites 4A and HS from natural materials. *Ann. De Chim. Sci. Des. Mater.* **2002**, *27*, 1–13. [CrossRef]
78. Shawabkeh, R.; Al-Harahsheh, A.; Hami, M.; Khlaifat, A. Conversion of oil shale ash into zeolite for cadmium and lead removal from wastewater. *Fuel* **2004**, *83*, 981–985. [CrossRef]
79. Shawabkeh, R.; Al-Harahsheh, A.; Al-Otoom, A. Production of zeolite from Jordanian oil shale ash and application for zinc removal from wastewater. *Oil Shale* **2004**, *21*, 125–136.
80. Fernandes-Machado, N.R.C.; Miotto-Bigatão, D.M.M. Synthesis of Na–A and –X zeolites from oil shale ash. *Fuel* **2005**, *84*, 2289–2294. [CrossRef]

81. Fernandes-Machado, N.R.C.; Miotto-Bigatão, D.M.M. Use of zeolites synthesized from oil shale ash for arsenic removal from polluted wáter. *Química Nova* **2007**, *30*, 1108–1114. [CrossRef]
82. Abdmeziem-Hamoudi, K.; Siffert, B. Synthesis of molecular sieve zeolites from a smectite-type clay material. *Appl. Clay Sci.* **1989** *4*, 1–9. [CrossRef]
83. Kovo, A.S.; Hernandez, O.; Holmes, S.M. Synthesis and characterization of zeolite Y and ZSM-5 from Nigerian Ahoko Kaolin using a novel, lower temperature, metakaolinization technique. *J. Mater. Chem.* **2009**, *19*, 6207–6212. [CrossRef]
84. Hamadi, A.; Nabih, K. Alkali Activation of Oil Shale Ash Based Ceramics. *E-J. Chem.* **2012**, *9*, 1373–1388. [CrossRef]
85. Ngoc, D.T.; Pham, T.H.; Nguyen, K.D.H. Synthesis, characterization and application of nanozeolite NaX from Vietnamese kaolin *Adv. Nat. Sci. Nanosci. Nanotechnol.* **2013**, *4*, 1–12. [CrossRef]
86. Bao, W.W.; Zou, H.F.; Gan, S.C.; Xu, X.C.; Ji, G.J.; Zheng, K.Y. Adsorption of heavy metal ions from aqueous solutions by zeolite based on oil shale ash: Kinetic and equilibrium studies. *Chem. Res. Chin. Univ.* **2013**, *29*, 126–131. [CrossRef]
87. Zhou, Z.; Jin, G.; Liu, H.; Wu, J.; Mei, J. Crystallization mechanism of zeolite A from coal kaolin using a two-step method. *Appl Clay Sci.* **2014**, *97–98*, 110–114. [CrossRef]
88. Tong, L.; Luo, H.; Zhang, L.; Zhan, H. Preparation of single phase Na–X zeolite from oil shale ash by melting hydrothermal method. *China's Refract.* **2014**, *23*, 12–17.
89. Doyle, A.M.; Alismaeel, Z.T.; Albayati, T.M.; Abbas, A.S. High purity FAU-type zeolite catalysts from shale rock for biodiesel production. *Fuel* **2017**, *199*, 394–402. [CrossRef]
90. Bai, S.-X.; Zhou, L.-M.; Chang, Z.-B.; Zhang, C.; Chu, M. Synthesis of Na–X zeolite from Longkou oil shale ash by alkaline fusion hydrothermal method. *Carbon Resour. Convers.* **2018**, *1*, 245–250. [CrossRef]
91. Akolekar, D.; Chaffee, A.; Howe, R.F. The transformation of kaolin to low-silica X zeolite. *Zeolites* **1997**, *19*, 359–365. [CrossRef]
92. Xu, M.; Cheng, M.; Tan, D.; Liu, X.; Bao, X. Growth of zeolite KSO1 on calcined kaolin microspheres. *J. Mater. Chem.* **1999**, *9* 2965–2966. [CrossRef]
93. Eze, K.A.; Nwadiogbu, J.O.; Nwankwere, E.T. Effect of Acid Treatments on the Physicochemical Properties of Kaolin Clay. *Arch Appl. Sci. Res.* **2012**, *4*, 792–794.
94. Murat, M.; Amokrane, A.; Bastide, J.P.; Montanaro, L. Synthesis of zeolites from thermally activated kaolinite. Some observations on nucleation and growth. *Clay Miner.* **1992**, *27*, 119–130. [CrossRef]
95. Chandrasekhar, S.; Pramada, P.N. Investigation on the Synthesis of Zeolite NaX from Kerala Kaolin. *J. Porous Mater.* **1999**, *6* 283–297. [CrossRef]
96. Novembre, D.; Di Sabatino, B.; Gimeno, D.; Pace, C.; Sabatino, D. Synthesis and characterization of Na–X, Na–A and Na–P zeolites and hydroxysodalite from metakaolinite. *Clay Miner.* **2011**, *46*, 339–354. [CrossRef]
97. Sanhueza, V.; Kelm, U.; Cid, R. Synthesis of molecular sieves from Chilean kaolinites: Synthesis of NaA type zeolites. *J. Chem Technol. Biotechnol.* **1999**, *74*, 358–363. [CrossRef]
98. Belviso, C.; Cavalcante, F.; Lettino, A.; Fiore, S. A and X-type zeolites synthesised from kaolinite at low temperatura. *Appl. Clay Sci.* **2013**, *80–81*, 162–168. [CrossRef]
99. Basaldella, E.I.; Kikot, A.; Tara, J.C. Effect of pellet pore size and synthesis conditions in the in situ synthesis of LSX zeolite. *Ind Eng. Chem. Res.* **1995**, *34*, 2990–2996. [CrossRef]
100. Covarrubias, C.; García, R.; Arriagada, R.; Yánez, J.; Garland, M.T. Cr(III) exchange on zeolites obtained from kaolin and natural mordenite. *Microporous Mesoporous Mater.* **2006**, *88*, 220–231. [CrossRef]
101. Chorover, J.; Choi, S.; Amistadi, M.K.; Karthikeyan, K.G.; Crosson, G.; Mueller, K.T. Linking Cesium and Strontium Uptake to Kaolinite Weathering in Simulated Tank Waste Leachate. *Environ. Sci. Technol.* **2003**, *37*, 2200–2208. [CrossRef]
102. Buhl, J.C.; Löns, J. Synthesis and crystal structure of nitrate enclathrated sodalite $Na_8[AlSiO_4]_6(NO_3)_2$. *J. Alloy. Compd.* **1996**, *235* 41–47. [CrossRef]
103. Healey, A.; Johnson, G.; Weller, M. The synthesis and characterization of JBW-type zeolites. *Microporous Mesoporous Mater.* **2000** *37*, 153–163. [CrossRef]
104. Lin, D.-C.; Xu, X.-W.; Zuo, F.; Long, Y.-C. Crystallization of JBW, CAN, SOD and ABW type zeolite from transformation of metakaolin. *Microporous Mesoporous Mater.* **2004**, *70*, 63–70. [CrossRef]
105. Ríos, C.A.; Williams, C.D.; Castellanos, O.M. Synthesis and characterization of zeolites by alkaline activation of kaolinite and industrial by-products (fly ash and natural clinker). *Bistua* **2006**, *4*, 60–71.
106. Ríos, C.A.; Williams, C.D.; Maple, M. Synthesis of zeolites and zeotypes by hydrothermal transformation of kaolinite and metakaolinite. *Bistua* **2007**, *5*, 15–26.
107. Ríos, C.A.; Williams, C.D.; Castellanos, O.M. Synthesis of zeolite LTA from thermally treated kaolinite. *Rev. Fac. Ing.* **2010**, *53* 30–41.
108. Ríos, C.A.; Williams, C.D. Hydrothermal transformation of kaolinite in the system K_2O-SiO_2-Al_2O_3-H_2O. *DYNA Univ. Nac Colomb. Medellin* **2010**, *77*, 55–63.
109. Ríos, C.A.; Williams, C.D.; Roberts, C.L. Synthesis and characterisation of SOD-, CAN- and JBW-type structures by hydrothermal reaction of kaolinite at 200 °C. *DYNA Univ. Nac. Colomb. Medellin* **2011**, *78*, 38–47.

Article

Characterization of Fibrous Mordenite: A First Step for the Evaluation of Its Potential Toxicity

Dario Di Giuseppe [1,2]

[1] Department of Chemical and Geological Sciences, University of Modena and Reggio Emilia, Via G. Campi 103, I-41125 Modena, Italy; dario.digiuseppe@unimore.it

[2] Department of Sciences and Methods for Engineering, University of Modena and Reggio Emilia, Via Amendola 2, I-42122 Reggio Emilia, Italy

Received: 4 August 2020; Accepted: 28 August 2020; Published: 31 August 2020

Abstract: In nature, a huge number of unregulated minerals fibers share the same characteristics as asbestos and therefore have potential adverse health effects. However, in addition to asbestos minerals, only fluoro-edenite and erionite are currently classified as toxic/pathogenic agents by the International Agency for Research on Cancer (IARC). Mordenite is one of the most abundant zeolites in nature and commonly occurs with a fibrous crystalline habit. The goal of this paper is to highlight how fibrous mordenite shares several common features with the well-known carcinogenic fibrous erionite. In particular, this study has shown that the morphology, biodurability, and surface characteristics of mordenite fibers are similar to those of erionite and asbestos. These properties make fibrous mordenite potentially toxic and exposure to its fibers can be associated with deadly diseases such as those associated with regulated mineral fibers. Since the presence of fibrous mordenite concerns widespread geological formations, this mineral fiber should be considered dangerous for health and the precautionary approach should be applied when this material is handled. Future in vitro and in vivo tests are necessary to provide further experimental confirmation of the outcome of this work.

Keywords: mordenite; mineral fibers; erionite; potential toxicity

---

1. Introduction

Occupational or environmental exposure to mineral dusts is one of the main causes in the development of pneumoconiosis and lung cancer [1]. Among airborne particles, the most notorious toxic/carcinogenic agent is asbestos [1,2]. Asbestos is a generic term which refers to six minerals exploited commercially for their outstanding physical properties, mostly due to their fibrous morphology [3]. Asbestos minerals are chrysotile and amphibole asbestos (i.e., amosite, crocidolite, anthophyllite, tremolite, and actinolite) [2,3]. It is generally recognized that asbestos is a toxic and carcinogenic agent associated with the induction of mesothelioma, lung tumours and other lung diseases [2,4]. Hence, chrysotile and amphibole asbestos have been classified by International Agency for Research on Cancer (IARC) as carcinogenic to humans (Group 1) [2]. Concurrently, starting from the 1980s, many countries imposed a ban on asbestos [5]. Nowadays, asbestos and asbestos containing materials (ACM) are gradually being replaced by safe alternatives for human health such as rock wool and plastic fibers. In recent years, a general concern is growing regarding other mineral fibers that are not classified (or not regulated) as asbestos, but assumed to show the same potential toxicity [4]. The term elongate mineral particles (EMPs) has been coined to include all such mineral fibers [5]. In addition to the six asbestos minerals, the following EMPs have raised global concern: non-regulated fibrous amphibole (e.g., winchite [6,7], richterite [6,7], fluoro-edenite [8], fibrous glaucophane [9]), fibrous antigorite [10], balangeroite [11] and fibrous zeolites (e.g., erionite [12], offretite [13] and ferrierite [14]). Within this context, the case of Turkish fibrous erionite is noteworthy [15]. As provided by long-term epidemiological studies and several animal carcinogenicity tests, fibrous erionite is responsible for epidemics of mesothelioma in Cappadocia (Turkey), where villages were built with erionite-bearing rocks [2,15]. Recent works highlighted that although there are differences in the crystal chemistry and genetic environment, several fibrous zeolites possess the same chemical-physical properties that are deemed to prompt adverse effects in vivo by fibrous erionite [13,14,16]. One of the most common natural fibrous zeolites is mordenite [17,18]. Mordenite is a zeolite with orthorhombic structure (space group *Cmcm*) and ideal chemical formula $(Na_2,K_2,Ca)[Al_8Si_{40}O_{96}]\cdot 28H_2O$ [19,20]. The crystal structure of mordenite was determined by Meier [19], while its crystal chemistry was studied extensively by Passaglia [19]. Mordenite displays a framework characterized by edge-sharing 5-member tetrahedral rings of tetrahedral (TO_4), built up from a combination of 5-1 secondary building units [20–23]. These building units are linked by edge-sharing into chains along *c*, which are in turn linked together by 4-rings to form a puckered sheet perforated with 8-ring holes. A unidimensional pore system, consisting of 12 membered rings of TO4, runs parallel to the [001] direction. The channel wall has side pockets, in the [010] direction, circumscribed by 8-membered rings. Each side pocket connects the 12-ring channel through 8-membered rings with a distorted 8-ring channel running parallel to the [001] direction. Mordenite occurs mainly in tuff deposits (diagenetic origin), as well as, is often found in the voids of volcanic rocks (hydrothermal origin) coexisting with other zeolites (e.g., clinoptilolite, heulandite, stilbite, and scolecite) [24]. In both hydrothermal and diagenetic occurrences, mordenite generally shows a fibrous crystal habit [24,25]. Sedimentary deposits of mordenite are present in several countries (e.g., Italy [26], Greece [27], Japan [28], and the United States [14]) where its content is often high enough for mineral exploitation. Rocks containing mordenite are known to have a high cation exchange capacity [24,27]. Consequently, mordenite has been increasingly used in various application areas such as gas separation processes, sorbent and molecular sieve, animal feed supplements and catalytic process [24,27]. Because, a large amount of mordenite fiber can eventually be mined and used for the aforementioned applications, some concerns may raise due to the close resemblance of mordenite with fibrous erionite which is known to be a carcinogenic agent. Currently, mordenite is listed by the IARC in the Group 3 (not classifiable as to its carcinogenicity to humans) [29]. However, as stated in the IARC Monographs Preamble 2019 [30], an evaluation in Group 3 is not a statement of non-carcinogenicity or overall safety. Moreover, as pointed out by the IARC Working Group in the Monographs 68, the main cause of difficulties in the classification of mordenite is the lack of sufficient literature data [29]. Indeed, no data are currently available on the toxicity and pathogenicity of mordenite in exposed humans. At the same time, the number of in vitro investigations and animal carcinogenicity studies related to mordenite are very few if compared to those relating to asbestos or erionite [2,29]. One of the first studies concerning biological activity of mordenite was done by Suzuki [31] who used two naturally occurring erionite and mordenite samples, to test their carcinogenic and fibrogenic effects on mice. The results of this study showed that both the erionite- and mordenite-treated mice developed fibrosis, but the effect was more pronounced in the case of erionite. In a later study by Suzuki et al. [32], a larger group of animals was subjected to intraperitoneal and intra-abdominal injection of mordenite and other mineral fibers types. In the mice exposed to mordenite, no peritoneal tumours were seen at the end of the test (7–23 months after injection), and there were no tumours in other organs. Peritoneal fibrosis was observed; however, in mordenite-treated mice. The fibrogenicity of mordenite was confirmed by the works of Tátrai et al. [33] and Adamis et al. [34]. Overall, the few studies on mordenite found in the literature states that fibrous erionite is more carcinogenic than fibrous mordenite and the health risk associated to this latter is relatively low or null [29].

As a matter of fact, the key factors in toxicity, inflammation and pathogenicity of mineral fibers are [4,12,35,36]: (a) morphology, (b) surface reactivity, and (c) biopersistence.

The length (L) and width (W) of mineral fibers (i.e., the main parameters of fibers morphology) are key factors in toxicity, inflammation and pathogenicity of mineral fibres [37–40], while the curvature of the fibers affects the binding process of proteins and influence cell adhesion [35,41]. According to

Stanton et al. [37] the optimal morphological parameters of a fiber to induce lung disease are L > 8 μm and W < 0.25 μm. The crystal habit of a fiber influences its depositional pathway in the respiratory tract [35,38,42]. At same time, the density determines its aerodynamic diameter and therefore it affects the deposition depth of inhaled fibre in the airways [35,43].

Surface reactivity of a mineral fiber is strongly influenced by the iron found at the surface of mineral fibers and the electric charges surrounding the fibres [35,44–47]. Iron promotes the formation of the reactive oxygen species (ROS), with cyto- and genotoxic effects through a Fenton-like chain reaction [35,46–48]. To be active, iron sites must be on the surface of the fiber where they can be in contact with the extracellular or intracellular environment [35,49]. The electric charges surrounding the fibers, measured as zeta potential, may correlate with a number of phenomena responsible for adverse effects [35,50].

One of the key parameters in the mineral fiber toxicity is the biopersistence as it influences the capability to induce chronic inflammation and pathogenicity in the lungs [35,36,38,51,52]. Biopersistence is the ability of an exogenous particle (such as a mineral fiber) to persist in the human body regardless of chemical dissolution (biodurability) and physical clearance mechanisms [35,36,38,51,52]. Mineral fibers carried by the air can pass the upper respiratory system (the first defence mechanism of the body against alien particles) and settle in the alveolar space where they promote the recruitment of phagocytic cells (e.g., macrophages) in charge of dissolution and clearance of the fibers [35,51,52]. This second defence mechanism of the body against respirable particles is effective for small spherical objects but fails for longer and thinner fibers [37,38]. When macrophages fail to completely internalize the longer fibers (i.e., frustrated phagocytosis), an inflammatory activity is prompted [36]. This inflammatory activity become chronic, because fibers can be biopersistent and resist reiterated attempts of phagocytosis [35–40]. Generally, if a mineral fibre rapidly dissolves in the lung (i.e., low biodurability), it is assumed that it has a low biopersistence and is less toxic than a mineral fiber with high biodurability [36,51,52].

The main aim of this study is to verify if fibrous mordenite possesses the above-mentioned "toxicity kay factors". Moreover, to objectively evaluate the hazard associated to the fibrous mordenite respect to the fibrous erionite, these two mineral fibers were compared from the morphological and chemical-physical point of view. Fibrous mordenite from Poona, India, was used as a representative member of fibrous mordenite zeolites. The studied mordenite sample was characterized using a suite of experimental techniques, and its characteristics were compared to those of fibrous erionite described in literature.

2. Materials and Methods

2.1. Mordenite Sample

The mordenite sample taken into consideration comes from the zeolite collection of the Chemical and Geological Sciences Department of the University of Modena and Reggio Emilia, Italy. This sample was recovered in the suburb of Pashan in the Poona district, Maharashtra, India. It is a hydrothermal origin mordenite and is an orthorhombic silica-rich zeolite. Crystal structure and crystal chemistry of this zeolite were identified and described by Passaglia. [19]. Cell parameters of this mordenite are a = 18.094(4) Å; b = 20.477(4) Å; c = 7.520(2) Å; V = 2786 $Å^3$ [19]. Generally, this mineral occurs in the voids of basalt flows belonging to the Deccan Traps. Mordenite crystals are extremely thin, almost capillary size, and either form radiating tufts or tangled woolly adhering to the walls of cavities [53]. Figure 1 shows an image of the sample. The specimen is approximately 3 cm long and approximately 1 cm wide. The mordenite fibers form radial fibrous tufts or cotton-like material. Small quartz crystals are perched on the fibers or hidden between them.

Figure 1. A stereomicroscopic image of the mordenite from Poona district, Maharashtra, India.

2.2. Electron Microscopy

Morphological observations were performed using an environmental scanning electron microscope (ESEM) FEI Quanta-200 equipped with energy-dispersive X-ray (EDX) spectrometer Oxford INCA-350 for microchemical analyses. Operating conditions were 15 kV accelerating voltage, 3.5 μA beam current 11 mm working distance and 0° tilt angle. Small amounts of powder from samples were mixed with 1 mL of ethanol. A drop of the suspension was laid on a carbon tape mounted on an Al stub and gold coated (10 nm thick). Images were collected using the signal of secondary electrons. The surface of the samples was investigated, working at different magnification levels, from 1000× up to 20,000×. The fiber size and morphometric parameters were determined on ~100 individual fibers, using 15 scanning electron microscope (SEM) images. Lengths and diameters were calculated using ImageJ image analysis software, version 1.52a [54].

The chemical composition of the mordenite was obtained by SEM-EDX following the procedure described by Mattioli et al. [13]. SEM-EDX chemical data were acquired using a low counting time to avoid an increase in temperature on the fiber's surface and minimize the migration of alkaline metals [13].

2.3. Fiber Density and Aerodynamic Diameter

Theoretical density of the mordenite fibers was obtained using the following formula: (molecular weight × number of molecules per unit cell)/(volume of unit cell × Avogadro's number). The density of zeolite it is a fundamental parameter for the calculation of the aerodynamic diameter (D_{ae}) of its fibers [55]. D_{ae} influences the deposition depth of inhaled particles in the airways [55]. Particles with $D_{ae} > 5$ μm are deposited in the nasal respiratory tract, whereas particle with $3 \leq D_{ae} \leq 15$ μm are deposited in the lower respiratory tract that extends from the trachea to the lungs [43,55,56], which is the main focus of respiratory diseases [57]. Particles with $D_{ae} \approx 2$–3 and <0.2 can easily settle in the alveolar space [55]. D_{ae} can be calculate using the equation by Gonda and Abd El Khalik [58]:

$$D_{ae} = d\sqrt{\left(\frac{1}{\frac{2}{9}\left(\frac{1}{(\ln 2\beta - 0.5)}\right) + \frac{8}{9}\left(\frac{1}{(\ln 2\beta + 0.5)}\right)}\right)\left(\frac{\rho}{\rho_0}\right)} \quad (1)$$

where d = fiber diameter; β = fiber length/d (aspect ratio); ρ = density; ρ_0 = unit density (1 g/cm^3).

2.4. Determination of the Zeta Potential

When a solid particle is suspended in a liquid medium, its charged surface is surrounded by ions of opposite sign from the liquid phase [40,49,50]. The arrangement of the mobile ions at the solid–liquid interface is called the electrical double layer (EDL). According to the Gouy–Chapman–Stern model [50,59–61], EDL is a thin film of ions with nonzero net charge composed of two layers: the first one is a compact and relatively immobile layer of ions adsorbed to the solid surface (Stern layer); the second is a diffuse layer of mobile ions outside the Stern layer, which penetrates to a certain extent in the liquid phase. The zeta potential is the electrostatic potential at the boundary dividing the two layers [50,59–61]. Knowing the zeta potential of mineral fibers is an important tool for understanding the reactions that may occur at the fiber-living matter interface [35,55]. The zeta potential of the mordenite sample was determined using a Zetasizer Nano Series instrument (Malvern, Worcestershire, UK). Analysis were performed using double-distilled water (resistivity of 18.2 MΩ.cm) as dispersants. Measurements were conducted at pH 4.5 and 7, a temperature of 37 °C and equilibration time of 120 s. A solution having pH 7 was chosen to reproduce the extracellular environment [35,60,61]. Subsequently, the solution was adjusted to pH 4.5 using diluted HCl, to replicate the intracellular chemical environment of the macrophage [35,60,61].

2.5. Biodurability

Biodurability of mineral fibers is their ability to resist to the chemical/biochemical alteration [41,44]. Biodurability of the mordenite fibers was determined by dissolutions batch test at 37 °C. Test was conducted by leaching 6 samples (10 mg) of sample into a batch reactor with a water and HCl solution (100 mL) at pH = 4. The degree of dissolution is determined by measuring the change of the sample mass after different times: 24 h, 48 h, 1 week, 2 weeks, 1 month, 2 months, and 3 months. The dissolved mass fraction (Dmf) of the mordenite was calculated using the following equation [62]:

$$Dmf = 1 - (m_t/m_0) \tag{2}$$

The dissolved mass fraction of the mineral fibers can be used as a useful parameter to identify the relatively biodurability of a fiber. According to the results of the tests carried out by Gualtieri et al. [51,62], during the 90 days of the dissolution test, the Dfm of biodurable fibers (i.e., erionite and amphibole asbestos) never exceeds the threshold of 0.4. In contrast, chrysotile, commonly recognized as a modest biodegradable mineral fiber, shows Dfm values above 0.4, just two weeks after the start of the dissolution test [51,62].

3. Results

A gallery of acquired SEM images of the sample is reported in Figure 2. Mordenite from Poona occurs as very elongated and prismatic crystals with average width ~7 μm and length up to 600 μm. Furthermore, in the sample, acicular and fibrous mordenite crystals with diameters <1 μm are also present, sometimes grouped in bundles of large size. SEM images also showed that mordenite fibers have flat surface, roughly prismatic morphology and a low flexibility. The summary statistics of the geometry of the fibers not grouped in bundles are reported in Table 1. The data display a wide range of length (L) values which ranging from 2.09 to 118 μm (average 20.8 μm), but the longer fibers predominate in the population. In fact, more than 75% of the fibers are longer than 11 μm. The widths of the fibers range overall between 0.11 and 3.34 μm, 50% of the fibers show a width of higher than 0.59 μm.

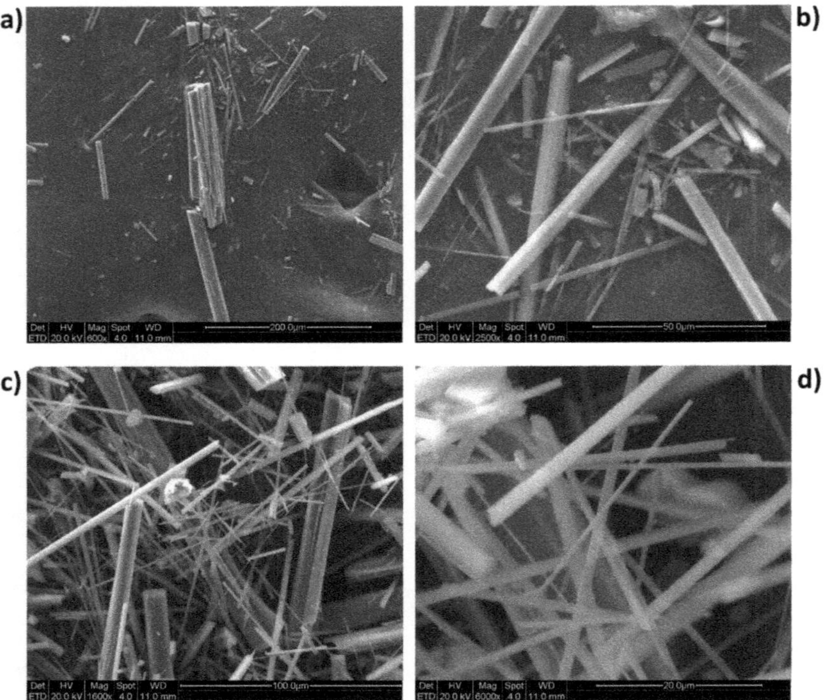

Figure 2. Selected examples of SEM microphotographs of the mordenite samples: (**a**,**b**) general overview of the sample; (**c**,**d**) close-up of very elongated and narrow mordenite crystals.

Table 1. Summary statistic of mordenite fibers geometry. L (length); W (width); Min (minimum); Max (maximum); σ (standard deviation). * Percentile is a value below which a given percentage of values in a dataset fall. Example: given a group of observations, the 25th percentile is the value that is greater or equal to 25% of the observations, i.e., the 25% of fibers have L ≤ 11.2 μm and W ≤ 0.61 μm.

		Percentiles *						
	Min	**5th**	**25th**	**50th**	**75th**	**95th**	**Max**	**σ**
L (μm)	2.09	3.52	11.2	20.8	36.7	56.1	118	20.5
W (μm)	0.11	0.18	0.61	0.95	1.24	1.80	3.34	0.58

According to the World Health Organization (WHO) guidelines, mineral fibers of respirable size with relevant biological activity (so-called "critical" fibers) are those equal to or longer than 5 μm and having diameters up to 3 μm, with a length/width ratio (aspect ratio) ≥ 3:1 [63]. 75% of the 200 observed fibers fit the WHO counting criteria concerning width and only 10% of the fibers are shorter than the length value set by the WHO. All of the investigated fibers showed an aspect ratio > 3.

As reported by Mattioli et al. [64], erionite fibers can exhibit two main morphologies. Fibrous erionite may occur with the typical "woolly" aspect, i.e., a thin and flexible fibrous habit that looks the same as wool. The fiber length may exceed 200 μm and the width ranging from 0.5 to 3 μm [64]. The fibers of the second type are acicular with diameter of 2–5 μm and lengths up to 30 μm [64]. They are generally grouped together in bundles up to 100 μm in length and variable width. As is the case for the mordenite, even the erionite bundles have great tendency to separate into straight and quite rigid thin fibrils (average diameter 0.5 μm) [64].

Applying the Equation (1) to a representative Poona mordenite fiber with theoretical density $\rho = 2.11/cm^3$, mean length 20.8 μm and average diameter (d) = 0.95 μm, the obtained Dae is 2.63 μm. Accordingly, it can be assumed that the mordenite fibers of the studied sample can easily penetrate through the respiratory tract and settle in the alveolar space [55]. Data concerning the aerodynamic diameter of erionite fibers are not available in the literature. However, since erionite is considered a positive carcinogen, its fibers are assumed to be classified as breathable/inhalable.

The results of the EDX analyses expressed as a percentage by weight with standard deviations are shown in Table 2. The data were obtained by averaging several point-analyses carried out on ten distinct mordenite crystals. Because EDX analyses of fibrous zeolite may be subject to critical errors, two criteria were adopted to assess the reliability of the data. Sum of the Si + Al atoms (47.79) close to the half of the oxygen atoms (48); total balance errors $E = [(Al - Al_{theor.})/Al_{theor.}] \times 100$ lower than 10%, where $Al_{theor.} = (Na + K) + 2(Ca + Mg)$.

Table 2. Elemental composition of Poona mordenite and erionite from Nevada and Karain [64]. * Chemical data are renormalized using a theoretical water content of 14.0 wt%. Standard deviation in brackets.

wt%	Mordenite Poona (India)	Erionite Nevada (USA)	Erionite Karain (Turkey)
SiO_2	68.82(1.57)	58.47(0.35)	56.66(0.32)
Al_2O_3	10.35(0.92)	14.30(0.21)	13.63(0.27)
Fe_2O_3	0.20(0.03)	0.13(0.05)	0.07(0.04)
MgO	0.08(0.01)	0.48(0.12)	0.02(0.10)
CaO	2.95(0.59)	3.35(0.23)	1.85(0.10)
Na_2O	3.17(0.98)	0.80(0.10)	2.48(0.16)
K_2O	0.21(0.09)	4.17(0.29)	5.49(0.45)
H_2O	14.0 *	18.46(0.45)	19.60(0.42)
Tot	99.98(1.84)	81.54	80.40

The chemical formula of mordenite from Poona, as determined from EDX spot analyses is:

$$(Na_{3.63}K_{0.16}Ca_{1.87}Mg_{0.07})[Al_{7.21}Si_{40.67}O_{96}]28\ H_2O$$

The unit formula was calculated on the basis of 96 oxygens. The Fe_2O_3 content was considered not incorporated in the zeolite. In Table 2, the chemical composition of fibrous erionite from Nevada (USA) and Karain (Turkey) is reported for comparison [64]. The mordenite has a content of tetrahedral atoms [Si/(Si + Al) = 0.85] within the known range of this mineral species (0.80–0.85). The Si/Al ratio of Poona mordenite (5.64) is significantly higher that the erionite from Nevada and Karain (3.47 and 3.52, respectively). The non-framework cation composition of mordenite is mostly Na-dominant, whereas, K is the dominant extra-framework cation in the structure of erionite samples.

The total Fe content is one of the most important factors for the fiber-induced patho-biological activity [35,48]. As far as erionite is concerned, several studies confirming the absence of iron in the structure of erionite [65–68]. Generally, iron is associated with iron-rich impurities (oxides, hydroxides or sulphates) present as particles or nanoparticles coating the surface of the erionite fibers [66]. Regarding mordenite, our investigations supporting the hypothesis that the Fe_2O_3 content as incorporated in the extra-zeolite phases. SEM micrographs and relative EDX analysis on the Poona mordenite sample revealing the presence of iron-rich particles at the surface of the fibers. Figure 3 displays a close-up of the fibers surface. On fibers surface, small particles and clusters, were observed. Relative EDX analysis showed that such particles are enriched in iron with respect to the bulk (Figure 4).

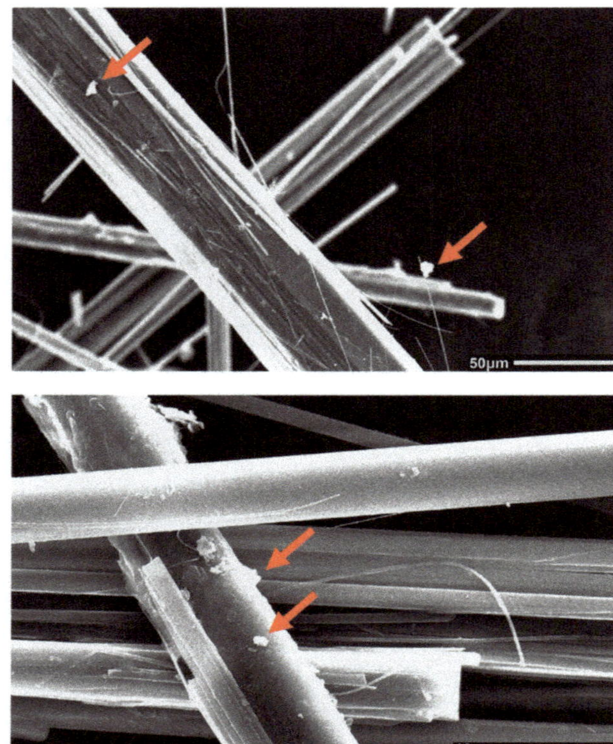

Figure 3. SEM images of mordenite fibers. Both images show mordenite fibers and particles on their surface (maybe hematite or goethite [65]). Some particles are pointed out with red arrows.

The zeta potential of fibrous zeolites was studied in depth by Pollastri et al. [50] and Gualtieri et al [16]. When double distilled water is chosen as the dispersion medium, the zeolite fibers show negative values of the zeta potential. According to the zeta potential model applied to zeolites [50], the presence of extra-framework cations determines a positive charge of the Stern layer, and a resulting attraction of negatively charged molecules to form the diffused corona. Consequently, the zeta potential will have negative values. The zeta potential of mordenite fibres under intracellular conditions (i.e., at pH 4.5) displays values in the range −10 to −20 mV; while in extracellular conditions (i.e., at pH 7.0) the values are in the range between −25 and −35 mV. These values are in agreement with those measured for erionite [50].

The raw dissolution curve of the investigated fibrous mordenite is reported in Figure 5. For comparison, the dissolution curves of UICC standard chrysotile and of fibrous erionite-Na from Jersey (Nevada, USA) is reported [62]. Chrysotile asbestos can be considered a representative member of mineral fibers with low biodurability [51,52,62], while erionite is considered a mineral fiber with high biodurability [51,62]. As showed in the Figure 5, the difference in the dissolution time of fibrous mordenite and erionite with respect to chrysotile asbestos is striking. After about 4000 h, the UICC chrysotile is completely dissolved and its Dmf value is greater than 0.8. Otherwise, after the same time interval, the Dmf of mordenite and erionite is lower than the threshold values of 0.4.

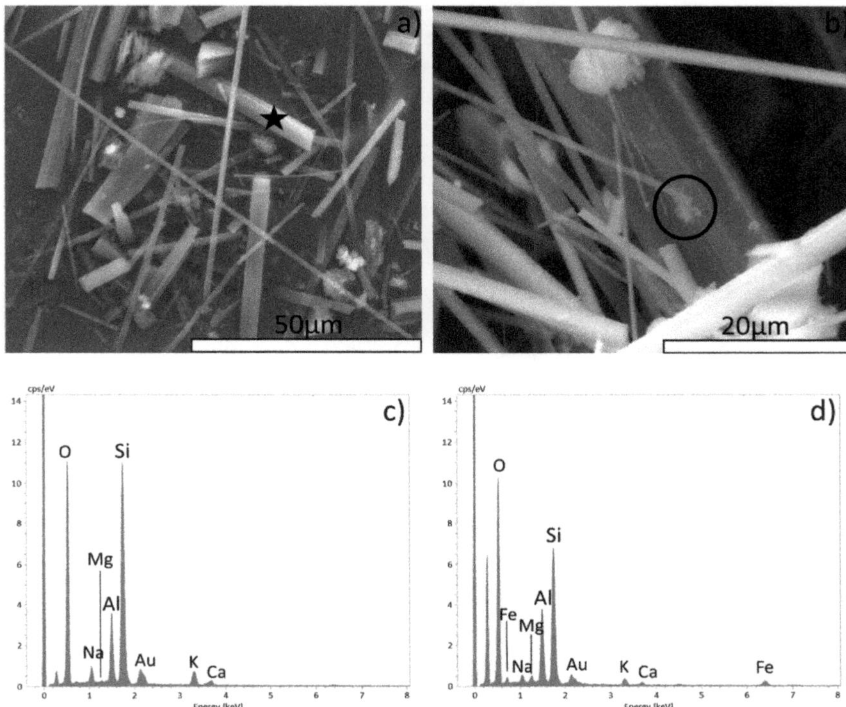

Figure 4. Gallery of SEM pictures and relative EDX spot analyses of the Poona sample. (**a**) Poona sample with acicular mordenite crystals: (**b**) Poona mordenite with microparticles enriched in iron present at the surface of the fibers. (**c**) Typical EDX spectrum recorded on mordenite surface free from the particles. No iron was found. (**d**) Representative EDX spectrum of the microparticles present in the fibers surface. A peak produced by iron is observed.

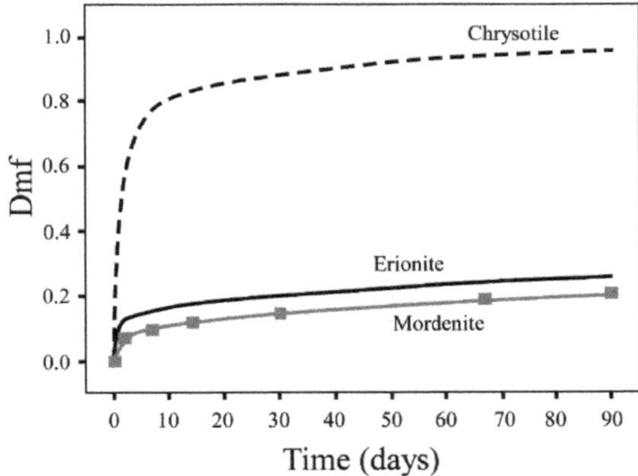

Figure 5. The raw dissolution curves of Poona mordenite. Trend of raw dissolution curves of chrysotile and fibrous erionite were reported for comparison [62].

4. Discussion

Despite the considerable efforts of the scientific community in the last decades, to date, a conclusive model explaining the toxicity mechanisms of asbestos has not been provided. Nevertheless, currently investigations based on in vitro and in vivo models suggested that a crucial role in the pathogenesis asbestos-related is played by the morphology of the fibers [35–38]. Chrysotile, asbestos amphibole, and fibrous erionite are known to possess the well-known "asbestos-like" habit, i.e., these minerals grow in a fibrous aggregate of curled, flexible, cylindrical, long, and thin crystals that readily separate. On the contrary, most of the Poona mordenite crystals have a stocky shape or acicular habit and null or low flexibility. However, in this context it is important to underline that natural mordenite zeolite group show a wide range of crystal habit. In particular, the diagenetic mordenite often has a morphology that resembles asbestos. Figure 6 shown a gallery of SEM image of mordenite from Ponza Island (Italy) [26] and Lovelock (Nevada, USA) [14]. Both mordenite samples occur in diagenetically altered rhyolitic rocks. Ponza mordenite occurs as an aggregate of very thin fibers that forms a "brush" texture (Figure 6a,b). Individual fibers have a slight degree of flexibility, W less than 0.5 μm and L in the range of 10–15 μm. Lovelock mordenite fibers appear with a fibrous crystal habit and its fibers have L ≈ 9 μm and W below 0.6 μm (Figure 6c,d).

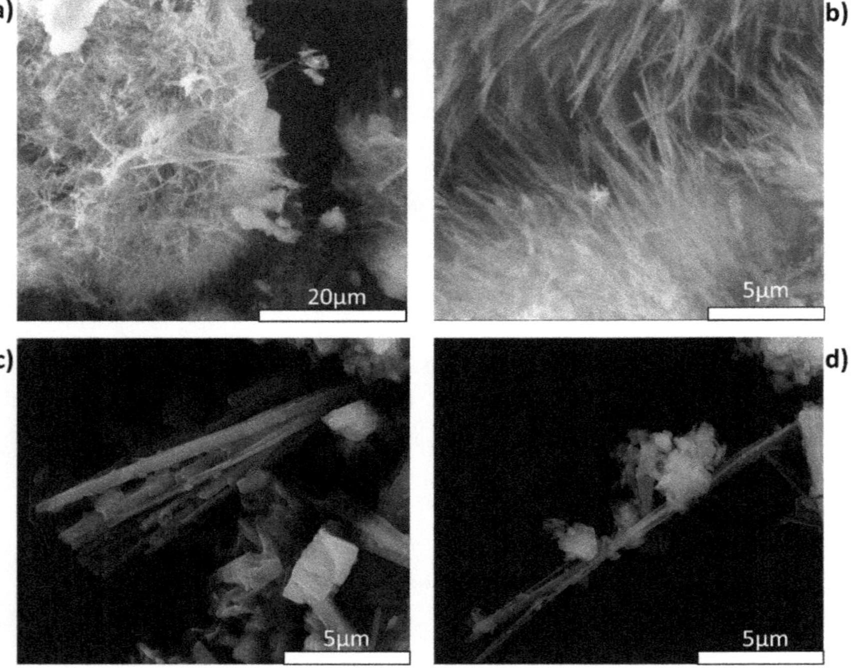

Figure 6. Gallery of SEM image of diagenetic mordenite. (**a**,**b**) Mordenite from Ponza Island (Italy). Fibers show a typical asbestiform habit. (**c**,**d**) Mordenite from Lovelock (Nevada, USA). The fibers bundles tend to separate into thin and quite rigid fibrils.

Although Poona mordenite does not show an asbestos-like habit and its morphology differs from that of fibrous erionite as described by Mattioli [64], its airborne fibers can be deposited in the alveolar space, where they high aspect ratio and biodurability may lead to the "frustrate phagocytosis" [38]. Since macrophages range in diameter from ~10 to 20 μm [37,39], they cannot completely engulf the long mordenite fibers (average length >20 μm). This leads to incomplete (frustrated) phagocytosis, which is characterized by the activation of the respiratory burst and the production of ROS and cytokines responsible for adverse effects at the cellular and subcellular levels. [4,35–39,48].

The zeta potential of mordenite fibres, as well as erionite, displays values in the range −10 to −35 mV. Such zeta potential values may affect several molecular and cellular biomechanics such as cross-talk and apoptosis mechanisms, fiber encapsulation by collagen and redox-activated iron-rich proteins, and the ROS generation [48,50].

As demonstrated by dissolution tests, mordenite possesses a high biodurability similar to that of erionite. According to Gualtieri et al. [16] the biodurability of zeolites is strictly related to the Si/Al framework ratio. In particular, the zeolite framework dissolution rate decreasing with increasing Si/Al ratio and zeolites such as erionite and mordenite with Si/Al ratio > 3, show little to null dissolution in an acid environment [69]. Given the close link between biodurability and biopersistence, mordenite fibers are expected to persist and accumulate in the lower respiratory tract after being inhaled. As reported by several studies, biopersistent fibers with a morphology similar to that of mordenite (L > 8 μm and W < 0.25 μm) are able to trigger a chronic inflammatory process and related adverse effects [36–38].

For both fibrous erionite and mordenite, iron is not structural but associated with mineral particles at the surface of the fibers. According to the model described for fibrous erionite in Gualtieri et al. [66], iron-rich particles may dissolve during phagocytosis when the zeolite fibers are engulfed in the phago-lysosome sacks at pH = 4–4.5. Dissolution may leave a residue of iron atoms at specific sites anchored to the surface windows of the zeolite channels. These surface sites can be potential catalysts for ROS generation. Therefore, as with erionite, mordenite has the potential to induce oxidative stress [36,70,71].

5. Conclusions

Among airborne pollutants, asbestos fibers have undoubtedly the worst reputation. However, recently, a new source of concern has arisen around fibrous zeolites and their potential health effect. Up to now, erionite is the only zeolite classified as carcinogen for humans by the IARC. In the present study we pointed out that, the fibrous mordenite has the same chemical-physical properties that are deemed to prompt adverse health effects. In particular, our investigations showed that mordenite, as with erionite, can display an asbestos-like shape and its fibers can be released as airborne particulate of breathable size. As demonstrated by the zeta potential analyses, the surface activity of mordenite is similar to erionite. Moreover, iron of mordenite is not structural but associated with secondary phases at the zeolite surface. The surface properties of fibrous mordenite, during fiber phagocytosis, may indirectly induce the formation of ROS with cyto-genotoxic effects. Based on the results of our dissolution tests is expected that mordenite is biodurable similar to erionite in the intracellular and extracellular environment. From this point of view, fibrous mordenite can be considered as a potential health hazard similar to fibrous erionite.

Currently, one of the main deposits of mordenite-rich exploded for industrial applications is the St. Cloud's Rome mordenite (a zeolite mineral) deposit in Rome, Oregon, in Malheur County (USA). Furthermore, mordenite occurs as the main phase in the zeolitized rocks of several countries such as Italy, Greece, Japan, and India. In this context, the implementation of a national/international policy for the elimination of potential exposure to dust containing mordenite fibers generated during extraction and grinding, as well as production and the use of material contains mordenite, should be taken into consideration by public authorities.

For a definitive and complete understanding of the potential toxicity of fibrous mordenite, in vitro toxicity test and animal carcinogenicity studies should be performed and compared to the outcome of our study to stimulate a critical evaluation and a possible reclassification of mordenite by the IARC.

Funding: This research was conducted under the project "Fibers a Multidisciplinary Mineralogical, Crystal-Chemical and Biological Project to Amend the Paradigm of Toxicity and Cancerogenicity of Mineral Fibers" (PRIN: PROGETTI DI RICERCA DI RILEVANTE INTERESSE NAZIONALE—Bando 2017—Prot. 20173X8WA4). The study was further supported in part by the project "CCIAARE—Attuazione di un progetto di accompagnamento delle imprese nell'ambito del progetto PID impresa 4.0" financed by the Camera di Commercio di Reggio Emilia (Italy).

Acknowledgments: Author warmly thank all the collaborators who contributed to the experimental part of the work: Alessandro Zoboli, Tommaso Giovanardi, Riccardo Fantini and Ilaria Baldini.

Conflicts of Interest: The author declares no conflict of interest. The funders had no role in the design of the study; in the collection, analyses, or interpretation of data; in the writing of the manuscript, or in the decision to publish the results.

References

1. Sisko, A.; Boffetta, P. *Occupational Cancers*; Springer Nature: London, UK, 2020; p. 640.
2. IARC (International Agency for Research on Cancer). Asbestos (chrysotile, amosite, crocidolite, tremolite, actinolite, and anthophyllite). *IARC Monogr. Eval. Carcinog Risks Hum. C* **2012**, *100*, 219–309.
3. Gualtieri, A.F. Mineral fibre-based building materials and their health hazards. In *Toxicity of Building Materials*; Woodhead Publishing: Cambridge, UK, 2012; pp. 166–195.
4. Carbone, M.; Kanodia, S.; Chao, A.; Miller, A.; Wali, A.; Weissman, D.; Adjei, A.; Baumann, F.; Boffetta, P.; Buck, B.; et al. Consensus Report of the 2015 Weinman International Conference on Mesothelioma. *J. Thorac. Oncol.* **2016**, *11*, 1246–1262. [CrossRef] [PubMed]
5. Middendorf, P.; Zumwalde, R.; Castellan, R.; Harper, M.; Wallace, W.; Stayner, L.; Castranova, V.; Hearl, F.; Sullivan, P.; Wallace, W.; et al. *Asbestos Fibers and Other Elongate Mineral Particles: State of the Science and Roadmap for Research*; NIOSH Current Intelligence Bulletin 62; National Institute for Occupational Safety and Health (NIOSH): Cincinnati, OH, USA, 2011; p. 174.
6. Baumann, F.; Buck, B.J.; Metcalf, R.V.; McLaurin, B.T.; Merkler, D.J.; Carbone, M. The Presence of Asbestos in the Natural Environment is Likely Related to Mesothelioma in Young Individuals and Women from Southern Nevada. *J. Thorac. Oncol.* **2015**, *10*, 731–737. [CrossRef] [PubMed]
7. Larson, T.C.; Antao, V.C.; Bove, F.J. Vermiculite worker mortality: Estimated effects of occupational exposure to Libby amphibole. *J. Occup. Environ. Med.* **2010**, *52*, 555–560. [CrossRef]
8. Comba, P.; Gianfagna, A.; Paoletti, L. Pleural mesothelioma cases in Biancavilla are related to a new fluoro-edenite fibrous amphibole. *Arch. Environ. Health* **2003**, *58*, 229–232. [CrossRef]
9. Di Giuseppe, D.; Harper, M.; Bailey, M.; Erskine, B.; Della Ventura, G.; Ardith, M.; Pasquali, L.; Tomaino, G.; Ray, R.; Mason, H.; et al. Characterization and assessment of the potential toxicity/pathogenicity of fibrous glaucophane. *Environ. Res.* **2019**, *178*, 108723. [CrossRef]
10. Petriglieri, J.R.; Laporte-Magoni, C.; Salvioli-Mariani, E.; Tomatis, M.; Gazzano, E.; Turci, F.; Cavallo, A.; Fubini, B. Identification and Preliminary Toxicological Assessment of a Non-Regulated Mineral Fiber: Fibrous Antigorite from New Caledonia. *Environ. Eng. Geosci.* **2020**, *26*, 89–97. [CrossRef]
11. Groppo, C.; Tomatis, M.; Turci, F.; Gazzano, E.; Ghigo, D.; Compagnoni, R.; Fubini, B. Potential toxicity of nonregulated asbestiform minerals: Balangeroite from the western Alps. Part 1: Identification and characterization. *J. Toxicol. Environ. Health A* **2005**, *68*, 1–19. [CrossRef]
12. Carbone, M.; Yang, H. Molecular pathways: Targeting mechanisms of asbestos and erionite carcinogenesis in mesothelioma. *Clin. Cancer Res.* **2012**, *18*, 598–604. [CrossRef]
13. Mattioli, M.; Giordani, M.; Arcangeli, P.; Valentini, L.; Boscardin, M.; Pacella, A.; Ballirano, P. Prismatic to asbestiform offretite from Northern Italy: Occurrence, morphology and crystal-chemistry of a new potentially hazardous zeolite. *Minerals* **2018**, *8*, 69. [CrossRef]
14. Zoboli, A.; Di Giuseppe, D.; Baraldi, C.; Gamberini, M.C.; Malferrari, D.; Urso, G.; Gualtieri, M.L.; Bailey, M.; Gualtieri, A.F. Characterisation of fibrous ferrierite in the rhyolitic tuffs at Lovelock, Nevada, USA. *Mineral. Mag.* **2019**, *83*, 577–586. [CrossRef]
15. Carbone, M.; Emri, S.; Dogan, A.U.; Steele, I.; Tuncer, M.; Pass, H.I.; Baris, Y.I. A mesothelioma epidemic in Cappadocia: Scientific developments and unexpected social outcomes. *Nat. Rev. Cancer* **2007**, *7*, 147–154. [CrossRef] [PubMed]

16. Gualtieri, A.F.; Gandolfi, N.B.; Passaglia, E.; Pollastri, S.; Mattioli, M.; Giordani, M.; Ottaviani, M.F.; Cangiotti, M.; Bloise, A.; Barca, D.; et al. Is fibrous ferrierite a potential health hazard? Characterization and comparison with fibrous erionite. *Am. Mineral.* **2018**, *103*, 1044–1055. [CrossRef]
17. Tschernich, R.W. *Zeolites of the World*; Geoscience Press: Tucson, AZ, USA, 1992; p. 563.
18. Stephenson, D.J.; Fairchild, C.I.; Buchan, R.M.; Dakins, M.E. A fiber characterization of the natural zeolite, mordenite: A potential inhalation health hazard. *Aerosol Sci. Technol.* **1999**, *30*, 467–476. [CrossRef]
19. Passaglia, E. The Crystal Chemistry of Mordenites. *Contrib. Mineral. Petrol.* **1975**, *50*, 65–77. [CrossRef]
20. Meier, W.M. The crystal structure of mordenite (ptilolite). *Z. Krist.* **1961**, *115*, 439–450. [CrossRef]
21. How, D.C.L. On mordenite, a new mineral from the trap of Nova Scotia. *J. Chem. Soc.* **1984**, *17*, 100–104.
22. Simoncic, P.; Armbruster, T. Peculiarity and defect structure of the natural and synthetic zeolite mordenite: A single-crystal X-ray study. *Am. Mineral.* **2004**, *89*, 421–431. [CrossRef]
23. Martucci, A.; Sacerdoti, M.; Cruciani, G.; Dalconi, C. In situ time resolved synchrotron powder diffraction study of mordenite. *Eur. J. Mineral.* **2003**, *15*, 485–493. [CrossRef]
24. Passaglia, E.; Sheppard, R.A. The crystal chemistry of zeolites. In *Natural Zeolites: Occurrence, Properties Application*; Bish, D.L., Ming, D.W., Eds.; Mineralogical Society of America and Geochemical Society: Washington, DC, USA, 2001; pp. 69–116.
25. Zhang, L.; Xie, S.; Xin, W.; Li, X.; Liu, S.; Xu, L. Crystallization and morphology of mordenite zeolite influenced by various parameters in organic-free synthesis. *Mater. Res. Bull.* **2011**, *46*, 894–900. [CrossRef]
26. Passaglia, E.; Artioli, G.; Gualtieri, A.; Carnevali, R. Diagenetic mordenite from Ponza, Italy. *Eur. J. Mineral.* **1995**, *7*, 429–438. [CrossRef]
27. Godelitsas, A.; Gamaletsos, P.; Roussos-Kotsis, M. Mordenite-bearing tuffs from Prassa quarry, Kimolos Island, Greece. *Eur. J. Mineral.* **2010**, *22*, 797–811. [CrossRef]
28. Negishi, T. Mordenite in the tuffs of the Shirasawa District, Miyagl Prefecture. *J. Jpn. Assoc. Mineral. Petrol. Econ. Geol.* **1972**, *67*, 29–34. [CrossRef]
29. IARC (International Agency for Research on Cancer). Silica, some silicates, coal dust and para-aramid fibrils. *IARC Monogr. Eval. Carcinog. Risks Hum. C* **1997**, *68*, 307–337.
30. IARC Monographs on the IDENTIFICATION of CARCINOGENIC Hazards to Humans. Available online: https://monographs.iarc.fr/wp-content/uploads/2019/07/Preamble-2019.pdf (accessed on 2 August 2020).
31. Suzuki, Y. Carcinogenic and fibrogenic effects of zeolites: Preliminary observations. *Environ. Res.* **1982**, *27*, 433–445. [CrossRef]
32. Suzuki, Y.; Kohyama, N. Malignant mesothelioma induced by asbestos and zeolite in the mouse peritoneal cavity. *Environ. Res.* **1984**, *35*, 277–292. [CrossRef]
33. Tátrai, E.; Bácsy, E.; Kárpáti, J.; Ungváry, G. On the examination of the pulmonary toxicity of mordenite in rats. *Pol. J. Occup. Med. Environ. Health* **1992**, *5*, 237–243.
34. Adamis, Z.; Tátrai, E.; Honma, K.; Six, É.; Ungváry, G. In Vitro and in Vivo Tests for Determination of the Pathogenicity of Quartz, Diatomaceous Earth, Mordenite and Clinoptilolite. *Ann. Occup. Hyg.* **2000**, *44*, 67–74. [CrossRef]
35. Gualtieri, A.F. Towards a quantitative model to predict the toxicity/pathogenicity potential of mineral fibers. *Toxicol. Appl. Pharm.* **2018**, *361*, 89–98. [CrossRef]
36. Carbone, M.; Adusumilli, P.S.; Alexander, H.R., Jr.; Baas, P.; Bardelli, F.; Bononi, A.; Bueno, R.; Felley-Bosco, E.; Galateu-Salle, F.; Jablons, D.; et al. Mesothelioma: Scientific clues for prevention, diagnosis, and therapy. *CA-Cancer J. Clin.* **2019**, *69*, 402–429. [CrossRef]
37. Stanton, M.F.; Layard, M.; Tegeris, A.; Miller, E.; May, M.; Morgan, E.; Smith, A. Relation of particle dimension to carcinogenicity in amphibole asbestoses and other fibrous minerals. *J. Natl. Cancer Inst.* **1981**, *67*, 965–975. [PubMed]
38. Donaldson, K.; Murphy, F.A.; Duffin, R.; Poland, C.A. Asbestos, carbon nanotubes and the pleural mesothelium: A review of the hypothesis regarding the role of long fibre retention in the parietal pleura, inflammation and mesothelioma. *Part. Fibre Toxicol.* **2010**, *7*, 5. [CrossRef] [PubMed]
39. Di Giuseppe, D.; Zoboli, A.; Vigliaturo, R.; Gieré, R.; Bonasoni, M.P.; Sala, O.; Gualtieri, A.F. Mineral fibres and asbestos bodies in human lung tissue: A case study. *Minerals* **2019**, *9*, 618. [CrossRef]
40. Padmore, T.; Stark, C.; Turkevich, L.A.; Champion, J.A. Quantitative analysis of the role of fiber length on phagocytosis and inflammatory response by alveolar macrophages. *BBA-Gen. Subj.* **2017**, *1861*, 58–67. [CrossRef] [PubMed]

41. Deng, Z.J.; Liang, M.L.; Tóth, I.; Monteiro, M.J.; Michin, R.F. Molecular interaction of poly (acrylic acid) gold nanoparticles with human fibrinogen. *ACS Nano* **2012**, *6*, 8962–8969. [CrossRef]
42. Harris, R.L.; Timbrell, V. Relation of alveolar deposition to the diameter and length of glass fibres. In Proceedings of the Inhaled Particles IV International Symposium Organized by the British Occupational Hygiene Society, Edinburgh, UK, 22–26 September 1977; Pergamon Press: Oxford, UK, 1977; p. 411.
43. Yeh, H.C.; Phalen, R.F.; Raabe, O.G. Factors influencing the deposition of inhaled particles. *Environ. Health Perspect.* **1976**, *15*, 147–156. [CrossRef]
44. Aust, A.E.; Cook, P.M.; Dodson, R.F. Morphological and chemical mechanisms of elongated mineral particle toxicities. *J. Toxicol. Environ. Health Part B* **2011**, *14*, 40–75. [CrossRef]
45. Bonneau, L.; Suquet, H.; Malard, C.; Pezerat, H. Studies on surface properties of asbestos: I. active sites on surface of chrysotile and amphiboles. *Environ. Res.* **1986**, *41*, 251–267. [CrossRef]
46. Turci, F.; Tomatis, M.; Lesci, I.G.; Roveri, N.; Fubini, B. The iron-related molecular toxicity mechanism of synthetic asbestos nanofibres: A model study for high-aspect-ratio nanoparticles. *Chem. Eur. J.* **2011**, *17*, 350–358. [CrossRef]
47. Hardy, J.A.; Aust, A.E. Iron in asbestos chemistry and carcinogenicity. *Chem. Rev.* **1995**, *95*, 97–118. [CrossRef]
48. Gualtieri, A.F.; Andreozzi, G.B.; Tomatis, M.; Turci, F. Iron from a geochemical viewpoint. Understanding toxicity/pathogenicity mechanisms in iron-bearing minerals with a special attention to mineral fibers. *Free Radic. Bio. Med.* **2019**, *133*, 21–37. [CrossRef] [PubMed]
49. Pollastri, S.; Gualtieri, A.F.; Ignatyev, K.; Strafella, E.; Pugnaloni, A.; Croce, A. Stability of mineral fibres in contact with human cell cultures. An in situ μXANES, μXRD and XRF iron mapping study. *Chemosphere* **2016**, *164*, 547–557. [CrossRef] [PubMed]
50. Pollastri, S.; Gualtieri, A.F.; Gualtieri, M.L.; Hanuskova, M.; Cavallo, A.; Gaudino, G. The zeta potential of mineral fibres. *J. Hazard. Mater.* **2014**, *276*, 469–479. [CrossRef] [PubMed]
51. Gualtieri, A.F.; Pollastri, S.; Gandolfi, N.B.; Gualtieri, M.L. In vitro acellular dissolution of mineral fibres: A comparative study. *Sci. Rep.* **2018**, *8*, 7071. [CrossRef] [PubMed]
52. Bernstein, D.M.; Chevalier, J.; Smith, P. Comparison of Calidria chrysotile asbestos to pure tremolite: Final results of the inhalation biopersistence and histopathology examination following short-term exposure. *Inhal. Toxicol.* **2005**, *17*, 427–449. [CrossRef] [PubMed]
53. Sukheswala, R.N.; Avasia, R.K.; Gangopadhyay, M. Zeolites and associated secondary minerals in the Deccan Traps of Western India. *Mineral. Mag.* **1974**, *39*, 658–671. [CrossRef]
54. National Institute of Mental Health. ImageJ. Available online: https://imagej.nih.gov/ij/ (accessed on 14 March 2020).
55. Gualtieri, A.F.; Mossman, B.T.; Roggli, V.L. Towards a general model for predicting the toxicity and pathogenicity of minerals fibres. In *Mineral Fibres: Crystal Chemistry, Chemical-Physical Properties, Biological Interaction and Toxicity*; Gualtieri, A.F., Ed.; European Mineralogical Union: London, UK, 2017; pp. 501–526.
56. Heyder, J.; Gebhart, J.; Rudolf, G.; Schiller, C.F.; Stahlhofen, W. Deposition of particles in the human respiratory tract in the size range 0.005–15 μm. *J. Aerosol Sci.* **1986**, *17*, 811–825. [CrossRef]
57. French, C.A. Respiratory tract. In *Cytology: Diagnostic Principles and Clinical Correlates;* Cibas, E.S., Ducatman, B.S., Eds.; Elsevier Saunders: Philadelphia, PA, USA, 2009; pp. 65–103.
58. Gonda, I.; Abd El Khalik, A.F. On the calculation of aerodynamic diameters of fibers. *Aerosol. Sci. Technol.* **1985**, *4*, 233–238. [CrossRef]
59. Gu, Y.; Li, D. The ζ-potential of glass surface in contact with aqueous solutions. *J. Colloid Interface Sci.* **2000**, *226*, 328–339. [CrossRef]
60. Sze, A.; Erickson, D.; Ren, L.; Li, D. Zeta-potential measurement using the Smoluchowski equation and the slope of the current–time relationship in electroosmotic flow. *J. Colloid Interface Sci.* **2003**, *261*, 402–410. [CrossRef]
61. Stern, O. The theory of the electrolytic double-layer. *Z. Elektrochem.* **1924**, *30*, 508–516.
62. Gualtieri, A.F.; Lusvardi, G.; Zoboli, A.; Di Giuseppe, D.; Gualtieri, M.L. Biodurability and release of metals during the dissolution of chrysotile, crocidolite and fibrous erionite. *Environ. Res.* **2019**, *171*, 550–557. [CrossRef] [PubMed]
63. WHO (World Health Organization). Determination of Airborne Fibre Number Concentrations. Available online: http://apps.who.int/iris/bitstream/10665/41904/1/9241544961.pdf (accessed on 4 August 2020).

64. Mattioli, M.; Giordani, M.; Dogan, M.; Cangiotti, M.; Avella, G.; Giorgi, R.; Dogan, U.A.; Ottaviani, M.F. Morpho-chemical characterization and surface properties of carcinogenic zeolite fibers. *J. Hazard. Mater.* **2016**, *306*, 140–148. [CrossRef] [PubMed]
65. Croce, A.; Allegrina, M.; Rinaudo, C.; Gaudino, G.; Yang, H.; Carbone, M. Numerous iron-rich particles lie on the surface of erionite fibers from Rome (Oregon, USA) and Karlik (Cappadocia, Turkey). *Microsc. Microanal.* **2015**, *21*, 1341–1347. [CrossRef]
66. Gualtieri, A.F.; Gandolfi, N.B.; Pollastri, S.; Pollok, K.; Langenhorst, F. Where is iron in erionite? A multidisciplinary study on fibrous erionite-Na from Jersey (Nevada, USA). *Sci. Rep.* **2016**, *6*, 37981. [CrossRef]
67. Ballirano, P.; Andreozzi, G.B.; Dogan, M.; Dogan, A.U. Crystal structure and iron topochemistry of erionite-K from Rome, Oregon, USA. *Am. Mineral.* **2009**, *94*, 1262–1270. [CrossRef]
68. Matassa, R.; Familiari, G.; Relucenti, M.; Battaglione, E.; Downing, C.; Pacella, A.; Cametti, G.; Ballirano, P. A deep look into erionite fibres: An electron microscopy investigation of their self-assembly. *Sci. Rep.* **2015**, *5*, 16757. [CrossRef]
69. Hartman, R.L.; Fogler, H.S. Understanding the dissolution of zeolites. *Langmuir* **2007**, *23*, 5477–5484. [CrossRef]
70. Urano, N.; Yano, E.; Evans, P.H. Reactive oxygen metabolites produced by the carcinogenic fibrous mineral erionite. *Environ. Res.* **1991**, *54*, 74–81. [CrossRef]
71. Pacella, A.; Cremisini, C.; Nardi, E.; Montereali, M.R.; Pettiti, I.; Giordani, M.; Mattioli, M.; Ballirano, P. Different erionite species bind iron into the structure: A potential explanation for fibrous erionite toxicity. *Minerals* **2018**, *8*, 36. [CrossRef]

Communication

Designer Synthesis of Ultra-Fine Fe-LTL Zeolite Nanocrystals

Fen Zhang [1,*], Yunhong Luo [1], Lei Chen [2], Wei Chen [1], Yin Hu [1], Guihua Chen [1], Shengyong You [1] and Weiguo Song [1,3]

1 Institute of Applied Chemistry, Jiangxi Academy of Sciences, Nanchang 330096, China; 201814060201@stu.ncst.edu.cn (Y.L.); chenwei@jxas.ac.cn (W.C.); huyin@jxas.ac.cn (Y.H.); chenguihua@jxas.ac.cn (G.C.); youshengyong@jxas.ac.cn (S.Y.); wsong@iccas.ac.cn (W.S.)
2 College of Chemistry and Chemical Engineering, Jiangxi Normal University, Nanchang 330022, China; chenlei1999@jxnu.edu.cn
3 Institute of Chemistry, Chinese Academy of Sciences, Beijing 100190, China
* Correspondence: zhangfen@jxas.ac.cn

Received: 28 August 2020; Accepted: 14 September 2020; Published: 15 September 2020

Abstract: Nanosized zeolites with larger external surface area and decreased diffusion pathway provide many potential opportunities in adsorption, diffusion, and catalytic applications. Herein, we report a designer synthesis of ultra-fine Fe-LTL zeolite nanocrystals under very mild synthesis conditions. We prepared Fe-LTL zeolite nanocrystals synthesized using L precursor. The precursor is aging at room temperature to obtain zeolite L nuclei. In order to investigate more details of Fe-LTL zeolite nanocrystals, various characterizations including X-ray diffraction (XRD), inductively coupled plasma (ICP), diffuse reflectance ultraviolet-visible (UV-Vis) spectroscopy, confirm the tetrahedral Fe^{3+} species in the zeolite framework. Besides, scanning electron microscope (SEM), Fourier transform infrared spectrometer (FT-IR), dynamic light scattering (DLS) indicate that the average particle size of Fe-LTL zeolite crystals is approximately 30 nm. Thus, ultra-fine Fe-LTL zeolite with large external surface area and shorter diffusion pathway to the active sites might have great potential in the near future.

Keywords: zeolite L precursor; Fe-LTL zeolite; ultra-fine

1. Introduction

Zeolites as promising adsorbents, ion-exchangers, and catalysts have been widely used in the fields of catalysis, separations, and adsorption [1–14]. In particular, they have accelerated the development of heterogeneous catalysis and separation processes because of their large internal surface area, unique channel systems, high pore volume, and adjustable active sites. The chemical composition and pore architecture of zeolites are important, however, the precise control of their crystallite morphology also has a significant influence on their applications in catalysis and adsorption. As the crystal size decreases below 100 nm, the zeolite external surface area, which is distinct from the internal pore surface and is negligible for micron-sized zeolites, increases dramatically, resulting in zeolites with over 25% of the total surface area on the external surface [15]. Shorter diffusion path length of nanosized zeolites is another advantage compared to micron-sized zeolites. The improved properties of nanosized zeolites for adsorption and diffusion provide many potential opportunities for their applications in environmental catalysis, environmental remediation, decontamination, and drug delivery [16]. Thus, zeolite nanocrystals have been paid much attention because of their widespread applications, the strategies for the preparation of zeolite nanocrystals are necessary and important [17–21]. Nonetheless, designer synthesis of pure zeolite nanocrystals is still challenging at present.

The structure of zeolite LTL is hexagonal (space group P6/mmm) with unit cell constants a = 18.4 Å and c = 7.5 Å [22]. The linkages of the cancrinite cages by double 6-rings (D6R) lead to the formation of columns in the c-direction and thus give rise to 12-membered rings with a free diameter of 7.1 Å [23]. Over past 20 years, zeolite L has drawn much attention for their applications in catalytic processes [24–27], ion-exchange and separations [28,29], and photonic devices [30,31]. Notably, iron catalysts, supported on different solids, have been widely applied in Fischer-Tropsch reaction for many years. Combining the advantages of iron and LTL zeolites, it is very meaningful to synthesize ultra-fine Fe-LTL zeolite nanocrystals, which are potentially significant for catalytic applications.

Herein, we report a designer synthesis of ultra-fine Fe-LTL zeolite nanocrystals under very mild synthesis conditions. We prepared Fe-LTL zeolite nanocrystals synthesized using L precursor. The precursor is aging at room temperature to obtain zeolite L nuclei. Notably, the average crystal size of Fe-LTL zeolite nanocrystals we obtained is approximately 30 nm, which is smaller than the previous reports. In order to investigate more details of Fe-LTL zeolite nanocrystals, various characterizations including X-ray diffraction (XRD), diffuse reflectance ultraviolet-visible (UV-Vis) spectroscopy, confirm the tetrahedral Fe^{3+} species in the zeolite framework. Besides, scanning electron microscope (SEM), Fourier transform infrared spectrometer (FT-IR), dynamic light scattering (DLS) indicate the nanoscale of zeolite crystals. Ultra-fine Fe-LTL zeolite with large external surface area and shorter diffusion pathway to the active sites might have great potential in the near future.

2. Materials and Methods

2.1. Materials

The following chemicals were utilized: sodium aluminate ($NaAlO_2$, 36.6% Na_2O, and 43.3% Al_2O_3, Sinopharm Chemical Reagent Co., Ltd., Shanghai, China), sodium hydroxide (NaOH, AR, 96%, Sinopharm Chemical Reagent Co., Ltd., Shanghai, China), potassium hydroxide (KOH, AR, 85%, Sinopharm Chemical Reagent Co., Ltd., Shanghai, China), silica sol (LUDOX HS-40, 40% SiO_2 in water, Sigma Aldrich, St. Louis, MO, USA), aluminum sulfate ($Al_2(SO_4)_3{\cdot}18H_2O$, AR, 99%, Sinopharm Chemical Reagent Co., Ltd., Shanghai, China), sodium silicate solution (253.8 g/L SiO_2, 77.8 g/L Na_2O of 1 L sodium silicate solution), ferric chloride hexahydrate ($FeCl_3{\cdot}6H_2O$, AR, Sinopharm Chemical Reagent Co., Ltd., Shanghai, China)

2.1.1. Synthesis of Zeolite L Precursor

As a typical run, zeolite L precursor was prepared by mixing 10 mL deionized H_2O, 0.472 g sodium aluminate and 3.76 g potassium hydroxide under stirring for 10 min, 8.4 g silica sol was added later. After the solution was stirred for 1 h, 0.24 g sodium hydroxide was finally added, followed by aging at room temperature for 72 h, giving a clear solution. The molar ratio of the precursor is $3.0Na_2O/14.3K_2O/Al_2O_3/28SiO_2/413H_2O$.

2.1.2. Synthesis of Fe-L Samples

Zeolite Fe-L nanocrystals were hydrothermally synthesized at the temperature of 80 °C for 3 days with $2.1Na_2O/3.4K_2O/Al_2O_3/15SiO_2/0.6FeCl_3/250H_2O$ molar ratios of initial synthesis gels in the presence of zeolite L precursor. As a typical run, 1.16 g potassium hydroxide was added to 10.9 g sodium silicate solution, followed by introducing 1.4 g deionized H_2O. After stirring for about 1 h, 1.665 g aluminum sulfate was added to the mixture, followed by the addition of 0.4 g $FeCl_3{\cdot}6H_2O$, finally the last addition of 2.0 mL zeolite L precursor. After being stirred for 2 h at room temperature, the mixture was transferred into an autoclave to crystallize at 80 °C for 3 days. The products were collected by filtration, washed with deionized H_2O, and dried in air, respectively. The H-form of the samples was prepared by triple ion-exchange with 1M of NH_4NO_3 solution at 80 °C for 1 h and calcination at 550 °C for 5 h.

2.1.3. Synthesis of Conventional Fe-L Zeolite

Conventional Fe-L crystal was hydrothermally synthesized at the temperature of 180 °C for 1 day with $1.56K_2O/Al/6.25SiO_2/0.25FeCl_3/90H_2O$ molar ratios of starting gels. The product was collected by filtration, washed with deionized H_2O, and dried in air. The H-form of the sample was prepared by triple ion-exchange with 1M of NH_4NO_3 solution at 80 °C for 1 h and calcination at 550 °C for 5 h.

2.2. Methods

Powder X-ray diffraction (XRD) (Rigaku, Tokoyo, Japan) data were experimented at room temperature with a Rigaku Ultimate VI X-ray diffractometer (40 kV, 40 mA) using CuKα (λ = 1.5406 Å) radiation. Crystallinity of samples is calculated by area of peaks at 5.7°, 19°, 22°, and 28° from the XRD patterns. Scanning electron microscopy (SEM) (Hitachi, Tokoyo, Japan) experiments were carried out on Hitachi SU-8010 and SU-1510 electron microscopes. Dynamic light scattering (DLS) (Malvern, Malvern City, UK) experiments were measured using a Malvern Instrument Zetasizer Nano ZS90. Nitrogen sorption experiments (Micromeritics, Atlanta, GA, USA) were performed on a Micromeritics TriStar II at −196 °C. The pore volume and surface area were calculated using the t-plot and BET methods. UV-Vis analysis, using $BaSO_4$ as the internal standard sample was performed on a Perkin-Elmer Lambda 20 spectrometer (Perkin Elmer, Waltham, MA, USA). Fourier transform infrared spectrometer (FT-IR) was measured on a FTIR 7600 Spectrometer (Lambda, Sydney, Australia). The sample composition was tested by ICP with a Perkin-Elmer 3300DV emission spectrometer.

3. Results and Discussion

Figure 1a–d shows the XRD patterns of as-synthesized and calcined Fe-LTL nanocrystals (Fe-LTL-N), conventional Fe-LTL (Fe-LTL-C), and calcined Fe-LTL-N, Fe-LTL-C samples. These samples give a series of characteristic peaks of the LTL framework structure. The peaks of Fe-LTL-N are broader than those of Fe-LTL-C, indicating the crystal size of Fe-LTL-N samples is nanoscale. After calcination, these peaks are basically remained, indicating their good thermal stability.

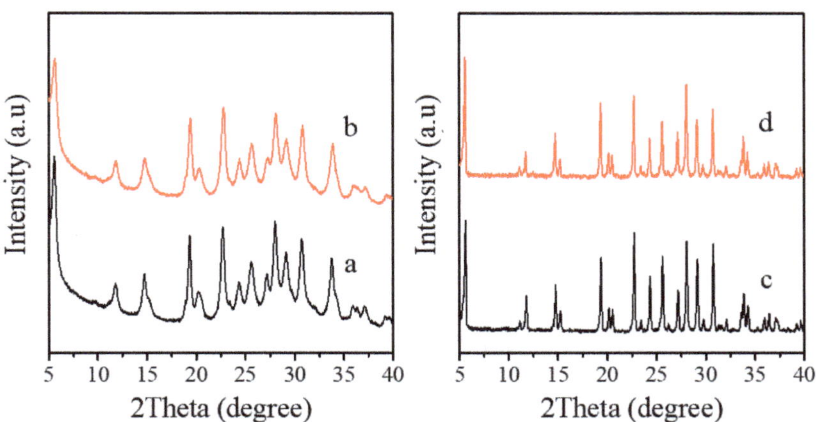

Figure 1. XRD patterns of (**a**) as-synthesized Fe-LTL-N, (**b**) calcined Fe-LTL-N, (**c**) as-synthesized Fe-LTL-C, (**d**) calcined Fe-LTL-C, respectively.

Figure 2 gives the SEM images of as-synthesized Fe-LTL-N, Fe-LTL-C samples. Figure 2a exhibits nanocylindrical morphology of Fe-LTL-N with size of 20–50 nm, while the morphology of Fe-LTL-C is micron cylindrical (Figure 2b). It is well consistent with the result from XRD testing of Fe-LTL-N and Fe-LTL-C samples. The particle size distribution (Figure 3) shows the average particle size of Fe-LTL-N is 30 nm. Notably, agglomeration of Fe-LTL-N is also exhibited in Figure 2a. Zeolite nanocrystals with smaller sizes have higher relative surface area and higher energy. To minimize its surface energy the nanocrystals create agglomeration.

Figure 2. SEM images of as-synthesized (**a**) Fe-LTL-N, (**b**) Fe-LTL-C.

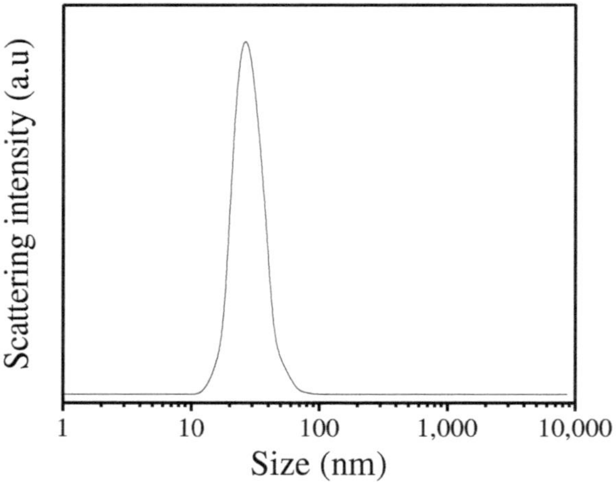

Figure 3. DLS curve of as-synthesized Fe-LTL-N.

The porosity and specific surface area of Fe-LTL-N and Fe-LTL-C samples were characterized by nitrogen adsorption measurement (Figure 4a,b), which displays the related parameters, as shown in Table 1. Fe-LTL-N sample shows a mixed isotherm curve of Type I and IV with a large H1-type hysteresis. As the Table 1 shows, the total pore volume of Fe-LTL-N is 0.38 cm^3g^{-1}, while the conventional Fe-LTL zeolite shows relatively low pore volume at only 0.19 cm^3g^{-1}. The results are consistent with that of XRD patterns and SEM images.

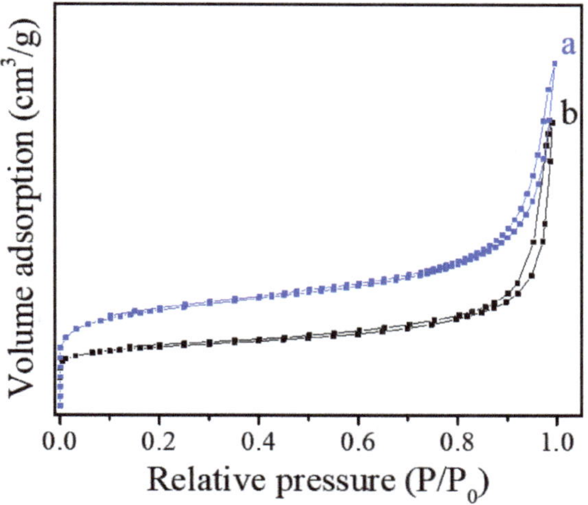

Figure 4. N_2 isotherm sorption of (**a**) Fe-LTL-N, (**b**) Fe-LTL-C.

Table 1. Textural parameters of Fe-LTL samples.

Sample	S_{BET} (m^2g^{-1})	S_{micro} (m^2g^{-1})	V_{total} (cm^3g^{-1})	V_{micro} (cm^3g^{-1})
Fe-LTL-N	285	228	0.38	0.10
Fe-LTL-C	142	128	0.19	0.12

Figure 5 displays the UV-vis spectrum of Fe-LTL-N, exhibiting a strong absorption band at 220 nm with a distinct shoulder at 245 nm and weak bands at 379, 419, and 446 nm, which is in good agreement with the previously reported results [32]. Both bands in the 220–250 nm range correspond to tetrahedral Fe in the zeolite, where the band at 220 nm may be due to a different specific environment of Fe within the zeolite framework. The absorption bands located at 379, 419, and 446 nm are assigned to tetrahedral Fe^{3+} in the zeolite framework [33]. There were no obvious absorption peaks at near 320 nm, indicating the absence of aggregated FeO_x species in the extra-framework.

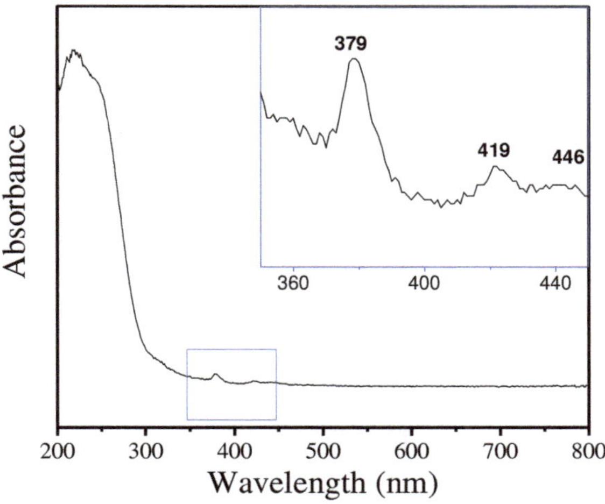

Figure 5. UV-vis spectrum of as-synthesized Fe-LTL-N.

Table 2 presents a systematic investigation on crystallization of Fe-LTL zeolite nanocrystals at 80 °C from aluminosilicate gels in the presence of L precursor. Interestingly, when the K_2O/SiO_2 in the gels is optimized from 0.122 to 0.300, products of amorphous phase are obtained such as, a mixture of amorphous phase with LTL structure, pure LTL, a mixture of LTL and amorphous (Figure 6A). However, as the K_2O/SiO_2 ratio in the synthesis gel increases, there is significant broadening and a decrease in height for peaks. Notably, sample obtained under Run 6 with poorer crystallinity shows decrease in height of resolved IR spectrum compared with Fe-LTL-N sample (Figure S1). Combined with the XRD pattern of sample (Run 6), it is probably a mixture of LTL and amorphous [34]. Moreover, the suitable K_2O/SiO_2 ratio for crystallization of pure LTL structure is changed from 0.229 to 0.265. It is found that the SiO_2/Al_2O_3 ratio has a great influence on the crystallization of LTL zeolite. When the SiO_2/Al_2O_3 ratio is less than 15, it is normally amorphous phase. When this ratio is distributed in the range of 15–25, a pure phase of LTL structure is usually formed (Figure 6B). Figure S1 also shows resolved IR spectra of samples obtained under conditions of Run 10 and 11. Obviously, sample (Run 10) gives similar IR spectrum as that of sample (Run 6), while sample (Run 11) has a further decrease in height for the peaks. Therefore, the sample synthesized under Run 10 and 11 are probably mixture of LTL and amorphous. Furthermore, the $FeCl_3/SiO_2$ ratio also influences the crystallization of Fe-LTL zeolite. The Fe content and Si/Al ratios of samples is listed in Table 3. Clearly, samples obtained with more iron content in the initial gel are likely to exhibit higher iron content and Si/Al ratios. As the Figure 6C shows, the increase of iron content up to a certain amount in the initial synthesis gels tends to produce samples with lower crystallinity. When this ratio in the initial gel reaches to 0.07, it is formed a mixture of LTL with amorphous. It has been claimed that the presence of greater concentrations of iron in the gel inhibited the formation of the cancrinite cage due to the larger Fe-O bond length compared to Al-O [34]. Moreover, cancrinite cage is crucial in the formation of LTL zeolites. Thus, the amount of iron in the synthesis gels really has a major impact on crystallinity of Fe-LTL zeolites. Moreover, crystallization temperature has a great effect on the crystallization of zeolite crystals. Figure 7 shows the XRD patterns of Fe-LTL zeolite nanocrystals synthesized with the same ratios of Run 4 at different crystallization temperatures. Reasonably, higher crystallization of Fe-LTL zeolites tends to be obtained at higher crystallization temperatures.

Table 2. Impact of synthesis conditions on crystallization of Fe-LTL zeolite nanocrystals (Na_2O/SiO_2 = 0.14, H_2O/SiO_2 = 250, at 80 °C for 3 d).

Run	K_2O/SiO_2	SiO_2/Al_2O_3	$FeCl_3/SiO_2$	Product [a]	Crystallinity [b] (%)	Yield [c] (%)
1	0.122	15	0.04	Amor	/	/
2	0.158	15	0.04	Amor	/	/
3	0.193	15	0.04	Amor+LTL	/	/
4	0.229	15	0.04	LTL	100	40
5	0.265	15	0.04	LTL	65	56
6	0.300	15	0.04	LTL + Amor	/	/
7	0.229	10	0.04	Amor	/	/
8	0.229	20	0.04	LTL	65	36
9	0.229	25	0.04	LTL	80	30
10	0.229	30	0.04	LTL + Amor	/	/
11	0.229	40	0.04	LTL + Amor	/	/
12	0.229	15	0.03	LTL	100	30
13	0.229	15	0.05	LTL	100	67
14	0.229	15	0.06	LTL	100	72
15	0.229	15	0.07	LTL + Amor	/	/
16	0.229	15	0.08	LTL + Amor	/	/

[a] The phase appearing first is dominant. [b] Run 4 was designated as 100% crystallinity. [c] yield (%) = quantity of product*100/quantity of SiO_2 in the synthesis gel.

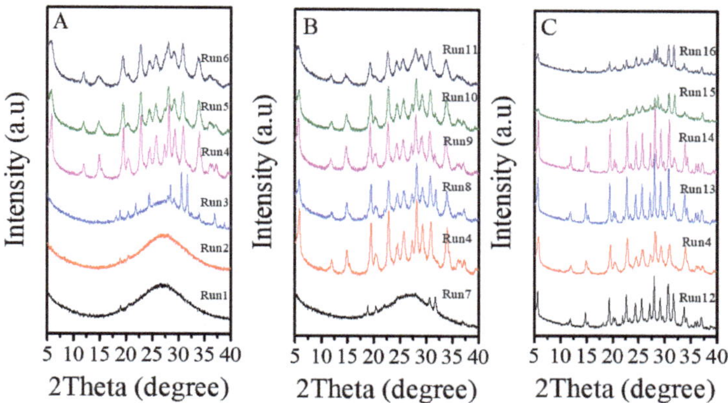

Figure 6. XRD patterns of Fe-LTL zeolite crystals synthesized under various conditions in Table 2. (**A**) XRD patterns of Fe-LTL zeolite crystals synthesized with different K_2O/SiO_2 ratios; (**B**) XRD patterns of Fe-LTL zeolite crystals synthesized with different SiO_2/Al_2O_3 ratios; (**C**) XRD patterns of Fe-LTL zeolite crystals synthesized with different $FeCl_3/SiO_2$ ratios.

Table 3. Fe/Si and Si/Al ratios of different samples.

Run	Fe/Si Ratio in the Initial Gel	Fe Content in the Obtained Sample [d]	Si/Al Ratio in the Obtained Sample [d]
12	0.03	2.07%	4.6
4	0.04	2.67%	4.7
13	0.05	3.33%	4.6
14	0.06	3.10%	5.3
15	0.07	3.74%	5.8
16	0.08	3.34%	5.9

[d] The iron content and Si/Al ratio was tested by ICP experiment.

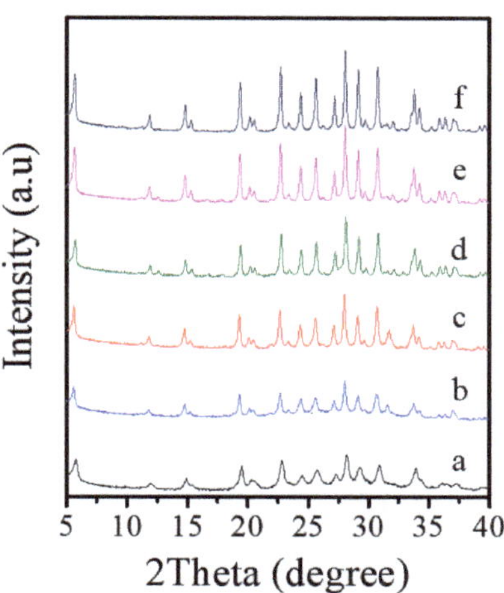

Figure 7. XRD patterns of Fe-LTL zeolite crystals synthesized at (**a**) 80 °C, (**b**) 100 °C, (**c**) 120 °C, (**d**) 140 °C, (**e**) 160 °C, (**f**) 180 °C, respectively.

After the systematic synthesis, aluminosilicate gel for crystallization of Fe-LTL zeolites at 80 °C with molar ratio of $2.1Na_2O/3.4K_2O/Al_2O_3/15SiO_2/0.6FeCl_3/250H_2O$ is suitable.

4. Conclusions

In summary, we have successfully synthesized ultra-fine Fe-LTL zeolite nanocrystals under very mild synthesis conditions. The average particle size of Fe-LTL zeolite nanocrystals is approximately 30 nm. Moreover, Fe-LTL-N zeolites that we obtained have advantages of larger external surface area, shorter diffusion pathway, good thermal stability, narrow particle distribution. Very importantly, tetrahedral Fe^{3+} is located in the zeolite framework of Fe-LTL-N. Designer synthesis of ultra-fine Fe-LTL zeolite nanocrystals in this work has significant potential applications for industrial catalysis in the near future.

Supplementary Materials: The following are available online at http://www.mdpi.com/2073-4352/10/9/813/s1, Figure S1: IR spectra of Fe-LTL zeolite crystals synthesized under various conditions in Table 2.

Author Contributions: Conceptualization and writing, F.Z.; methodology, Y.L., L.C. and G.C.; investigation, W.C., Y.H., S.Y., and W.S. All authors have read and agreed to the published version of the manuscript.

Funding: This research was funded by the Jiangxi Provincial Natural Science Foundation (20202ACB203003), the Science and Technology Program of Jiangxi Academy of Sciences (2020-YZD-3, 2019-XTPH1-09, 2019-YYB-09, 2018-YDHZ-01).

Conflicts of Interest: The authors declare no conflict of interest.

References

1. Davis, M.E.; Lobo, R.F. Zeolite and molecular sieve synthesis. Chemistry of Materials. *Chem. Mater.* **1992**, *4*, 756–768. [CrossRef]
2. Gies, H.; Marler, B. The structure-controlling role of organic templates for the synthesis of porosils in the system SiO_2/template/H_2O. *Zeolite* **1992**, *12*, 42–49. [CrossRef]
3. Davis, M.E. Ordered porous materials for emerging applications. *Nature* **2002**, *417*, 813–821. [CrossRef]
4. Cundy, C.S.; Cox, P.A. The hydrothermal synthesis of zeolites: History and development from the earliest days to the present time. *Chem. Rev.* **2003**, *103*, 663–702. [CrossRef] [PubMed]
5. Zou, X.; Conradsson, T.; Klingstedt, M.; Dadachov, M.S.; O'Keeffe, M. A mesoporous germanium oxide with crystalline pore walls and its chiral derivative. *Nature* **2005**, *437*, 716–719. [CrossRef] [PubMed]
6. Ma, S.Q.; Sun, D.F.; Wang, X.S.; Zhou, H.C. A mesh-adjustable molecular sieve for general use in gas separation. *Angew. Chem. Int. Ed.* **2007**, *46*, 2458–2462. [CrossRef] [PubMed]
7. Shin, J.; Camblor, M.A.; Woo, H.C.; Miller, S.R.; Wright, P.A.; Hong, S.B. PST-1: A synthetic small-pore zeolite that selectively adsorbs H_2. *Angew. Chem. Int. Ed.* **2009**, *48*, 6647–6650. [CrossRef] [PubMed]
8. Lee, J.H.; Park, M.B.; Lee, J.K.; Min, H.K.; Song, M.K.; Hong, S.B. Synthesis and characterization of ERI-type UZM-12 zeolites and their methanol-to-olefin performance. *J. Am. Chem. Soc.* **2010**, *132*, 12971–12982. [CrossRef]
9. Na, K.; Jo, C.; Kim, J.; Cho, K.; Jung, J.; Seo, Y.; Messinger, R.J.; Chmelka, B.F.; Ryoo, R. Directing zeolite structures into hierarchically nanoporous architectures. *Science* **2011**, *333*, 328–332. [CrossRef] [PubMed]
10. Zhang, X.Y.; Liu, D.X.; Xu, D.D.; Asahina, S.; Cychosz, K.A.; Agrawal, K.V.; Al Wahedi, Y.; Bhan, A.; Al Hashimi, S.; Terasaki, O.; et al. Synthesis of self-pillared zeolite nanosheets by repetitive branching. *Science* **2012**, *336*, 1684–1687. [CrossRef]
11. Goel, S.; Wu, Z.J.; Zones, S.I.; Iglesia, E. Synthesis and catalytic properties of metal clusters encapsulated within small-pore (SOD, GIS, ANA) zeolites. *J. Am. Chem. Soc.* **2012**, *134*, 17688–17695. [CrossRef]
12. Jo, C.; Jung, J.; Shin, H.S.; Kim, J.; Ryoo, R. Capping with Multivalent Surfactants for Zeolite Nanocrystal Synthesis. *Angew. Chem. Int. Ed.* **2013**, *52*, 10014–10017. [CrossRef]
13. Moller, K.; Bein, T. Mesoporosity-a new dimension for zeolites. *Chem. Soc. Rev.* **2013**, *42*, 3689–3707. [CrossRef] [PubMed]
14. Bian, C.Q.; Wang, X.; Yu, L.; Zhang, F.; Zhang, J.; Fei, Z.X.; Qiu, J.P.; Zhu, L.F. Generalized methodology for inserting metal heteroatoms into the layered zeolite precursor RUB-36 by interlayer expansion. *Crystals* **2020**, *10*, 530. [CrossRef]
15. Song, W.; Justice, R.E.; Jones, C.A.; Grassian, V.H.; Larsen, S.C. Size-dependent properties of nanocrystalline silicalite synthesized with systematically varied crystal sizes. *Langmuir* **2004**, *20*, 4696–4702. [CrossRef] [PubMed]

16. Larsen, S.C. Nanocrystalline zeolites and zeolite structures: Synthesis, characterization, and applications. *J. Phys. Chem. C* **2007**, *111*, 18464–18474. [CrossRef]
17. Choi, M.; Na, K.; Kim, J.; Sakamoto, Y.; Terasaki, O.; Ryoo, R. Stable single-unit-cell nanosheets of zeolite MFI as active and long-lived catalysts. *Nature* **2009**, *461*, 246–249. [CrossRef]
18. Eng-Poh, N.; Chateigner, D.; Bein, T.; Valtchev, V.; Mintova, S. Capturing ultrasmall EMT zeolite from template-free systems. *Science* **2012**, *335*, 70–73.
19. Feng, G.D.; Cheng, P.; Yan, W.F.; Boronat, M.; Li, X.; Su, J.H.; Wang, J.Y.; Li, Y.; Corma, A.; Xu, R.R. Accelerated crystallization of zeolites via hydroxyl free radicals. *Science* **2016**, *351*, 1188–1191. [CrossRef]
20. Chen, W.; Fan, Z.L.; Pan, X.L.; Bao, X.H. Effect of confinement in carbon nanotubes on the activity of Fischer-Tropsch iron catalyst. *J. Am. Chem. Soc.* **2008**, *130*, 9414–9419. [CrossRef]
21. Awala, H.; Gilson, J.-P.; Retoux, R.; Boullay, P.; Goupil, J.-M.; Valtchev, V.; Mintova, S. Template-free nanosized faujasite-type zeolites. *Nat. Mater.* **2015**, *14*, 447–451. [CrossRef] [PubMed]
22. Ohsuna, T.; Horikawa, Y.; Hiraga, K.; Terasaki, O. Surface Structure of Zeolite L Studied by High-Resolution Electron Microscopy. *Chem. Mater.* **1998**, *10*, 688–691. [CrossRef]
23. Laulus, O.; Valtchev, V.P. Crystal morphology control of LTL-type zeolite crystals. *Chem. Mater.* **2004**, *16*, 3381–3389.
24. Davis, R.J. Aromatization on zeolite L-supported Pt clusters. *Heterogen. Chem. Rev.* **1994**, *1*, 41–53.
25. Joshi, P.N.; Bandyopadhyay, R.; Awate, S.V.; Shiralkar, V.P.; Rao, B.S. Influence of physicochemical parameters on n-hexane dehydrocyclization over Pt/LTL zeolites. *React. Kinet. Catal. Lett.* **1994**, *53*, 231–236. [CrossRef]
26. Jentoft, R.E.; Tsapatsis, M.; Davis, M.E.; Gates, B.C. Platinum clusters supported in zeolite LTL: Influence of catalyst morphology on performance in n-hexane reforming. *J. Catal.* **1998**, *179*, 565–580. [CrossRef]
27. Miller, J.T.; Agrawal, N.G.B.; Lane, G.S.; Modica, F.S. Effect of pore geometry on ring closure selectivities in platinum L-zeolite dehydrocyclization catalysts. *J. Catal.* **1996**, *163*, 106–116. [CrossRef]
28. Barrer, R.M.; Galabova, I.M. Ion-exchanged forms of zeolite L, erionite, and offretite and sorption of inert-gases. *Adv. Chem. Ser.* **1973**, 356–373. [CrossRef]
29. Lee, T.P.; Saad, B.; Ng, E.P.; Salleh, B. Zeolite linde type L as micro-solid phase extraction sorbent for the high performance liquid chromatography determination of ochratoxin A in coffee and cereal. *J. Chromatogr. A* **2012**, *1237*, 46–54. [CrossRef]
30. Calzaferri, G.; Huber, S.; Maas, H.; Minkowski, C. Host–Guest Antenna Materials. *Angew. Chem. Int. Ed.* **2003**, *42*, 3732–3735. [CrossRef]
31. Gigli, L.; Arletti, R.; Tabacchi, G.; Fabbiani, M.; Vitillo, J.G.; Martra, G.; Devaux, A.; Miletto, I.; Quartieri, S.; Calzaferri, G. Structure and host-guest interactions of perylene-diimide dyes in zeolite L nanochannels. *J. Phys. Chem. C.* **2018**, *122*, 3401–3418. [CrossRef]
32. Goldfarb, D.; Bernado, M.; Strohmaier, K.G.; Vaughan, D.E.W.; Thomann, H. Characterization of iron in zeolites by X-band and Q-band ESR, pulsed ESR, and UV-visible spectroscopies. *J. Am. Chem. Soc.* **1994**, *116*, 6344–6353. [CrossRef]
33. Zhang, H.; Chu, L.; Xiao, Q.; Zhu, L.; Yang, C.; Meng, X.; Xiao, F.-S. One-pot synthesis of Fe-Beta zeolite by an organotemplate-free and seed-directed route. *J. Mater. Chem. A* **2013**, *1*, 3254–3257. [CrossRef]
34. Camblor, M.A.; Corma, A.; Mifsud, A.; Perez-Pariente, J. Valencia, Synthesis of nanocrystalline zeolite Beta in the absence of alkali metal cations. *Stud. Surf. Sci. Catal.* **1997**, *105*, 341–348.

Article

Guided Crystallization of Zeolite Beads Composed of ZSM-12 Nanosponges

Kassem Moukahhal [1,2,3], Ludovic Josien [1,2], Habiba Nouali [1,2], Joumana Toufaily [3], Tayssir Hamieh [3], T. Jean Daou [1,2,*] and Bénédicte Lebeau [1,2,*]

1 Université de Haute Alsace (UHA), CNRS, Axe Matériaux à Porosité Contrôlée (MPC), IS2M UMR 7361, F-68100 Mulhouse, France; kassem.moukahhal@uha.fr (K.M.); ludovic.josien@uha.fr (L.J.); habiba.nouali@uha.fr (H.N.)

2 Université de Strasbourg, F-67000 Strasbourg, France

3 Laboratory of Materials, Catalysis, Environment and Analytical Methods Faculty of Sciences, Section I, Lebanese University Campus Rafic Hariri, Hadath, Lebanon; joumana.toufaily@ul.edu.lb (J.T.); tayssir.hamieh@ul.edu.lb (T.H.)

* Correspondence: jean.daou@uha.fr (T.J.D.); benedicte.lebeau@uha.fr (B.L.); Tel.: +33-389336739 (T.J.D.); +33-389336882 (B.L.)

Received: 24 August 2020; Accepted: 15 September 2020; Published: 17 September 2020

Abstract: The direct route using a bifunctional amphiphilic structuring agent for the synthesis of hierarchical nanozeolites coupled with pseudomorphic transformation was used for the crystallization of hierarchized zeolite beads/hollow spheres composed of ZSM-12 (MTW structural-type) with nanosponge morphology. These beads/hollow spheres have the same average diameter of 20 μm as their counterpart amorphous mesoporous silica beads used as precursor in the starting synthesis mixture. The effects of synthesis parameters, such as stirring and treatment time at 140 °C, on the morphology, structure, and texture of the materials have been investigated using X-ray diffraction (XRD), N_2 sorption, scanning electronic microscopy (SEM), and transmission electronic microscopy (TEM) techniques. Static conditions were found necessary to maintain the morphology of the starting amorphous silica beads. An Ostwald ripening phenomenon was observed with the increase in hydrothermal treatment time leading to the dissolution of the interior of some beads to form core shell beads or hollow spheres with larger crystals on the outer surface. These ZSM-12 beads/hollow spheres possess higher porous volume than conventional ZSM-12 zeolite powder and can be used directly for industrial applications.

Keywords: zeolite beads; hierarchical zeolite; pseudomorphic transformation; ZSM-12; shaping; nanosponges

1. Introduction

Zeolites constitute a category of aluminosilicate crystals with microporous frameworks. They are good candidates for adsorption, catalysis and separation due to their shape selectivity and high thermal stability [1–6]. However, the diffusion of molecules is limited in micropores and thus the performance of zeolites in catalytic reactions or adsorption often depends strongly on the size, morphology and porosity (pore size and pore topology) of the crystals [7–10]. In catalysis, diffusional limitations often lead to coke formation or reduce the accessibility to active sites. For molecular decontamination, adsorption capacity and kinetics are mainly governed by pore size and porous volume. Zeolites having a one-dimensional (1D) pore system, such as MTW, TON, or MOR type, and generally suffer from a long diffusion path in crystals with high aspect ratios (length/width) and pore channels oriented along the length [11]. In recent years, a particular attention has been brought to rationally design the crystal morphology and/or the porosity of a given zeolite in order to reduce diffusion path in microporosity [12]. In this context, the reduction of crystal size or introducing larger pores appeared the best strategies. Post-treatments consisting in dealumination or desilication are the conventional methods for generating mesopores within zeolite crystals [13–16]. The crystallization of zeolites with hierarchical micro/mesoporous structures by using bifunctional amphiphilic structure directing agent was also identified as a performant alternative [9,17–26]. Moreover, the addition of additives, such as crystal growth modifiers [27–29] or seed crystals (salt-aided seed-induced route) [8,30], also makes it possible to tailor the size and the morphology of the crystals by inhibiting the crystal growth along certain specific network planes or by controlling the process of dissolution-recrystallization, respectively.

ZSM-12 zeolite of MTW-type is a highly silicic 1D-pore system with a framework Si/Al molar ratio always higher than 20, and has shown interesting features for acid catalysis such as high coking resistance and high chemical stability [31]. It was first synthesized by Rosinski and Rubin [32] in a reaction medium containing diethyl sulphate and triethylamine. In 1985, LaPierre et al. resolved the structure using X-ray powder diffraction [33]. The unit cell is monoclinic, the space group is C2/m and the unit cell parameters are: a = 24.88, b = 5.02 Å, c = 12.15 Å, and $\beta = 107.7°$. The structure contains rings with four, five, and six T atoms (T = Si or Al) and has a one-dimensional system of channels with 12 T atoms with pore openings of ~5.6 × 6.1 Å. Looking at the structure in a direction perpendicular to the axis of the channels we can observe that the walls are formed only by six-atom rings T with an opening of about 2.5 Å. This small opening prevents the passage of adsorbed molecules between the channels [34]. As a consequence, ZSM-12 zeolite is characterized by non-interconnecting tubular-like linear channels that were identified to be responsible for not allowing the accumulation of coking precursors [31].

The particular micropores dimensions of the ZSM-12 zeolite, make it performant catalyst in reactions involving cyclic hydrocarbons. Pazzucconi et al. used ZSM-12 zeolite for alkylation of naphthalene in liquid phase at high pressure in the presence of 1,2,4-trimethylbenzene as a solvent, and they reported a high activity and high selectivity of ZSM-12 zeolite to 2,6-dimethylnaphthalene [35,36]. Higher stability and higher activity on the reaction time in the catalytic cracking of naphthalene were observed on ZSM-12 zeolite, which can provide insight into the catalytic performance for many reactions. However, ZSM-12 zeolite also suffers diffusional limitations in micropores and the hierarchization of the pore network was investigated and revealed beneficial to improve its catalytic performance.

Dugkhuntod et al. recently reported the synthesis of hierarchical ZSM-12 nanolayers via a dual template method [37]. Hierarchical ZSM-12 nanolayers exhibit an improvement of catalytic activity in terms of levulinic acid conversion compared to other hierarchical zeolite nanosheets, such as MFI (ZSM-5) and FAU. The better catalytic performance was related to the enhanced mesoporosity and the presence of a large one-dimensional 12-membered ring network, which can promote the accessibility of bulky-molecules to active sites of zeolite.

Zeolites or hierarchized zeolites synthesized with conventional hydrothermal routes leads to micrometer-sized crystals or nanocrystal aggregates, respectively, forming fine powder products. However, the use of these materials in powder form can be disadvantageous for certain industrial applications such as those based on continuous flow processes. Indeed, they can generate too much pressure drop and cause clogging of the cartridges/columns/filters used on an industrial scale. Moreover, these zeolite powders can be a source of secondary contamination and often fine powders causes problems of handling (due to particles spreading). Therefore, the development of methods for shaping these materials into objects of controllable shape and size is of great technological importance. However, shaping methods of zeolite powder usually require several steps such as compacting, grinding and sieving for obtaining grains of similar size, or assembling the zeolite particles by extrusion or mixing in presence of organic (e.g., polymers) or inorganic (e.g., clays, sodium silicate) binders, which can reduce the performance of the zeolite materials [38–40], or removal of sacrificial macrotemplates (resin, …) [41,42]. A one-step and binderless process such as pseudomorphic transformation is an interesting alternative for shaping zeolite powder. This strategy consists by controlling the dissolution/crystallization process to transform a shaped material into another one having different chemical composition but the same shape. This concept was successfully used by Anne Galarneau's team to synthesize micrometric beads and zeolite monoliths of the LTA, SOD and FAU type with low Si/Al ratios [43–46]. Our group has recently reported the direct synthesis of hierarchical ZSM-5 beads [47] and monoliths with nanosheets morphology [48], and also plain and hollow purely silica beads of silicalite-1 nanosheets by using the pseudomorphic transformation method [49]. Here, we report the use of the pseudomorphic transformation concept to elaborate hierarchical ZSM-12 zeolite beads with nanosponges morphology thanks to the use of a dual-porogenic surfactant as a template and growth inhibitor, which was used for zeolite *BEA nanosponge synthesis [9,21,25].

2. Materials and Methods

2.1. Preparation of ZSM-12 Beads

2.1.1. Synthesis of the Dual-Porogenic Surfactant Template

The dual-porogenic surfactant $C_{22}H_{45}$-$N^+(CH_3)_2$-C_6H_{12}-$N^+(CH_3)_2$-CH_2-*Phe*-CH_2-$N^+(CH_3)_2$-C_6H_{12}-$N^+(CH_3)_2$-CH_2-*Phe*-CH_2-$N^+(CH_3)_2$-C_6H_{12}-$N^+(CH_3)_2$-$C_{22}H_{45}$, $(Br^-)_2(Cl^-)_4$, (abbreviated as N_6-Diphe, Phe = C_6H_4) have been used to generate ZSM-12 nanosponges beads and hollow spheres. This surfactant has been synthetized in three steps following the procedure described by Na et al. [21]. 16.2 g (0.01 mol) of 1-bromodocosane (TCI EUROPE N.V., Zwijndrecht, Belgium) and 68.8 g (0.1 mol) of N,N,N′,N′-tetramethyl-1,6-diaminohexane (Sigma-Aldrich Chemie S.a.r.l., Saint-Quentin Fallavier, France) were dissolved in 400 mL of acetonitrile/toluene mixture (1:1 vol/vol) and heated under reflux at 70 °C for 12 h with magnetic stirring. After cooling to room temperature, the solvent was evaporated by rotary evaporator, and the product was filtered, washed with diethyl ether, and dried under vacuum at 50 °C. The product obtained was denoted $C_{22\text{-}6\text{-}0}$.

In a second step, 8 g of $C_{22\text{-}6\text{-}0}$ (0.01 mol) and 23.56 g (0.1 mol) of α,α′- dichloro-p-xylene (Sigma-Aldrich Chemie S.a.r.l., Saint-Quentin Fallavier, France) were dissolved in 100 mL of acetonitrile, which were heated to 60 °C for 48 h with magnetic stirring. After evaporation of the organic solvent by rotary evaporator, the solid product of the formula $C_{22}H_{45}$-$N^+(CH_3)_2$-C_6H_{12}-$N^+(CH_3)_2$-CH_2-(p-C_6H_4)-CH_2-$Cl(Br^-)(Cl^-)$, designated $C_{22\text{-}6\text{-}phe\text{-}Cl}$, was precipitated. This product was washed with diethyl ether, filtered, and dried under vacuum at 50 °C. Finally, 10 g (0.02 mol) of $C_{22\text{-}6\text{-}phe\text{-}Cl}$ and 1.12 g (0.01 mol) of *N,N,N′,N′*-tetramethyl-1,6-diaminohexane (Sigma-Aldrich Chemie S.a.r.l., Saint-Quentin Fallavier, France) were dissolved in 100 mL of chloroform and refluxed for 24 h under magnetic agitation. The solvent was evaporated by rotary evaporator and a final solid product (N6-diphe) was recovered. The solid was washed with diethyl ether, filtered, and dried under vacuum at 50 °C. The yield was 80%.

Purity was checked by ^1H liquid nuclear magnetic resonance (NMR), the chemical shifts obtained are as follows: ^1H NMR (CDCl3, 400 MHz, 25 °C): δ (ppm) [C3] 0.83 (m, 6H); 1.3 (m, 41H); 1.59 (s, 4H); 1.71 (s, 4H); 2.01 (s, 3H); [C2] 3.36 (s, 12H); [C1] 3.47 (m, 4H); 3.72 (m, 4H).

2.1.2. Synthesis of ZSM-12 Nanosponges Beads

First, 0.3 g of sodium hydroxide (NaOH 99%, Carlo Erba Reagents, Val-de-Reuil, France) and 0.037 g sodium aluminate ($NaAlO_2$ 92%, Strem Chemicals, Inc., Bischeim, France) were dissolved in 21.3 g of distilled water. Then, 6.14 g of pure ethanol (ETOH, 99%, VWR International S.A.S, Fontenay-sous-Bois, France) and 1.6 g of the dual-porogenic surfactant the N6-Diphe were added to the previous mixture. In a next step, the obtained mixture was mixed with 1 g of amorphous mesoporous silica beads with an average size of 20 μm (data given by the supplier SiliCycle [Inc.,] Quebec, Canada) to set the molar composition to: 1 SiO_2: 0.22 Na_2O: 0.0125 Al_2O_3: 0.05 N_6-Diphe: 8 EtOH: 71 H_2O. The resulting precursor gel was introduced in a PTFE®-lined stainless-steel autoclave of 45 mL for crystallization at 140 °C for 4 days under tumbling conditions at 60 rpm or under static mode for several days (4, 5, 7, 14 and 21 days). The final product is recovered by filtration, then washed with distilled water before drying for 12 h at 80 °C. The dual porogenic surfactant was eliminated by calcination under air at 550 °C for 8 h.

The synthesized samples are denoted t-S, or t-NS, where "t" stands for the hydrothermal treatment time at 140 °C, and "S" or "NS" means that the samples was synthetized under tumbling stirring or not (static mode), respectively

2.2. Characterization of Zeolite Beads

The purity and crystallinity of the obtained samples was checked by X-ray diffraction (XRD). The samples were introduced in glass capillaries and the X-ray patterns were recorded using a STOE STADI-P (STOE & Cie GmbH, Darmstadt, Germany) diffractometer using Cu $K\alpha_1$ radiation (λ = 0.15406 nm) in the range $3° < 2\theta < 50°$.

Scanning and transmission electron microscopy (SEM and TEM) were performed to analyze morphology, homogeneity, and particle sizes. Philips XL30 FEG (Field Emission Gun) (Philips, Verdun, France) and JEOL JSM-7900F (JEOL, Val-de-Reuil, France) scanning electron microscopes (SEM) were used. The later was equipped with a BRUKER QUANTAX spectrometer (BRUKER, Champs sur Marne, France) for the energy dispersive X-ray analyses (EDX). The transmission electron microscopy (TEM) was carried out on a JEOL (Val-de-Reuil, France) model ARM 200, operating at 200 kV, with a point-to point resolution of 80 pm. For EDX analyses, samples were previously embedded in an epoxy resin for cold mounting (Epofix from Struer S.A.S. Champigny sur Marne, France), which was grinded with various grinding papers and then polished until a soft surface was obtained.

Nitrogen sorption isotherms were recorded at −196 °C using a Micromeritics (Merignac, France) sorptometer (model ASAP 2420). In prior analyses, the samples were out-gassed under vacuum at 300 °C for 15 h. The BET method was used for determining the specific surface area. The total pore volume was determined from the total amount of N_2 adsorbed at saturation (i.e., at $p/p° = 0.99$ in absence of textural porosity). The Density Functional Method (DFT) method was applied on the adsorption branch in order to obtain the pore size distributions [50]. The microporous volumes were calculated by using a modified t-plot method.

The amounts of dual porogenic agent occluded within the porosity of sample were determined by thermogravimetric analysis (METTLER TOLEDO, Viroflay, France, model TG/DSC STARe) from 20 to 800 °C under air with a heating rate of 5 $°C/min^{-1}$.

The Si/Al molar ratios were determined using X-ray fluorescence spectrometry (XRF) performed on a PANalytical (Limeil-Brévannes, France) equipment model Zetium and from EDX analyses.

3. Results and Discussion

The purity and crystallinity of the synthesized zeolites were investigated by X-ray diffraction (Figure 1). The X-ray patterns of the all obtained materials present diffraction peaks, which correspond to a pure phase of ZSM-12 zeolite in agreement with the corresponding patterns available in the literature [51]. The diffraction peaks are broader for samples hydrothermally treated under static conditions for 4, 5 and 7 days (4-NS, 5-NS and 7-NS, respectively), which indicate a small crystal size and/or a low crystallinity degree. The thinner diffraction peaks with higher intensity observed after 4 days of hydrothermal treatment with mechanical stirring (4-S) and after 14 and 21 days in static mode (14-NS and 21-NS, respectively) indicate the presence of larger crystal sizes and/or high crystallinity degree.

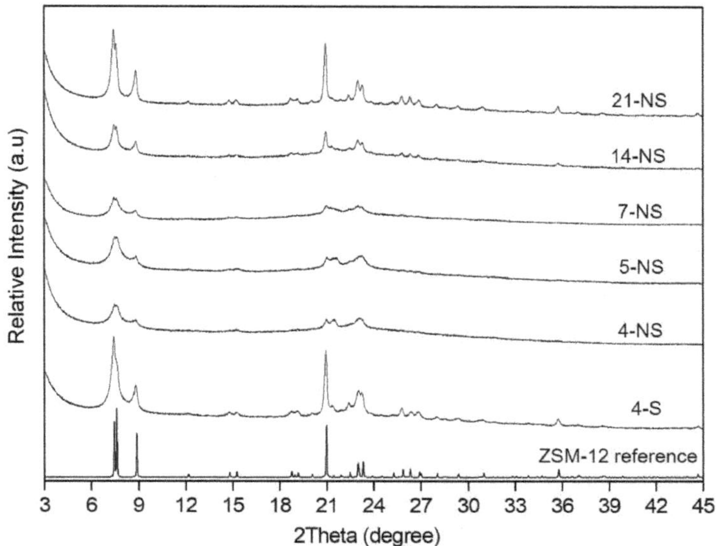

Figure 1. XRD patterns of the calcined ZSM-12 zeolites obtained by pseudomorphic transformation and XRD reference pattern of ZSM-12 zeolite [51].

SEM images displayed in Figure 2 show that the morphology of the parent silica beads (Figure S1) has been totally lost for the sample 4-S synthesized under stirring (Figure 2a). However, the nanosponge morphology was observed by TEM (Figure 3a). Indeed, TEM image show nano-sized zeolite crystals delimiting mesopores with narrow pore size distribution as a result of the self-assembly of the long hydrocarbon chains of the multiammonium surfactant used as structure directing agent. This, in complement of XRD data, indicates that stirring favor a fast crystallization of ZSM-12 nanosponges but induce the loss of the parent bead morphology. On the other hand, the spherical morphology of the parent amorphous silica beads was maintained when the hydrothermal treatment was carried out in static mode for all investigated durations (Figure 2b–f). The breaking of the beads upon stirring allowed a faster diffusion of reactants present in the liquid phase (aluminum source, bifunctional structuring agent N_6-Diphe) to the silica backbone compared to the plain beads. SEM images at high magnification displayed in Figure 2 show that beads are constituted of spherical agglomerates of ZSM-12 nanocrystals, with an average of 400 ± 90 nm diameter after four and five days of hydrothermal treatment. The size of the agglomerates of ZSM-12 nanocrystals seems to increase slightly to reach about 1 ± 0.2 μm by increasing the duration of the hydrothermal treatment to seven and 14 days, and reaches about 1.7 ± 0.2 μm after 21 days. In parallel, the size ZSM-12 nanocrystals composing these agglomerates increases from 4 nm (4-S and 4-NS samples) to reach a maximum of 80 nm after 21 days of hydrothermal treatment as shown in Figure 3a,b. The nanocrystal size distribution seems also to increase while increase the hydrothermal treatment time, especially for 7-NS, 14-NS and 21-NS samples (Figure 3c–e). These results are consistent with the sharper X-ray diffractions peaks when increasing the time of the hydrothermal treatment. According to the TEM images of samples 4-S, 4-NS, 7-NS, 14-NS, and 21-NS displayed in Figure 3, nanosponges are composed by randomly agglomerated nanocrystals interconnected by mesopores, which are characteristic of nanosponge morphology [9,25]. Since beads have been crushed for TEM observations, it is difficult to identify the location of the crystals of different sizes in the obtained beads. However, by correlation with SEM observations these big crystals are probably located at the outer surface of the beads while small ones are inside as already observed in our previous work on the synthesis of silicalite-1 zeolite beads and hollow spheres, where the crystallization seems to begin on the external surface of the beads and to diffuse with the time inside the beads [49].

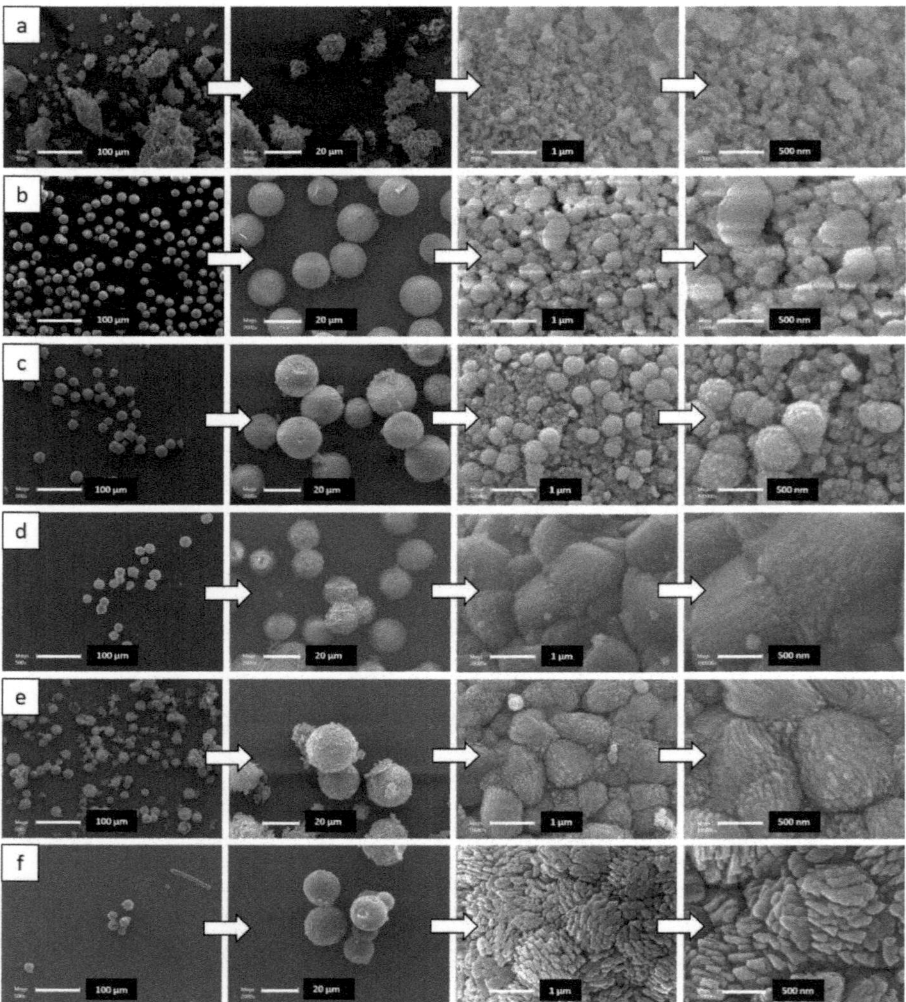

Figure 2. SEM images of all samples (**a**) 4-S (**b**) 4-NS (**c**) 5-NS (**d**) 7-NS (**e**) 14-NS (**f**) 21-NS obtained after the pseudomorphic transformation of the amorphous silica beads.

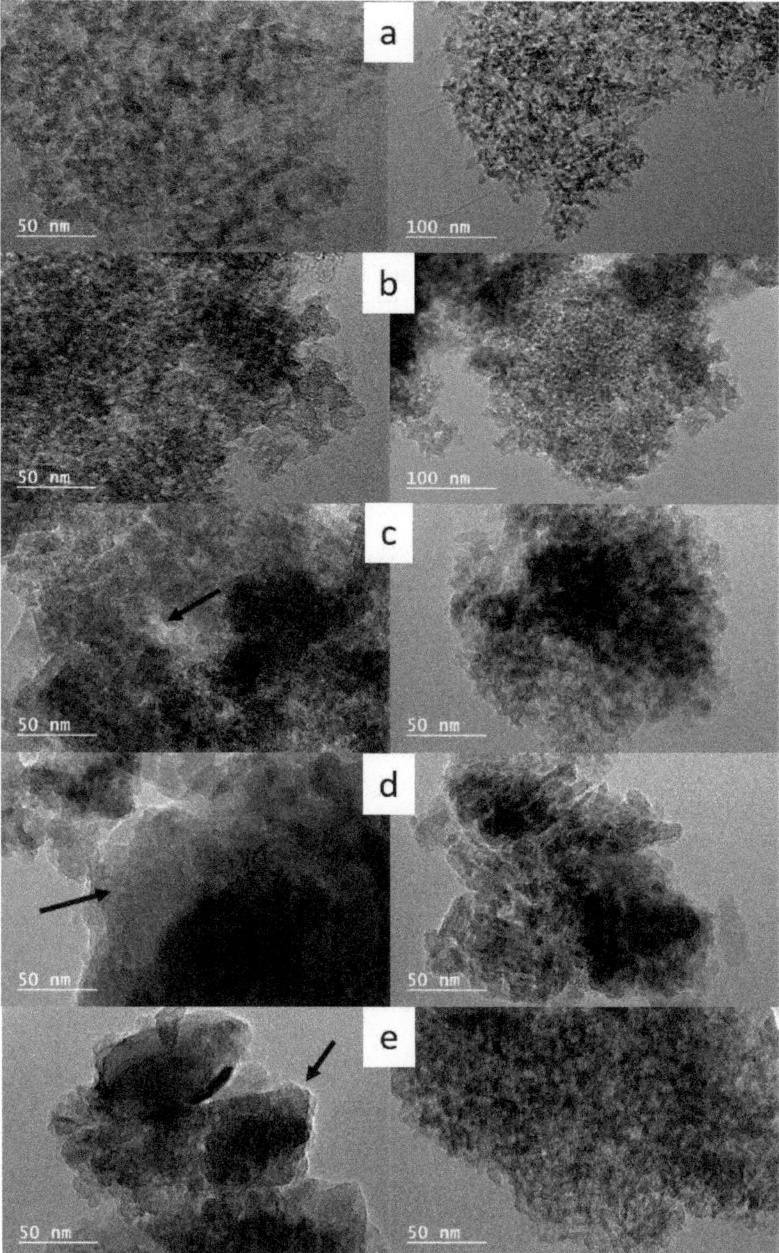

Figure 3. TEM images of (**a**) 4-S (**b**) 4-NS (**c**) 7-NS (**d**) 14-NS (**e**) 21-NS samples showing the morphology of the obtained hierarchized ZSM-12 zeolites (Black arrows indicate the presence of larger crystals).

The nitrogen adsorption-desorption isotherms at 77 K of the calcined materials are shown in Figure 4a. The nitrogen adsorption–desorption isotherm of the parent silica beads is of type IV (according to an IUPAC classification) (Figure S2) [52]. For the samples obtained by pseudomorphic transformation, isotherms are a mixture of type I (low relative pressure p/p°) and type IV (high p/p°). All isotherms present a H3 hysteresis in the $0.4 < p/p^0 < 1$ range, whichis characteristic of mesoporosity induced by the alkyl chains and the phenyl groups of the dual-porogenic surfactant (N6-Diphe). However, for 4-S and 4-NS samples, some interparticle mesoporosity is also clearly observed (type II at high p/p°). The type I shape of the isotherms is coherent with the presence of the micropores. The textural properties of all materials determined from the nitrogen adsorption-desorption isotherms are presented in Table 1. The microporous volumes have been determined by the t-plot method with the correction proposed by Galarneau et al. for hierarchical zeolites [53,54]. The microporous volume of the sample 4-S, where the morphology of the starting beads is not conserved, is 0.19 cm^3/g. A lower microporous volume (0.15 cm^3/g) has been measured for the sample synthesized with the same duration (4 days) but under static conditions. These microporous volumes are higher than the one expected for conventional ZSM-12 zeolite (0.05–0.11 cm^3/g) [37,55,56] because of the presence of secondary micropores with a diameter of 1.5 nm revealed by the pore size distribution obtained by applying DFT Method [50] (Figure 4b). The volume of these secondary micropores seem to decrease by increasing the hydrothermal treatment time. It is assumed that they were generated by the presence of dual-porogenic surfactant N6-Diphe. Indeed, such secondary microporosity generated by this kind of bifunctional amphiphilic structuring agent was already observed in hierarchized ZSM-5 [47]. It is noteworthy that secondary micropores of the same size are also present in the parent amorphous silica beads. However, no porosity was detected on as-made materials, suggesting that original secondary micropores (due to amorphous silica beads) disappeared upon pseudomorphic transformation. The total microporous volume decreases to about 0.11–0.13 cm^3/g with the increase of the hydrothermal treatment time (5 days and more). This phenomenon can be explained by "Ostwald ripening" when small crystals are used as nutriment to give birth to bigger ones. This assumption is consistent with SEM and TEM observations, and with the thinner diffraction peaks obtained by XRD. Figure 4b shows the presence of mesopores of about 5 nm in average diameter with a wide mesopore size distribution going from 2 to 15 nm for the obtained samples. These mesopores do not come from the parent amorphous silica beads since these latter have also a large pore size distribution but centered at 10 nm (Figure S1). They are formed by the micellization of the hydrophobic long chains of dual structure directing agents and are responsible of the nanosponge morphology. It is noteworthy that mesoporous volumes (without interparticle porosity for 4-NS sample) increase linearly with the hydrothermal treatment time (Figure 5).

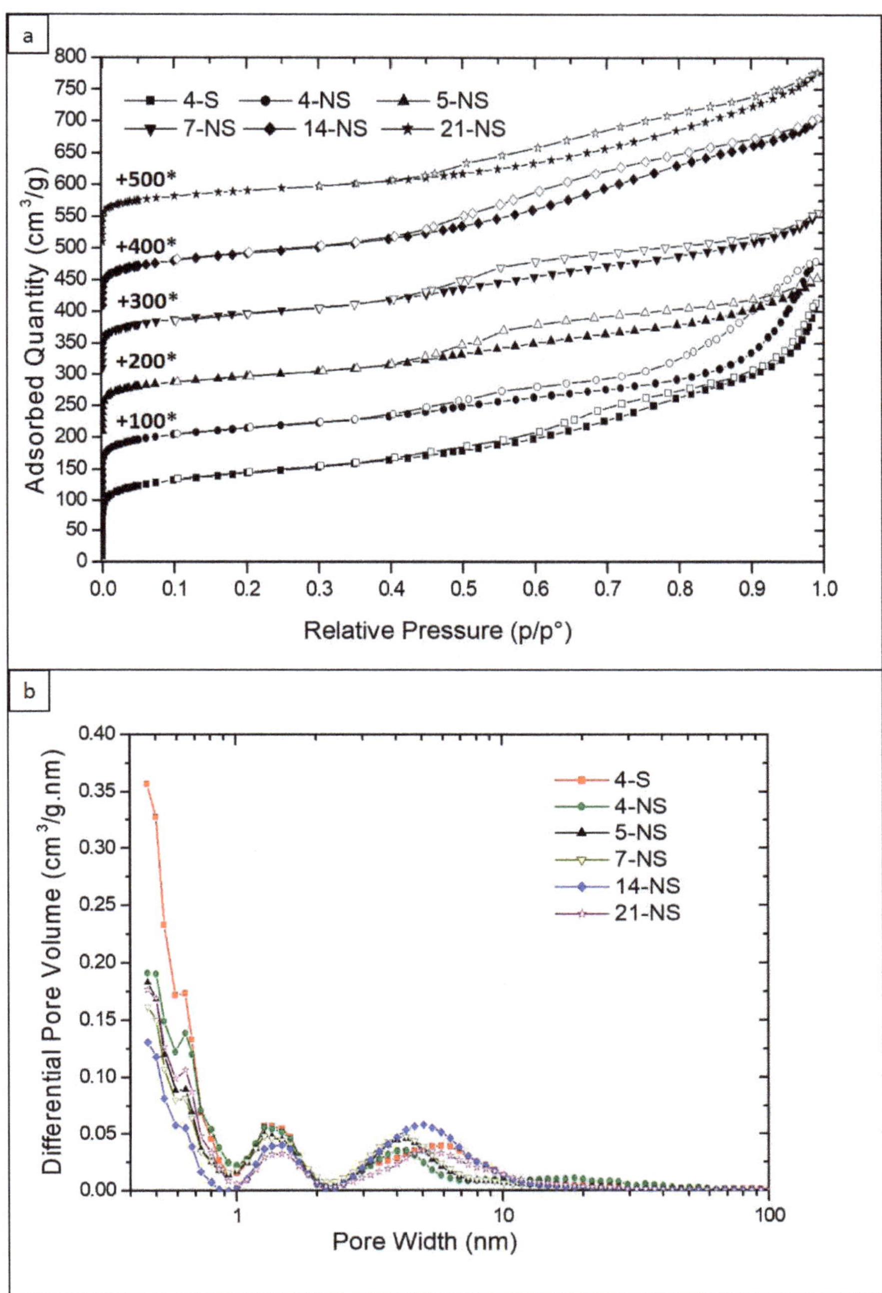

Figure 4. (**a**) N_2 isotherms sorption at 77 K and (**b**) pore size distributions for all the calcined samples obtained after pseudomorphic transformation. (*) For clarity reason, isotherms have been shifted along *y*-axis and these values (in cm^3/g) are the used increments.

Table 1. Textural properties of the calcined obtained samples and weight losses determined by TGA on as-made samples.

	SBET [a] (m^2/g)	Vtot [b] (cm^3/g)	Vmicro [c] (cm^3/g)	Vmeso [d] (cm^3/g)	Mesopores Diameter [e] (nm)	Physisorbed Water (%) [f]	Organic Matter (%) [g]
4-S	498	0.51 (0.42) *	0.19	0.32 (0.23) **	5.5	2.6	26
4-NS	417	0.59 (0.37) *	0.15	0.44 (0.22) **	5.8	3.8	22
5-NS	351	0.39	0.12	0.27	4.2	3.3	24.5
7-NS	347	0.40	0.13	0.27	4.1	3.6	24.8
14-NS	328	0.43	0.11	0.32	5.6	4.2	34.1
21-NS	327	0.47	0.12	0.35	4.5	2.6	38.5

[a] Determined by using BET (Brunauer-Emmet-Teller) method. [b] Determined at p/p° = 0.99. * value corresponding to the total pore volume determined on the plateau after hysteresis at p/p° = 0.91 without interparticle porosity. [c] Determined with corrected t-plot method [53,54]. [d] Vmeso = Vtot − Vmicro. ** value calculated with Vtot determined at p/p° = 0.91. [e] Determined from pore size distributions. [f] Weight loss determined by TGA in the temperature range 30–120 °C. [g] Weight loss determined by TGA in the temperature range 120–700 °C.

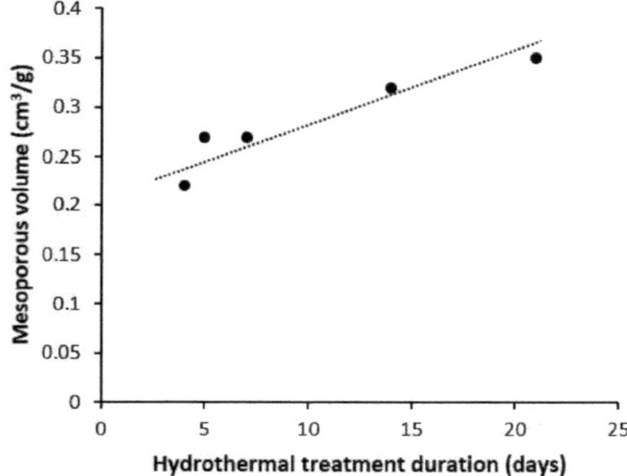

Figure 5. Evolution of the mesoporous volume of 4-NS, 5-NS, 7-NS, 14-NS and 21-NS samples (without interparticle porosity for 4-NS sample) with the hydrothermal treatment duration.

TGA analysis under air done on the as-synthesized samples shows 2 weight losses: a small one between 30 and 120 °C corresponding to the physisorbed water and a large one between 120 and 700 °C corresponding to the oxidation of the dual-porogenic surfactant used for the synthesis (Table 1). As well as the mesoporous volume, the amount of dual-porogenic surfactant determined by TGA increases linearly with the hydrothermal treatment time (Figure 6). This indicates that the incorporation of the dual porogenic surfactant N6-Diphe increases with the hydrothermal treatment time, resulting in a higher internal mesopore volume.

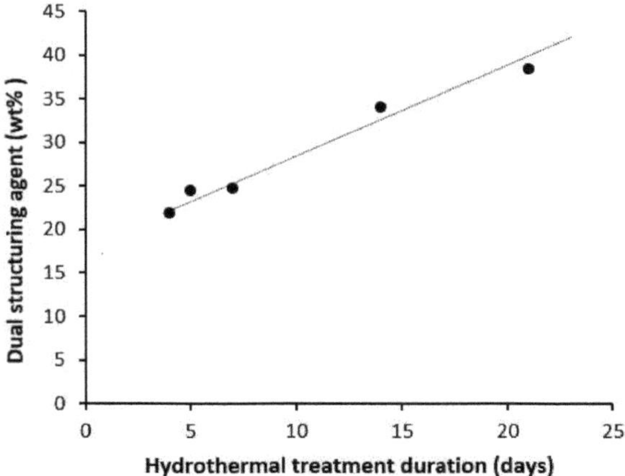

Figure 6. Evolution of the weight percentage of the dual structuring agent determined by TGA of 4-NS, 5-NS, 7-NS, 14-NS and 21-NS samples with the hydrothermal treatment duration.

In order to study the evolution of the inner part of the spheres and the silica and aluminum distribution as function of the hydrothermal treatment time, SEM observations and EDX analyses (Figure 7) were performed on three of the synthesized samples. For that, it was necessary to prepare cross section of the beads by embedding the samples in resin and then polish them. After four days of hydrothermal treatment in static mode, beads are plain, and the distribution of silica and aluminum is homogeneous on the whole of the bead indicating the nanocrystals formation in the whole beads. When increasing the duration of treatment to seven days, core-shell beads are mainly observed with a more important silica content in the shell than in the core. Few hollow spheres are also observed indicating the partial dissolution of the bead center. After 21 days, core-shell beads are still observed in coexistence with a higher number of hollow spheres. These observations are consistent with the assumption that the crystallization begins from the outer surface and propagates to the center, and with time the interior of the beads dissolve to feed the growth of the crystals located at the surface of the shell. From EDX analyses performed on all ZSM-12 nanosponges beads and hollow spheres, an average value of 30 was determined for the Si/Al ratio, which is close to the average value of 35 determined by XRF. These values are coherent with the initial Si/Al molar ratio of the precursor gel that was 40.

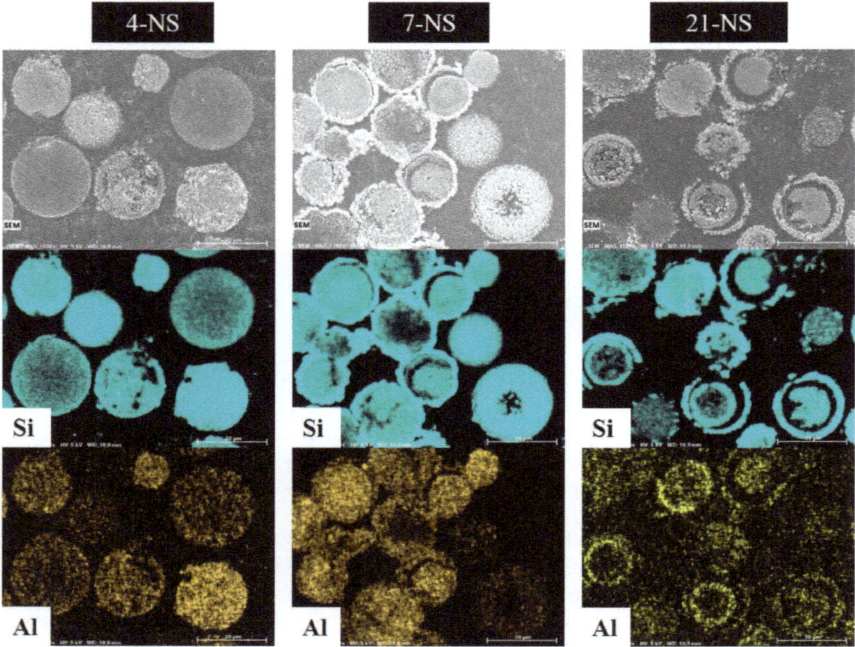

Figure 7. SEM images and corresponding EDX element mapping (Si and Al) for the samples 4-NS, 7-NS and 21-NS.

4. Conclusions

The pseudomorphic transformation pathway was applied for the synthesis of 20-µm beads composed of ZSM-12 zeolite nanosponges with MTW structure type. When hydrothermal treatment was performed with mechanical stirring (4-S), the starting amorphous silica spheres were broken and the diffusion of reactants within mesopores of the starting amorphous silica beads was faster. The crystallization of the amorphous silica walls into ZSM-12 was thus faster compared to the case when hydrothermal treatment was performed under static conditions for the same duration (4-NS). In both cases, ZSM-12 nanosponges were formed as observed by TEM images but the crystallization rate is lower for 4-NS, in agreement with broader XRD peaks for this latter sample. With one additional day of hydrothermal treatment under static conditions (5-NS), the crystallization rate increases in accordance with XRD data that showed thinner peaks. After seven days of hydrothermal treatment at 140 °C under static conditions an Ostwald ripening phenomenon has started, which results in growth of the crystals located at the outer surface of the beads. These bigger crystals are responsible for the thinner peaks observed by XRD. As observed by SEM and EDX, they form a shell rich in silicon. Beads and hollow spheres are composed of ZSM-12 zeolite nanocrystals (4 to 80 nm in size) interconnected to each other through a mesoporous network giving rise to a nanosponge morphology. This morphology generates an additional porosity (mesoporosity and secondary microporosity) to the microporosity of the zeolites (total pore volume greater than the one of conventional ZSM-12 zeolite) and improves the diffusion within this material, which should result in improving the kinetics as well as the adsorption capacities and catalytic properties of this material.

Supplementary Materials: The following are available online at http://www.mdpi.com/2073-4352/10/9/828/s1, Figure S1. Scanning electron microscopy (SEM) images of the parent amorphous 20 μm silica spheres. An average size of 23 ± 3 μm was determined from measurements on about 100 beads, Figure S2. (**a**) Nitrogen sorption isotherm at 77 K and (**b**) DFT pore size distribution determined from the adsorption branch of the N_2 isotherm for the parent amorphous silica beads.

Author Contributions: Conceptualization, T.J.D. and B.L.; methodology, K.M., T.J.D., J.T., and B.L.; validation, L.J. and H.N.; investigation, K.M.; data curation, K.M., T.J.D., and B.L.; writing—original draft preparation, K.M.; writing—review and editing, T.J.D. and B.L.; supervision, T.J.D., J.T., T.H., and B.L.; project administration, T.J.D. and B.L. All authors have read and agreed to the published version of the manuscript.

Funding: This research received funding from Institut Universitaire de France (IUF).

Acknowledgments: The XRD: adsorption and Electronic microscopies platforms of IS2M are acknowledged. Laure Michelin, Habiba Nouali, Ludovic Josien, and Loïc Vidal are warmly thanked for their help in XRD, N_2 adsorption–desorption, SEM and TEM acquisition data, respectively.

Conflicts of Interest: The authors declare no conflict of interest.

References

1. Dusselier, M.; Van Wouwe, P.; Dewaele, A.; Jacobs, P.A.; Sels, B.F. Shape-selective zeolite catalysis for bioplastics production. *Science* **2015**, *349*, 78–80. [CrossRef] [PubMed]
2. Zhang, R.; Liu, N.; Lei, Z.; Chen, B. Selective transformation of various nitrogen-containing exhaust gases toward N2 over zeolite catalysts. *Chem. Rev.* **2016**, *116*, 3658–3721. [CrossRef] [PubMed]
3. Valtchev, V.; Majano, G.; Mintova, S.; Pérez-Ramírez, J. Tailored crystalline microporous materials by post-synthesis modification. *Chem. Soc. Rev.* **2013**, *42*, 263–290. [CrossRef] [PubMed]
4. Smit, B.; Maesen, T.L.M. Towards a molecular understanding of shape selectivity. *Nature* **2008**, *451*, 671–678. [CrossRef]
5. Deroche, I.; Daou, T.J.; Picard, C.; Coasne, B. Reminiscent capillarity in subnanopores. *Nat. Commun.* **2019**, *10*, 4642. [CrossRef] [PubMed]
6. Huve, J.; Ryzhikov, A.; Nouali, H.; Lalia, V.; Augé, G.; Daou, T.J. Porous sorbents for the capture of radioactive iodine compounds: A review. *RSC Adv.* **2018**, *8*, 29248–29273. [CrossRef]
7. Reddy, J.K.; Motokura, K.; Koyama, T.; Miyaji, A.; Baba, T. Effect of morphology and particle size of ZSM-5 on catalytic performance for ethylene conversion and heptane cracking. *J. Catal.* **2012**, *289*, 53–61. [CrossRef]
8. Zhang, H.; Hu, Z.; Huang, L.; Zhang, H.; Song, K.; Wang, L.; Shi, Z.; Ma, J.; Zhuang, Y.; Shen, W.; et al. dehydration of glycerol to acrolein over hierarchical ZSM-5 zeolites: Effects of mesoporosity and acidity. *ACS Catal.* **2015**, *5*, 2548–2558. [CrossRef]
9. Kabalan, I.; Lebeau, B.; Nouali, H.; Toufaily, J.; Hamieh, T.; Koubaissy, B.; Bellat, J.-P.; Daou, T.J. New generation of zeolite materials for environmental applications. *J. Phys. Chem. C* **2016**, *120*, 2688–2697. [CrossRef]
10. Astafan, A.; Pouilloux, Y.; Patarin, J.; Bats, N.; Bouchy, C.; Daou, T.J.; Pinard, L. Impact of extreme downsizing of *BEA-type zeolite crystals on n -hexadecane hydroisomerization. *New J. Chem.* **2016**, *40*, 4335–4343. [CrossRef]
11. Larlus, O.; Valtchev, V.P. Crystal morphology control of LTL-type zeolite crystals. *Chem. Mater.* **2004**, *16*, 3381–3389. [CrossRef]
12. Drews, T.O.; Tsapatsis, M. Progress in manipulating zeolite morphology and related applications. *Curr. Opin. Colloid Interface Sci.* **2005**, *10*, 233–238. [CrossRef]
13. Wakihara, T.; Ihara, A.; Inagaki, S.; Tatami, J.; Sato, K.; Komeya, K.; Meguro, T.; Kubota, Y.; Nakahira, A. Top-down tuning of nanosized ZSM-5 zeolite catalyst by bead milling and recrystallization. *Cryst. Growth Des.* **2011**, *11*, 5153–5158. [CrossRef]
14. Tao, Y.; Kanoh, H.; Abrams, L.; Kaneko, K. Mesopore-modified zeolites: Preparation, characterization, and applications. *Chem. Rev.* **2006**, *106*, 896–910. [CrossRef]
15. Pérez-Ramírez, J.; Verboekend, D.; Bonilla, A.; Abelló, S. Zeolite catalysts with tunable hierarchy factor by pore-growth moderators. *Adv. Funct. Mater.* **2009**, *19*, 3972–3979. [CrossRef]
16. Zhang, B.; Zhang, Y.; Hu, Y.; Shi, Z.; Azhati, A.; Xie, S.; He, H.; Tang, Y. Microexplosion under microwave irradiation: A facile approach to create mesopores in zeolites. *Chem. Mater.* **2016**, *28*, 2757–2767. [CrossRef]

17. Choi, M.; Na, K.; Kim, J.; Sakamoto, Y.; Terasaki, O.; Ryoo, R. Stable single-unit-cell nanosheets of zeolite MFI as active and long-lived catalysts. *Nature* **2009**, *461*, 246–249. [CrossRef]
18. Xu, D.; Ma, Y.; Jing, Z.; Han, L.; Singh, B.; Feng, J.; Shen, X.; Cao, F.; Oleynikov, P.; Sun, H.; et al. π–π interaction of aromatic groups in amphiphilic molecules directing for single-crystalline mesostructured zeolite nanosheets. *Nat. Commun.* **2014**, *5*, 4262. [CrossRef]
19. Singh, B.K.; Xu, D.; Han, L.; Ding, J.; Wang, Y.; Che, S. Synthesis of single-crystalline mesoporous ZSM-5 with three-dimensional pores via the self-assembly of a designed triply branched cationic surfactant. *Chem. Mater.* **2014**, *26*, 7183–7188. [CrossRef]
20. Kore, R.; Srivastava, R.; Satpati, B. ZSM-5 zeolite nanosheets with improved catalytic activity synthesized using a new class of structure-directing agents. *Chem. Eur. J.* **2014**, *20*, 11511–11521. [CrossRef]
21. Na, K.; Jo, C.; Kim, J.; Cho, K.; Jung, J.; Seo, Y.; Messinger, R.J.; Chmelka, B.F.; Ryoo, R. Directing zeolite structures into hierarchically nanoporous architectures. *Science* **2011**, *333*, 328–332. [CrossRef] [PubMed]
22. Dhainaut, J.; Daou, T.J.; Bidal, Y.; Bats, N.; Harbuzaru, B.; Lapisardi, G.; Chaumeil, H.; Defoin, A.; Rouleau, L.; Patarin, J. One-pot structural conversion of magadiite into MFI zeolite nanosheets using mononitrogen surfactants as structure and shape-directing agents. *Cryst. Eng. Comm* **2013**, *15*, 3009–3015. [CrossRef]
23. Rioland, G.; Albrecht, S.; Josien, L.; Vidal, L.; Daou, T.J. The influence of the nature of organosilane surfactants and their concentration on the formation of hierarchical FAU-type zeolite nanosheets. *New J. Chem.* **2015**, *39*, 2675–2681. [CrossRef]
24. Kabalan, I.; Rioland, G.; Nouali, H.; Lebeau, B.; Rigolet, S.; Fadlallah, M.-B.; Toufaily, J.; Hamiyeh, T.; Daou, T.J. Synthesis of purely silica MFI-type nanosheets for molecular decontamination. *RSC Adv.* **2014**, *4*, 37353–37358. [CrossRef]
25. El Hanache, L.; Lebeau, B.; Nouali, H.; Toufaily, J.; Hamieh, T.; Daou, T.J. Performance of surfactant-modified *BEA-type zeolite nanosponges for the removal of nitrate in contaminated water: Effect of the external surface. *J. Hazard. Mater.* **2019**, *364*, 206–217. [CrossRef]
26. El Hanache, L.; Sundermann, L.; Lebeau, B.; Toufaily, J.; Hamieh, T.; Daou, T.J. Surfactant-modified MFI-type nanozeolites: Super-adsorbents for nitrate removal from contaminated water. *Microporous Mesoporous Mater.* **2019**, *283*, 1–13. [CrossRef]
27. Lupulescu, A.I.; Rimer, J.D. Tailoring silicalite-1 crystal morphology with molecular modifiers. *Angew. Chem. Int. Ed.* **2012**, *51*, 3345–3349. [CrossRef]
28. Lupulescu, A.I.; Kumar, M.; Rimer, J.D. A facile strategy to design zeolite L crystals with tunable morphology and surface architecture. *J. Am. Chem. Soc.* **2013**, *135*, 6608–6617. [CrossRef] [PubMed]
29. Gaona Gomez, A.; de Silveira, G.; Doan, H.; Cheng, C.-H. A facile method to tune zeolite L crystals with low aspect ratio. *Chem. Commun.* **2011**, *47*, 5876–5878. [CrossRef]
30. Zhang, H.; Zhao, Y.; Zhang, H.; Wang, P.; Shi, Z.; Mao, J.; Zhang, Y.; Tang, Y. Tailoring zeolite ZSM-5 crystal morphology/porosity through flexible utilization of silicalite-1 seeds as templates: Unusual crystallization pathways in a heterogeneous system. *Chem. Eur. J.* **2016**, *22*, 7141–7151. [CrossRef]
31. Zhang, W.; Smirniotis, P.G. On the exceptional time-on-stream stability of HZSM-12 zeolite: Relation between zeolite pore structure and activity. *Catal. Lett.* **1999**, *60*, 223–228. [CrossRef]
32. Rosinski, E.J.; Rubin, M.K. Preparation of ZSM-12 Type Zeolites. U.S. Patent 4,391,785, 5 July 1983.
33. LaPierre, R.B.; Rohrman, A.C.; Schlenker, J.L.; Wood, J.D.; Rubin, M.K.; Rohrbaugh, W.J. The framework topology of ZSM-12: A high-silica zeolite. *Zeolites* **1985**, *5*, 346–348. [CrossRef]
34. Kokotailo, G.T.; Fyfe, C.A.; Feng, Y.; Grondey, H.; Gies, H.; Marler, B.; Cox, D.E. Powder X-ray diffraction and solid state NMR techniques forzeolite structure determination. In *Studies in Surface Science and Catalysis*; Beyer, H.K., Karge, H.G., Kiricsi, I., Nagy, J.B., Eds.; Catalysis by Microporous Materials Elsevier: Amsterdam, The Netherlands, 1995; Volume 94, pp. 78–100.
35. Pazzuconi, G.; Terzoni, G.; Perego, C.; Bellussi, G. Selective alkylation of naphthalene to 2,6-dimethylnaphthalene catalyzed by MTW zeolite. In *Studies in Surface Science and Catalysis*; Elsevier: Amsterdam, Netherlands, 2001; Volume 135, p. 152. ISBN 978-0-444-50238-4.
36. Pazzucconi, G.; Perego, C.; Millini, R.; Frigerio, F.; Mansani, R.; Rancati, D. Process for the Preparation of 2,6-dimethylnaphthalene Using a MTW Zeolitic Catalyst. U.S. Patent 6,147,270, 14 November 2000.
37. Dugkhuntod, P.; Imyen, T.; Wannapakdee, W.; Yutthalekha, T.; Salakhum, S.; Wattanakit, C. Synthesis of hierarchical ZSM-12 nanolayers for levulinic acid esterification with ethanol to ethyl levulinate. *RSC Adv.* **2019**, *9*, 18087–18097. [CrossRef]

38. Itani, L.; Valtchev, V.; Patarin, J.; Rigolet, S.; Gao, F.; Baudin, G. Centimeter-sized zeolite bodies of intergrown crystals: Preparation, characterization and study of binder evolution. *Microporous Mesoporous Mater.* **2011**, *138*, 157–166. [CrossRef]
39. Rioland, G.; Bullot, L.; Daou, T.J.; Simon-Masseron, A.; Chaplais, G.; Faye, D.; Fiani, E.; Patarin, J. Elaboration of FAU-type zeolite beads with good mechanical performances for molecular decontamination. *RSC Adv.* **2015**, *6*, 2470–2478. [CrossRef]
40. Rioland, G.; Daou, T.J.; Faye, D.; Patarin, J. A new generation of MFI-type zeolite pellets with very high mechanical performance for space decontamination. *Microporous Mesoporous Mater.* **2016**, *221*, 167–174. [CrossRef]
41. Fawaz, E.G.; Salam, D.A.; Nouali, H.; Deroche, I.; Rigolet, S.; Lebeau, B.; Daou, T.J. Synthesis of binderless ZK-4 zeolite microspheres at high temperature. *Molecules* **2018**, *23*, 2647. [CrossRef]
42. Tosheva, L.; Valtchev, V.; Sterte, J. Silicalite-1 containing microspheres prepared using shape-directing macro-templates. *Microporous Mesoporous Mater.* **2000**, *35–36*, 621–629. [CrossRef]
43. Mańko, M.; Vittenet, J.; Rodriguez, J.; Cot, D.; Mendret, J.; Brosillon, S.; Makowski, W.; Galarneau, A. Synthesis of binderless zeolite aggregates (SOD, LTA, FAU) beads of 10, 70μm and 1mm by direct pseudomorphic transformation. *Microporous Mesoporous Mater.* **2013**, *176*, 145–154. [CrossRef]
44. Didi, Y.; Said, B.; Cacciaguerra, T.; Nguyen, K.L.; Wernert, V.; Denoyel, R.; Cot, D.; Sebai, W.; Belleville, M.-P.; Sanchez-Marcano, J.; et al. Synthesis of binderless FAU-X (13X) monoliths with hierarchical porosity. *Microporous Mesoporous Mater.* **2019**, *281*, 57–65. [CrossRef]
45. Said, B.; Cacciaguerra, T.; Tancret, F.; Fajula, F.; Galarneau, A. Size control of self-supported LTA zeolite nanoparticles monoliths. *Microporous Mesoporous Mater.* **2016**, *227*, 176–190. [CrossRef]
46. Didi, Y.; Said, B.; Micolle, M.; Cacciaguerra, T.; Cot, D.; Geneste, A.; Fajula, F.; Galarneau, A. Nanocrystals FAU-X monoliths as highly efficient microreactors for cesium capture in continuous flow. *Microporous Mesoporous Mater.* **2019**, *285*, 185–194. [CrossRef]
47. Moukahhal, K.; Daou, T.J.; Josien, L.; Nouali, H.; Toufaily, J.; Hamieh, T.; Galarneau, A.; Lebeau, B. Hierarchical ZSM-5 beads composed of zeolite nanosheets obtained by pseudomorphic transformation. *Microporous Mesoporous Mater.* **2019**, *288*, 109565. [CrossRef]
48. Moukahhal, K.; Le, N.H.; Bonne, M.; Toufaily, J.; Hamieh, T.; Daou, T.J.; Lebeau, B. Controlled crystallization of hierarchical monoliths composed of nanozeolites. *Cryst. Growth Des.* **2020**, *20*, 5413–5423. [CrossRef]
49. Moukahhal, K.; Lebeau, B.; Josien, L.; Galarneau, A.; Toufaily, J.; Hamieh, T.; Daou, T.J. Synthesis of hierarchical zeolites with morphology control: Plain and hollow spherical beads of silicalite-1 nanosheets. *Molecules* **2020**, *25*, 2563. [CrossRef] [PubMed]
50. Landers, J.; Gor, G.Y.; Neimark, A.V. Density functional theory methods for characterization of porous materials. *Colloids Surf. A* **2013**, *437*, 3–32. [CrossRef]
51. Mintova, S.; Ristić, A.; Rangus, M.; Novak Tušar, N. *Verified Syntheses of Zeolitic Materials*; Synthesis Commission of the International Zeolite Association: Caen, France, 2016; ISBN 978-0-692-68539-6.
52. Thommes, M.; Cychosz, K.A. Physical adsorption characterization of nanoporous materials: Progress and challenges. *Adsorption* **2014**, *20*, 233–250. [CrossRef]
53. Galarneau, A.; Mehlhorn, D.; Guenneau, F.; Coasne, B.; Villemot, F.; Minoux, D.; Aquino, C.; Dath, J.-P. Specific surface area determination for microporous/mesoporous materials: The case of mesoporous FAU-Y zeolites. *Langmuir* **2018**, *34*, 14134–14142. [CrossRef]
54. Galarneau, A.; Villemot, F.; Rodriguez, J.; Fajula, F.; Coasne, B. Validity of the t-plot method to assess microporosity in hierarchical micro/mesoporous materials. *Langmuir* **2014**, *30*, 13266–13274. [CrossRef]
55. Carvalho, K.T.G.; Urquieta-Gonzalez, E.A. Microporous–mesoporous ZSM-12 zeolites: Synthesis by using a soft template and textural, acid and catalytic properties. *Catal. Today* **2015**, *243*, 92–102. [CrossRef]
56. Dimitrov, L.; Mihaylov, M.; Hadjiivanov, K.; Mavrodinova, V. Catalytic properties and acidity of ZSM-12 zeolite with different textures. *Microporous Mesoporous Mater.* **2011**, *143*, 291–301. [CrossRef]

Article

Synthesis of Pure Phase NaP2 Zeolite from the Gel of NaY by Conventional and Microwave-Assisted Hydrothermal Methods

Pimwipa Tayraukham [1], Nawee Jantarit [1], Nattawut Osakoo [1,2] and Jatuporn Wittayakun [1,*]

[1] School of Chemistry, Institute of Science, Suranaree University of Technology, Nakhon Ratchasima 30000, Thailand; pimwipatrk@gmail.com (P.T.); nawee1995@gmail.com (N.J.); nattawut.o@g.sut.ac.th (N.O.)

[2] Institute of Research and Development, Suranaree University of Technology, Nakhon Ratchasima 30000, Thailand

* Correspondence: jatuporn@sut.ac.th; Tel.: +66-44-224-256

Received: 29 September 2020; Accepted: 15 October 2020; Published: 19 October 2020

Abstract: The gel of zeolite NaY has potential as a precursor of other zeolites. The particular interest in this work is to convert the gel of NaY to NaP2. We found that the pure phase NaP2 can be produced simply by the conventional hydrothermal (CH) method at 150 °C for 24 h. This NaP2 sample, named CH150, has an average particle size of 10.3 µm and an Si/Al ratio of 1.82. In the case of single crystallization via microwave-assisted hydrothermal (MH) method, various parameters were studied, including the crystallization temperature (90, 150, 175 °C) and time (15, 30, 45, 60 min). The samples were analyzed by X-ray diffraction and scanning electron microscopy. However, mixed phases of P1 and P2 or ANA were obtained from all samples. Another attempt was made by a double crystallization via MH method as followed: at 90 °C for 1 h, quickly cooled down to room temperature in the microwave chamber and aged for 23 h, and finally at 150 °C for 1 h. The sample, named MH90A150, has an average crystal size of 16.45 µm and an Si/Al ratio of 1.85. The high Al content of NaP2 in both samples (CH150 and MH90A150) could lead to interesting applications.

Keywords: NaP2 zeolite; NaY gel; microwave-assisted hydrothermal; conventional hydrothermal

1. Introduction

Zeolite P has a gismondine (GIS) framework type, composed from secondary building units (SBU) containing four- and eight-membered rings of T atoms (T = Si or Al linked together by sharing oxygen atoms). Its pore diameter is 2.8 × 4.8 Å [010] [1] which makes it useful as molecular sieve [2]. Zeolite P is useful for many applications such as the separation of small gases [3–5] or heavy metals [6,7] and application in sensing material [8].

Baerlocher et al. and Mccusker et al. [9,10] have found the differences and flexibility in unit cell structures of zeolite P. The distortion of the SBU planes generates two isotypes, P1 and P2, which have different pore sizes and shapes. Likewise, Oleksiak et al. [11] have reported that the eight-membered pore of P2 is narrower than that of P1. Consequently, the P2 type has a higher adsorption equilibrium constant for small gases such as H_2 [11].

Despite the interesting advantage, the synthesis of P2 requires a long crystallization period and produces impurity phases. Table 1 compares the synthesis of zeolite P with various parameters. The standard procedure from the International Zeolite Association Synthesis Commission (IZA) is the conventional hydrothermal method (CH) which has a long crystallization time, i.e., 60 days (Table 1, entry 1) [12]. Several researchers have improved the synthesis by extending the nucleation time (Table 1, entry 2) [13], employing a structure-directing agent (SDA) (Table 1, entry 3) [14] and seeding (Table 1, entry 4) [15]. Although the synthesis time is shortened, it still takes several days. An approach for the fast synthesis is the microwave-assisted hydrothermal method (MH (Table 1, entry 5–6) [16,17] due to its rapid and homogeneous heating [18].)

Table 1. Gel composition and synthesis parameters of zeolite P from the literature.

Gel Composition	Main Product	Crystallization Time	Crystallization Temperature (°C)	Method	References
2.2 SiO_2: Al_2O_3: 5.28 NaF: 105.6 H_2O	P1	60 days	85	CH [1]	[12]
$Na_8Al_8Si_8O_{32} \cdot 15.2\ H_2O$	P2	5 days	100	CH [1]	[13]
3.75 SiO_2: 0.2 Al_2O_3: 4.3 Na_2O: 220 H_2O [4]	P1	42 h	100	CH [1]	[14]
0.0065 $NaAlO_2$: 0.023 $Na_2Si_3O_7$ (seed) [3] 0.04 $NaAlO_2$: 0.147 $Na_2Si_3O_7$ (feed)	P1	24 h	100	CH [1]	[15]
1 SiO_2: 0.25 Al_2O_3: 3 Na_2O: 410H_2O [5,6]	P1	3 h	110	MH [2]	[16]
17 Na_2O: Al_2O_3: 17 SiO_2: 345 H_2O (seed) [3] 3 Na_2O: Al_2O_3: 8 SiO_2: 209 H_2O (Overall)	P2	2 h	150	MH [2]	[17]

[1] CH = conventional hydrothermal method, [2] MH = microwave-assisted hydrothermal method, [3] use of seeding for nuclei preparation, use of structure-directing agent: [4] D-methionine, [5] triethanolamine, [6] triisopropanolamin.

The impurity phases from the conventional synthesis method can be minimized by recrystallization, also called double crystallization. For example, Sousa et al. [19] recrystallized ZSM-22 (structure type TON) and ZSM-35 (structure type FER) by basic treatment to produce a pure phase of zeolite NaP1. Nevertheless, there are no reports about the synthesis of zeolite NaP2 from recrystallization. Although zeolite NaP2 has been synthesized from NaY gel, the mixed zeolite phases were obtained [20]. Consequently, this study aimed to synthesize the high purity of zeolite NaP2 from the gel of zeolite NaY through double crystallization without the separation of solid product after the first heating. Seeding and microwave radiation will be used to decrease hydrothermal time and avoid uneven temperature.

2. Materials and Methods

2.1. Materials

Chemicals in this study were sodium hydroxide anhydrous pellets (NaOH, 97%, Carlo Erba), silicon dioxide (SiO_2, 99%, Carlo Erba) and sodium aluminate ($NaAlO_2$, 95%, Sigma Aldrich). Sodium silicate (Na_2SiO_3), which was a silicon source precursor, was prepared according to the following chemical composition: 28.7 wt% SiO_2, 8.9 wt% Na_2O. Moreover, nitric acid (HNO_3, 69%, ANaPURE), hydrochloric acid (HCl, 37%, ACI Labscan), hydrofluoric acid (HF, 49% QRëC) and boric acid (H_3BO_3, 99.5%, Merck) were used in the digestion step.

2.2. Preparation of Zeolite Synthesis Gel

The overall gel with the molar ratio of $10SiO_2$:Al_2O_3:$4.62Na_2O$:$180H_2O$ was prepared by the method from that of zeolite NaY [21] with the scale on-fourth, from a seed gel ($10SiO_2$:Al_2O_3:$10.67Na_2O$:$180H_2O$) and feedstock gel ($10SiO_2$:Al_2O_3:$4.30Na_2O$:$180H_2O$). It was aged for 24 h before crystallization. All the gel ratios were calculated from the weight of each component.

2.3. Single Crystallization

The overall gel was crystallized by conventional and microwave-assisted hydrothermal methods (CH and MH, respectively). In the CH method, the overall gel was poured into a Teflon-lined autoclave (250 mL) and heated with a rate of 5 °C/min to 150 °C and held for 24 h in a muffle furnace (Carbolite, CWF 12/23) and cooled down outside the furnace to room temperature. This sample was named CH150. In the MH method, the overall gel was transfused into two microwave tubes (100 mL) with equal amount and heated by microwave (Anton Paar, Multiwave 3000) with XF-100 rotor and the power of 900 W to 150 °C within 4 min, kept for the desired period of 1 h and cooled down quickly in the microwave chamber. This sample was named MH150.

To understand the temperature effect, the samples MH90 and MH175 were prepared by microwave-assisted method and crystallized for 1 h at 90 °C and 175 °C, respectively. The result of the time was investigated by crystallization at 150 °C and 175 °C, held for 15, 30, 45 and 60 min.

2.4. Double Crystallization via MH Method

There were three steps to investigate the phase change from each step. Firstly, the overall gel was transferred into the microwave tubes with an equal amount for evenness. The sample was heated by microwave from room temperature up to 90 °C and crystallized for 1 h before cooling down. This sample was named MH90. Secondly, the sample was kept unopened (aged) for 23 h at room temperature in the same microwave tubes. This sample was named MH90A. Finally, the sample after aging was crystallized again at 150 °C for 1 h. This sample was named MH90A150.

After crystallization, all solid products were separated by centrifugation, washed several times with deionized (DI) water until the pH of the filtrates was 7–8, and dried at 90 °C for 18 h to obtain white powder samples.

2.5. Characterization of Solid Products

Phases and structures of crystalline products were analyzed by powder X-ray diffraction (XRD) on a Bruker (AXS D8, Billerica, MA, USA) diffractometer using Cu Kα radiation with a current of 40 mA and a potential of 40 kV. All the samples were analyzed on the same day for a confident comparison of crystallinity. The functional groups of the samples were determined by Fourier-transform infrared spectroscopy (FTIR) on a Bruker (Tensor 27, Billerica, MA, USA), using attenuated total reflectance (ATR) mode with a resolution of 4^{-1}. Morphology was studied by a field emission scanning electron microscope (FESEM), JEOL (JSM-7800F, Tokyo, Japan). The silicon–aluminum (Si/Al) mole ratios in the solid products were analyzed by inductively coupled plasma optical emission spectrometry (ICP–OES) using an Perkin Elmer (Optima 8000, Waltham, MA, USA) with argon carrier gas. The samples were digested [22] as follows. Fifty milligrams of each sample was mixed in 0.5 mL aqua regia, $1HNO_3$:3HCl volumetric ratio, and 3.0 mL of HF. The mixture in a 250 mL polypropylene (PP) bottle was heated at 110 °C for 1 h and cooled down to room temperature. Then, 2.8 g of H_3BO_3 and 10 mL of DI water were added into the mixture and the volume was adjusted with DI water in a 100 mL PP volumetric flask. The emission wavelength of silicon and aluminum in solid samples were 251.611 and 396.153 nm, respectively.

3. Results and Discussion

3.1. Comparison of Conventional and Microwave-Assisted Hydrothermal Methods of Single Crystallization

Figure 1 compares the XRD patterns of zeolite products from single crystallization by the conventional hydrothermal method at 150 °C for 24 h (CH150) and microwave-assisted hydrothermal method at 150 °C for 1 h (MH150). The pattern of CH150 corresponds to the characteristic phase of zeolite NaP2. The splitting of peaks refers to two rotating types of secondary building units (SBUs) planes or four-membered ring (4MR) in GIS structure [11]. Other zeolite phases are not observed. In comparison with Khabuanchalad et al. [20], the crystallization of the gel of NaY at 100 °C for 5 days produced zeolite NaP2 with a trace of NaY. This work successfully synthesizes the pure phase NaP2 from the gel of zeolite NaY by the conventional hydrothermal method through crystallization at 150 °C for 24 h.

The XRD pattern of MH150 shows the mixed zeolite phases. The main phase is NaP1 and the minor phase is NaP2. At the same crystallization temperature, Le et al. [17] obtained NaP2 as the main phase with the trace of NaY. However, they used a different gel composition, microwave power (1600 W) and longer crystallization time (2 h). The crystallization at 100 °C and 120 °C only produced zeolite NaY [17]. When they used the gel with Si/Al = 5 and a crystallization time of 0.5 h, the product was pure phase NaP2.

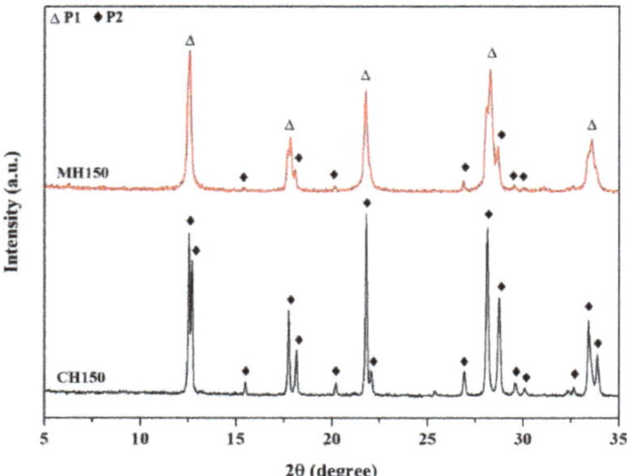

Figure 1. Powder X-ray diffraction (XRD) patterns of zeolite products from conventional and microwave-assisted hydrothermal methods with single crystallization at 150 °C for 24 h (CH150) and 1 h (MH150).

Figure 2a shows the SEM image of CH150. This sample has polycrystals with polygons, some have a diamond-like shape and some have trisecting cracks. The particle size is between 4 and 14 μm with an average of 10.3 μm (calculated from 50 particles). From the ICP analysis, the Si/Al ratio of this sample is 1.82. The morphology of CH150 is similar to zeolite P2 in the literature [11,17]. Although the sample has various shapes and morphology, the XRD results only indicate the pure phase of NaP2. Consequently, one can conclude that NaP2 has various morphology. In the case of MH150, the SEM image (see Figure 2b) shows larger particles with a different morphology from CH150. The average particle size of MH150 is 17.4 μm and the Si/Al ratio was 1.77. The major phase has a cheese ball-like morphology of NaP1 similar to those in the literature [23,24]. Those zeolite microcrystals are well grown on the microsphere surface and the size of each particle is below one micrometer. The observation of mixed phases from SEM is consistent with the XRD results.

Figure 2. A field emission scanning electron microscopy (FESEM) images of the zeolite products from conventional and microwave-assisted hydrothermal methods with single crystallization at 150 °C for 24 h (**a**, CH150) and 1 h (**b**, MH150).

From the XRD and SEM results, the phase of products depends on the heating methods. The faster heating rate by MH method provides the faster crystal growth rate, producing the larger particle size compared with the CH method. This behavior was also observed in the synthesis of zeolite NaY by CH and MH methods [25]. Moreover, the homogeneous heating from microwave results in the homogeneous growth of nuclei that affect the narrow particle size distribution of MH150. Therefore, the crystallization of MH is about 24 times faster than CH at the same temperature. Microwave radiation significantly reduced the time of crystallization due to fast spreading heat to all positions [26]. However, the fast heating rate results in mixed phases.

3.2. Effect of Temperature and Time on Single Crystallization by Microwave-Assisted Hydrothermal Methods

Figure 3a shows the XRD patterns of MH90, MH150 and MH175 which are the products from crystallization by microwave heating at 90, 150 and 175 °C, respectively. The MH90 sample has pure phase NaY zeolite [27]. This result is consistent with the work by Le et al. [17] that NaY was obtained at 100 and 120 °C. The sample MH150 shows mixed phases of NaP1 and NaP2 zeolites. Finally, the MH175 sample displays the mixed phases of NaP2 as the major phase and Anacime (ANA) zeolite as the minor phase [28]. The densities of faujasite (FAU) [29], GIS [1] and ANA [30] zeolites are 13.3, 16.4 and 19.2 T/1000 Å^3, respectively. The results confirm that the higher crystallization temperature produces zeolite with a higher framework density [17]. Although the exact densities of the two GIS zeolites are not clear, our results imply that the density of NaP2 is higher than NaP1 due to the smaller pore size [11]. From these results, the suitable crystallization temperature for NaP2 is 175 °C. However, the product still contains ANA zeolite as an impurity. Consequently, we investigated the shorter crystallization time and lower temperature to improve purity.

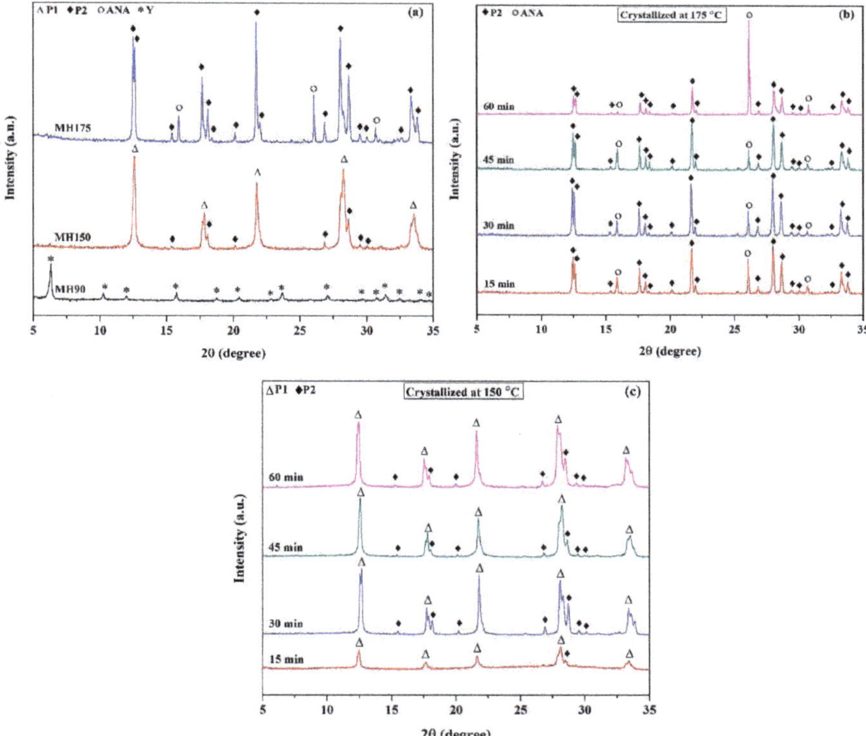

Figure 3. Powder XRD patterns of zeolites from single crystallization via MH (**a**) for 1 h at 90 °C, 150 °C and 175 °C, (**b**) at 175 °C, and (**c**) 150 °C for various times: 15, 30, 45 and 60 min.

Figure 3b displays the XRD patterns of zeolites from single crystallization by MH at 175 °C for 15, 30, 45 and 60 min. All samples still contain an ANA phase and the sample from 60 min has more ANA phase than the others. The observation agrees with the literature that the longer crystallization time generates a product with high density [31] with a competitive phase. Consequently, we investigated the crystallization at 150 °C to reduce the ANA phase. Figure 3c shows the XRD patterns of the samples crystallized for 15, 30, 45 and 60 min. The main phase of all samples was NaP1 [11] along with NaP2.

Figure 4 presents the SEM images of zeolites from single crystallization via MH at 150 °C for 15, 30, 45 and 60 min. The sample from 15 min has an amorphous phase which is from a short crystallization time. Moreover, other samples have a mixed phase of NaP1 (cheese ball like) [23,24] and NaP2 (polygonal) [11,17] consistent with the XRD patterns. The presence of the mixed-phase suggests that a longer aging time is necessary.

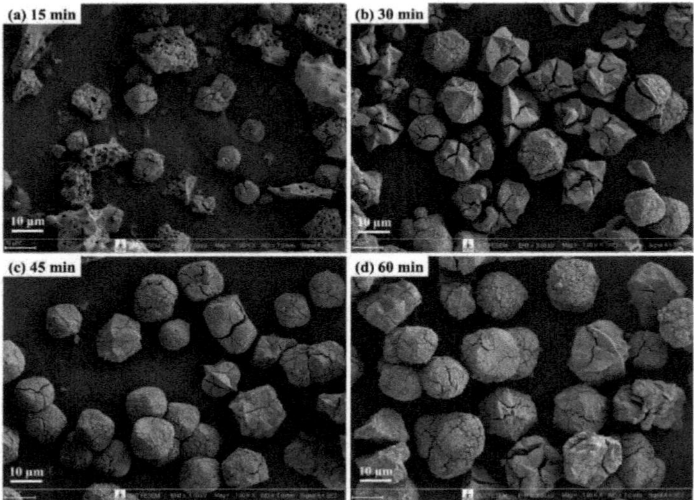

Figure 4. FESEM images of zeolite products from microwave-assisted hydrothermal methods with single crystallization at 150 °C for various times: (**a**) 15, (**b**) 30, (**c**) 45 and (**d**) 60 min.

3.3. *Effect of Double Crystallization by Microwave-Assisted Hydrothermal Methods*

Figure 5 compares the XRD patterns of MH90 with the further aged sample (MH90A) and crystallized at 150 °C (MH90A150). The patterns of MH90 and MH90A correspond to NaY [27]. The peaks from MH90A are stronger than those of MH90, indicating that the longer aging time improves crystallinity. After crystallization at 150 °C, the pattern of MH90A150 exhibits the pure phase of NaP2 zeolite. These results suggest that the synthesis condition of MH90A150 is suitable to transform the gel of zeolite NaY to NaP2. Concurrently, the second crystallization under microwave heating causes NaY to depolymerize the small building units [19] of NaP2 which has the denser structure.

Figure 6 compares the FTIR spectra of MH90 and MH90A150 and the interpretations are summarized in Table 2. The spectra of MH90 and MH90A150 correspond to the functional groups of zeolite NaY [32–34] and NaP [34–37], consistent with the XRD results.

Figure 7 exhibits the SEM images of MH90 and MH90A150. The morphology of MH90 (Figure 7a) is a polygon shape similar to the NaY zeolite in the literature [38]. The average particle size of MH90 is 230 nm and the Si/Al ratio is 1.88. The MH90A150 particles (Figure 7b) have the polygonal morphology, similar to NaP2 zeolite in the literature [11,17]. To ensure the reproducibility of pure phase NaP2 zeolite from this method, we repeated the synthesis with the same manners. The similar XRD patterns (Figure 8) indicate that pure NaP2 could be produced by our synthesis method. The average particle size of MH90A150 is 16.45 µm and the Si/Al ratio is 1.85. The results are consistent with the XRD and FTIR data. Comparing between NaP2 from CH150 and MH90A150, the particle size distribution of MH90A150 is narrower than that of CH150 (standard deviation of 7.5% vs. 27.6%).

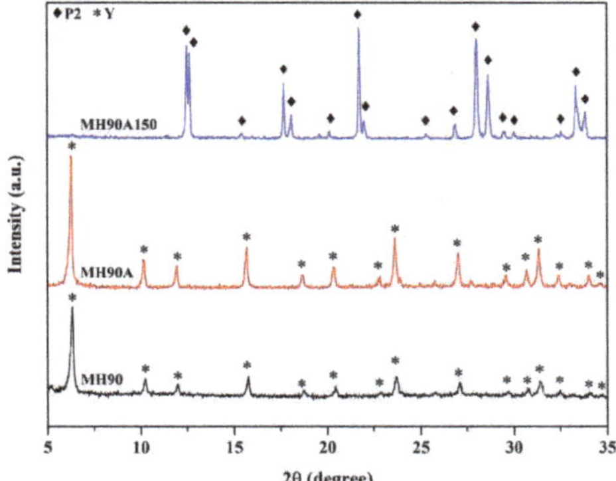

Figure 5. XRD patterns of zeolites from microwave-assisted hydrothermal method at 90 °C for 1 h (MH90), aged at room temperature for 23 h (MH 90A) and crystallized at 150 °C for 1 h (MH90A150).

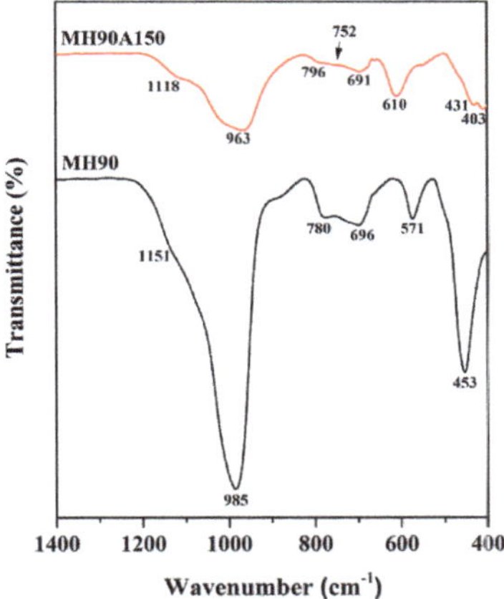

Figure 6. Fourier-transform infrared spectroscopy (FTIR)–attenuated total reflectance (ATR) spectra of zeolite products: MH90 from the microwave-assisted hydrothermal method at 90 °C for 1 h and MH90A150 from the microwave-assisted hydrothermal method at 90 °C for 1 h followed by aging at room temperature for 23 h and recrystallization at 150 °C for 1 h.

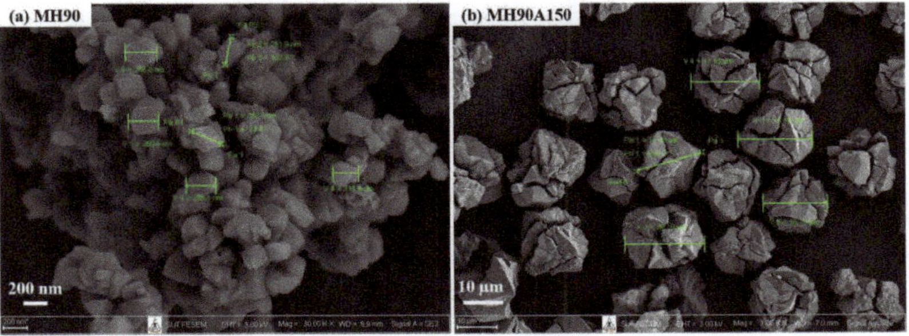

Figure 7. FESEM images of zeolites from the microwave-assisted hydrothermal method (**a**) at 90 °C for 1 h (MH90), followed by aging at room temperature for 23 h and recrystallized (**b**) at 150 °C for 1 h (MH90A150).

Table 2. Band assignment of FTIR spectra of MH90 (FAU) and MH90A150 (GIS) samples.

Wavenumber (cm^{-1})	Assignment	References
FAU zeolite		[32–34]
1151	Asymmetric stretching vibration of external linkage TO_4 (T = Si, Al)	
985	Asymmetric stretching vibration of internal tetrahedral TO_4	
780	Symmetric stretching vibration of external linkage TO_4	
696	Symmetric stretching vibration of internal linkage TO_4	
571	Vibration of double-6-membered ring (D6R)	
453	Bending vibration of T–O bond in internal tetrahedron	
GIS zeolite		[34–37]
1118	Asymmetric stretching vibration of external linkage TO_4	
963	Asymmetric stretching vibration of internal tetrahedral TO_4	
796	Symmetric stretching vibrations of external linkage TO_4	
752	Vibration of 4-membered ring (4MR)	
691	Symmetric stretching vibration of internal linkage TO_4	
610	Double ring vibration in GIS zeolite framework	
431, 403	Bending vibration of T–O	

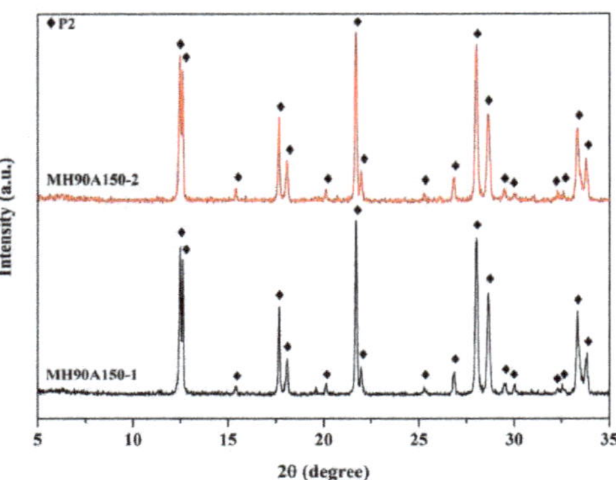

Figure 8. Powder XRD patterns of NaP2 zeolites from microwave-assisted double crystallization for repetition at the first time (MH90A150–1) and the second time (MH90A150–2).

All the conditions and results from this work are summarized in Table 3. The pure phase NaP2 can be synthesized by two methods: CH method at 150 °C for 24 h and double crystallization via MH method at 90 °C and 150 °C. Single crystallization via MH method at 90 °C produced pure phase NaY but at 150 and 175 °C produced mixed phases of NaP2 and NaP1 or ANA.

Table 3. Gel composition, synthesis parameters and phase of the products in this study.

Sample Name	Crystallization Conditions				Phase of Product [4]	
	Method	Synthesis Process	Temperature (°C)	Time (min)	Major	Minor
CH150	CH [1]	Single	150	1440	NaP2	-
MH90	MH [2]	Single	90	60	NaY	-
MH150	MH [2]	Single	150	15–60	NaP1	NaP2
MH175	MH [2]	Single	90	15–60	NaP2	ANA
MH90A	MH [2]	Single	90 and aged [3]	60	NaY	-
MH90A150	MH [2]	Double	90, 150	60, 60	NaP2	-

[1] CH = conventional hydrothermal method, [2] MH = microwave-assisted hydrothermal method, [3] crystallization at 90 °C followed by aging at room temperature for 23 h, [4] measurement by XRD technique.

4. Conclusions

The synthesis gel of zeolite NaY with the molar ratio $10SiO_2{:}Al_2O_3{:}4.62Na_2O{:}180H_2O$ was prepared and used as a precursor of zeolite NaP. The pure phase zeolite NaP2 was obtained after aging the synthesis gel of zeolite NaY for 24 h followed by the crystallization by conventional hydrothermal (CH) method at 150 °C for 24 h. The sample (CH150) has polycrystals with polygons, some have a diamond-like shape, and some have trisecting cracks. The average size was 10.3 μm and the Si/Al ratio was 1.82.

From the single crystallization via microwave-assisted hydrothermal (MH) method, the pure phase NaY, the mixed phase of NaP1 (major) and NaP2 (minor), and the mixed phase of NaP2 (major) and ANA (minor) were produced after crystallization at 90 °C, 150 °C and 175 °C for 1 h, respectively. The crystallization for various times (15–60 min) at 150 °C and 175 °C also produced the mixed phases.

From double crystallization via microwave-assisted hydrothermal (MH) method, the pure phase zeolite NaP2 was obtained by three steps: (1) crystallization at 90 °C for 1 h and quickly cooling down to room temperature in the microwave chamber, (2) aging at room temperature for 23 h, and (3) crystallization at 150 °C for 1 h. This NaP2 sample has a similar morphology to the CH150 sample with an average crystal size of 16.45 μm and Si/Al ratio 1.85.

Author Contributions: Conceptualization, P.T., N.O., and J.W., methodology, P.T. and N.J., validation, J.W., formal analysis, P.T. and N.J., investigation, P.T. and N.J., data curation, P.T., writing—original draft preparation, P.T., writing—review and editing, P.T., N.J., N.O., and J.W., supervision, J.W., project administration, P.T., funding acquisition, P.T. All authors have read and agreed to the published version of the manuscript.

Funding: P. Tayruakham is supported by the Science Achievement Scholarship of Thailand (SAST). N. Jantarit is supported by the scholarship from the Thai government under the Development and Promotion of Science and Technology Talents (DPST) Project. N. Osakoo is supported by full-time Doctoral Researcher Grants from Suranaree University of Technology.

Conflicts of Interest: The authors declare no conflict of interest. The funders had no role in the design of the study; in the collection, analyses, or interpretation of the data; in the writing of the manuscript, or in the decision to publish the results.

References

1. Baerlocher, C.; McCusker, L.B.; Olson, D.H. GIS—I41/amd. In *Atlas of Zeolite Framework*, 6th ed.; Baerlocher, C., McCusker, L.B., Olson, D.H., Eds.; Elsevier Science B.V.: Amsterdam, The Netherlands, 2007; pp. 146–147. [CrossRef]
2. Flanigen, E.M. Chapter 2 Zeolites and molecular sieves: An historical perspective. *Stud. Surf. Sci. Catal.* **2001**, *35*, 137. [CrossRef]
3. Yin, X.; Zhu, G.; Wang, Z.; Yue, N.; Qiu, S. Zeolite P/NaX composite membrane for gas separation. *Microporous Mesoporous Mater.* **2007**, *105*, 156–162. [CrossRef]
4. Dong, J.; Lin, Y. In Situ Synthesis of P-Type Zeolite Membranes on Porous α-Alumina Supports. *Ind. Eng. Chem. Res.* **1998**, *37*, 2404–2409. [CrossRef]
5. Fischer, M.; Bell, R.G. Influence of Zeolite Topology on CO_2/N_2 Separation Behavior: Force-Field Simulations Using a DFT-Derived Charge Model. *J. Phys. Chem. C* **2012**, *116*, 26449–26463. [CrossRef]
6. Ali, I.O.; El-Sheikh, S.M.; Salama, T.M.; Bakr, M.F.; Fodial, M.H. Controllable synthesis of NaP zeolite and its application in calcium absorption. *Sci. China Mater.* **2015**, *58*, 621–633. [CrossRef]
7. Nery, J.G.; Mascarenhas, Y.P.; Cheetham, A.K. A study of the highly crystalline, low-silica, fully hydrated zeolite P ion exchanged with (Mn^{2+}, Cd^{2+}, Pb^{2+}, Sr^{2+}, Ba^{2+}) cations. *Microporous Mesoporous Mater.* **2003**, *57*, 229–248. [CrossRef]
8. Arrigo, I.; Caprì, M.; Corigliano, F.; Bonavita, A.; Rizzo, G.; Neri, G.; Donato, N. An Exploratory Study on the Potential of Zeolite P-Based Materials for Sensing Applications. In *Sensors and Microsystems*; Springer: Berlin/Heidelberg, Germany, 2011; Volume 91, pp. 67–71. [CrossRef]
9. Baerlocher, C.; Meier, W.M. The crystal structure of synthetic zeolite Na-P 1, an isotype of gismondine. *Zeitschrift für Kristallographie—Cryst. Mater.* **1972**, *135*, 339–354. [CrossRef]

10. McCusker, L.; Baerlocher, C.; Nawaz, R. Rietveld refinement of the crystal structure of the new zeolite mineral gobbinsite. *Zeitschrift für Kristallographie—Cryst. Mater.* **1985**, *171*, 281–289. [CrossRef]
11. Oleksiak, M.D.; Ghorbanpour, A.; Conato, M.T.; McGrail, B.P.; Grabow, L.C.; Motkuri, R.K.; Rimer, J.D. Synthesis Strategies for Ultrastable Zeolite GIS Polymorphs as Sorbents for Selective Separations. *Chem. A Eur. J.* **2016**, *22*, 15961. [CrossRef]
12. Rees, L.V.C.; Chandrasekhar, S. Chapter 51—GIS zeolite P Si(54), Al(46). In *Verified Syntheses of Zeolitic Materials*, 2nd ed.; Robson, H., Lillerud, K.P., Eds.; Elsevier Science: Amsterdam, The Netherlands, 2001; pp. 169–170. [CrossRef]
13. Albert, B.; Cheetham, A.; Stuart, J.; Adams, C. Investigations on P zeolites: Synthesis, characterisation, and structure of highly crystalline low-silica NaP. *Microporous Mesoporous Mater.* **1998**, *21*, 133–142. [CrossRef]
14. Azizi, S.N.; Tilami, S.E. Recrystallization of Zeolite Y to Analcime and Zeolite P with D-Methionine as Structure-Directing Agent (SDA). *Zeitschrift für Anorg. und Allg. Chem.* **2009**, *635*, 2660–2664. [CrossRef]
15. Yamni, K.; Tijani, N.; Lamsiah, S. Influence of the crystallization temperatureon the synthesis of Y zeolite. *IOSR J. Appl. Chem.* **2014**, *7*, 58–61. [CrossRef]
16. Sathupunya, M.; Gulari, E.; Wongkasemjit, S. ANA and GIS zeolite synthesis directly from alumatrane and silatrane by sol-gel process and microwave technique. *J. Eur. Ceram. Soc.* **2002**, *22*, 2305–2314. [CrossRef]
17. Le, T.; Wang, Q.; Pan, B.; Ravindra, A.; Ju, S.; Peng, J. Process regulation of microwave intensified synthesis of Y-type zeolite. *Microporous Mesoporous Mater.* **2019**, *284*, 476–485. [CrossRef]
18. Yang, G.; Park, S.-J. Conventional and Microwave Hydrothermal Synthesis and Application of Functional Materials: A Review. *Materials* **2019**, *12*, 1177. [CrossRef] [PubMed]
19. Sousa, L.V.; Silva, A.O.S.; Silva, B.J.; Quintela, P.H.; Barbosa, C.B.; Frety, R.; Pacheco, J.G.A. Preparation of zeolite P by desilication and recrystallization of zeolites ZSM-22 and ZSM-35. *Mater. Lett.* **2018**, *217*, 259–262. [CrossRef]
20. Khabuanchalad, S.; Khemthong, P.; Prayoonpokarach, S.; Wittayakun, J. Transformation of zeolite NaY synthesized from rice husk silica to NaP during hydrothermal synthesis. *Suranaree J. Sci. Technol.* **2008**, *15*, 225–231.
21. Ginter, D.M.; Bell, A.T.; Radke, C.J. Chapter 46—FAU Linde type Y Si(71), Al(29). In *Verified Syntheses of Zeolitic Materials*, 2nd ed.; Robson, H., Lillerud, K.P., Eds.; Elsevier Science: Amsterdam, The Netherlands, 2001; pp. 156–158. [CrossRef]
22. Bunmai, K.; Osakoo, N.; Deekamwong, K.; Rongchapo, W.; Keawkumay, C.; Chanlek, N.; Prayoonpokarach, S.; Wittayakun, J. Extraction of silica from cogon grass and utilization for synthesis of zeolite NaY by conventional and microwave-assisted hydrothermal methods. *J. Taiwan Inst. Chem. Eng.* **2018**, *83*, 152–158. [CrossRef]
23. Azizi, S.N.; Daghigh, A.A.; Abrishamkar, M. Phase Transformation of Zeolite P to Y and Analcime Zeolites due to Changing the Time and Temperature. *J. Spectrosc.* **2013**, *2013*, 1–5. [CrossRef]
24. Sharma, P.; Song, J.-S.; Han, M.H.; Cho, C.-H. GIS-NaP1 zeolite microspheres as potential water adsorption material: Influence of initial silica concentration on adsorptive and physical/topological properties. *Sci. Rep.* **2016**, *6*, 22734. [CrossRef]
25. Panzarella, B.; Tompsett, G.A.; Yngvesson, K.S.; Conner, W.C. Microwave Synthesis of Zeolites. 2. Effect of Vessel Size, Precursor Volume, and Irradiation Method. *J. Phys. Chem. B* **2007**, *111*, 12657–12667. [CrossRef]
26. Li, Y.; Yang, W. Microwave synthesis of zeolite membranes: A review. *J. Membr. Sci.* **2008**, *316*, 3–17. [CrossRef]
27. Bunmai, K.; Osakoo, N.; Deekamwong, K.; Kosri, C.; Khemthong, P.; Wittayakun, J. Fast synthesis of zeolite NaP by crystallizing the NaY gel under microwave irradiation. *Mater. Lett.* **2020**, *272*, 127845. [CrossRef]
28. Gu, B.; Bai, J.; Yang, W.; Li, C. Synthesis of ANA-zeolite-based Cu nanoparticles composite catalyst and its regularity in styrene oxidation. *Microporous Mesoporous Mater.* **2019**, *274*, 318–326. [CrossRef]
29. Baerlocher, C.; McCusker, L.B.; Olson, D.H. FAU—Fd3¯m. In *Atlas of Zeolite Framework*, 6th ed.; Baerlocher, C., McCusker, L.B., Olson, D.H., Eds.; Elsevier Science B.V.: Amsterdam, The Netherlands, 2007; pp. 140–141. [CrossRef]
30. Baerlocher, C.; McCusker, L.B.; Olson, D.H. ANA—Ia3¯d. In *Atlas of Zeolite Framework*, 6th ed.; Baerlocher, C., McCusker, L.B., Olson, D.H., Eds.; Elsevier Science B.V.: Amsterdam, The Netherlands, 2007; pp. 46–47. [CrossRef]
31. Maldonado, M.; Oleksiak, M.D.; Chinta, S.; Rimer, J.D. Controlling Crystal Polymorphism in Organic-Free Synthesis of Na-Zeolites. *J. Am. Chem. Soc.* **2013**, *135*, 2641–2652. [CrossRef]

32. Kulawong, S.; Chanlek, N.; Osakoo, N. Facile synthesis of hierarchical structure of NaY zeolite using silica from cogon grass for acid blue 185 removal from water. *J. Environ. Chem. Eng.* **2020**, *8*, 104114. [CrossRef]
33. Shariatinia, Z.; Bagherpour, A. Synthesis of zeolite NaY and its nanocomposites with chitosan as adsorbents for lead(II) removal from aqueous solution. *Powder Technol.* **2018**, *338*, 744–763. [CrossRef]
34. Mozgawa, W.; Król, M.; Barczyk, K. FT-IR studies of zeolites from different structural groups. *Chemik* **2011**, *65*, 671–674.
35. Sharma, P.; Yeo, J.-G.; Han, M.-H.; Cho, C.H. Knobby surfaced, mesoporous, single-phase GIS-NaP1 zeolite microsphere synthesis and characterization for H_2 gas adsorption. *J. Mater. Chem. A* **2013**, *1*, 2602–2612. [CrossRef]
36. Chandrasekhar, S.; Pramada, P. Kaolin-based zeolite Y, a precursor for cordierite ceramics. *Appl. Clay Sci.* **2004**, *27*, 187–198. [CrossRef]
37. Flanigen, E.M.; Khatami, H.; Szymanski, H.A. Infrared Structural Studies of Zeolite Frameworks. In *Advances in Chemistry*; American Chemical Society (ACS): Washington, DC, USA, 1974; Volume 101, pp. 201–229.
38. Qamar, M.; Baig, I.; Azad, A.-M.; Ahmed, M.; Qamaruddin, M. Synthesis of mesoporous zeolite Y nanocrystals in octahedral motifs mediated by amphiphilic organosilane surfactant. *Chem. Eng. J.* **2016**, *290*, 282–289. [CrossRef]

MDPI

Article

Thermophysical Properties of Kaolin–Zeolite Blends up to 1100 °C

Ján Ondruška [1], Tomáš Húlan [1], Ivana Sunitrová [1], Štefan Csáki [2], Grzegorz Łagód [3], Alena Struhárová [4] and Anton Trník [1,5,*]

1 Department of Physics, Faculty of Natural Sciences, Constantine the Philosopher University in Nitra, Tr. A. Hlinku 1, 94974 Nitra, Slovakia; jondruska@ukf.sk (J.O.); thulan@ukf.sk (T.H.); ivka.sunitrova@gmail.com (I.S.)
2 Institute of Plasma Physics, Czech Academy of Sciences, Za Slovankou 3, 18200 Prague, Czech Republic; csaki@ipp.cas.cz
3 Faculty of Environmental Engineering, Lublin University of Technology, Nadbystrzycka 40B, 20-618 Lublin, Poland; g.lagod@pollub.pl
4 Department of Materials Engineering and Physics, Faculty of Civil Engineering, Slovak University of Technology, Radlinskeho 11, 810 05 Bratislava, Slovakia; alena.struharova@stuba.sk
5 Department of Materials Engineering and Chemistry, Faculty of Civil Engineering, Czech Technical University in Prague, Thákurova 7, 166 29 Prague, Czech Republic
* Correspondence: atrnik@ukf.sk; Tel.: +421-37-6408-616

Abstract: In this study, the thermophysical properties such as the thermal expansion, thermal diffusivity and conductivity, and specific heat capacity of ceramic samples made from kaolin and natural zeolite are investigated up to 1100 °C. The samples were prepared from Sedlec kaolin (Czech Republic) and natural zeolite (Nižný Hrabovec, Slovakia). Kaolin was partially replaced with a natural zeolite in the amounts of 10, 20, 30, 40, and 50 mass%. The measurements were performed on cylindrical samples using thermogravimetric analysis, a horizontal pushrod dilatometer, and laser flash apparatus. The results show that zeolite in the samples decreases the values of all studied properties (except thermal expansion), which is positive for bulk density, porosity, thermal diffusivity, and conductivity. It has a negative effect for thermal expansion because shrinkage increases with the zeolite content. Therefore, the optimal amount of zeolite in the sample (according to the studied properties) is 30 mass%.

Keywords: kaolin; zeolite; kaolinite; clinoptilolite; thermal expansion; thermal diffusivity; thermal conductivity; specific heat capacity

Citation: Ondruška, J.; Húlan, T.; Sunitrová, I.; Csáki, Š.; Łagód, G.; Struhárová, A.; Trník, A. Thermophysical Properties of Kaolin–Zeolite Blends up to 1100 °C. *Crystals* **2021**, *11*, 165. https://doi.org/10.3390/cryst11020165

Academic Editor: Magdalena Król
Received: 4 January 2021
Accepted: 3 February 2021
Published: 7 February 2021

1. Introduction

Ceramic production is known as one of the oldest sectors of human activity. Commonly used materials in the production of traditional ceramics are kaolin, illitic clays, feldspars, quartz, and Al_2O_3. Traditional ceramics are usually used in the building industry, such as bricks and tiles. The partial substitution of raw materials with waste or new materials can improve the properties of ceramic products and also reduce the cost of their production [1]. Nowadays, many published studies deal with partial substitution of traditional input raw materials for production of ceramics by waste materials such as fly or bottom ash [2–12], waste glass [13–16], waste calcite [17], etc. Húlan et al. [7] determined that a higher Young's modulus was reached after sintering with a lower amount of fly ash. The Young's modulus and the flexural strength decreased linearly with the amount of fly ash. Hasan et al. [14] observed an increase in the compressive strength and a decrease in water absorption of the samples with the addition of waste glass. They also found that the partial replacement of natural clay in a brick with waste soda-lime glass made the brick production sustainable and eco-friendly. Kováč et al. [17] showed that a high content of waste calcite may double the energy consumption during the creation of anorthite at a temperature of 950 °C and also that the waste calcite had a slight positive effect on the final contraction of samples.

In this study, kaolin and natural zeolite were used for the preparation of blends. Kaolin is the most commonly used material in the ceramic and paper industries, cosmetics, medicine, etc. It mainly consists of mineral kaolinite ($Si_2Al_2O_5(OH)_4$) with about 80 mass% depending on its origin and type [18]. Kaolinite belongs to the group of phyllosilicates and it crystalizes in the triclinic crystal system. Kaolinite has a 1:1 sheet structure composed of tetrahedral $[Si_2O_5]^{2-}$ sheets and octahedral $[Al_2(OH)_4]^{2+}$ sheets. It has a pseudo-hexagonal symmetry with the lattice parameters a = 0.515 nm, b = 0.895 nm, c = 0.740 nm, α = 91.68°, β = 104.87°, and γ = 89.9° [19–21]. The distance between layers is 0.72 nm [22]. During heating up to 1100 °C, three important thermal reactions take place in kaolinite. The first is the dehydration, which occurs in the temperature interval from 35 to 250 °C, where physically bound water is removed from the surface of crystals and pores [23]. The second reaction is the dehydroxylation of kaolinite [19,21,24–26], where the chemically bound water escapes from its structure and kaolinite is transformed into a new phase metakaolinite. It may be described by the following equation:

$$Al_2O_3 \cdot 2SiO_2 \cdot 2H_2O \rightarrow Al_2O_3 \cdot 2SiO_2 + 2H_2O_{(g)}. \quad (1)$$

This reaction takes place in the temperature interval from 450 to 700 °C. Metakaolinite ($Al_2Si_2O_7$) has a similar structure to kaolinite, but the lattice parameter c is changed to 0.685 nm (it is smaller by about 0.055 nm). The structure of metakaolinite does not contain OH^- ions, and the distance between the layers is shorter than in kaolinite. In addition, metakaolinite is more defective and less stable than kaolinite [24,25,27]. The third and last process begins above 925 °C and is connected with the transformation of metakaolinite to an Al-Si spinel, γ–Al_2O_3, and amorphous SiO_2, according to the following equation [28,29]:

$$2(Al_2O_3 \cdot 2SiO_2) \rightarrow 2Al_2O_3 \cdot 3SiO_2 + SiO_{2(amorphous)}. \quad (2)$$

Moreover, above the temperature of 700 °C, solid-state sintering occurs, which is a high-temperature technological process that transforms individual ceramic particles into a compact polycrystalline body. For traditional, kaolin-based ceramics (earthenware, stoneware or pottery), solid-state sintering occurs when the powder compact is densified entirely in the solid state [24,25,30,31].

Zeolites are microporous, hydrated crystalline aluminosilicates which are porous and widely used due to their structure and absorption properties. Many studies deal with natural or synthetic zeolites [32–41]. Usually, zeolites contain alkaline metals or metals of alkaline earth and frequently (e.g., in the case of clinoptilolite) crystallize in a monoclinic crystal system [32]. The three-dimensional structure of zeolites consists of tetrahedral silicate and aluminum, which are interconnected by oxygen atoms. The charge of their structure is negative and this charge can balance between monovalent and divalent cations [33,34]. Zeolites are porous and they are widely used due to their absorption properties (for example, in agriculture, ecology, the rubber industry, the building industry, households, and medicine [35–38]). Several important processes occur in zeolites during heating. Above the temperature of 100 °C, physically bound water escapes from the crystal surface and pores (capillaries) [39]. During heating up to 900 °C, the infrared spectral features attributed to the Si (Al)-O stretching and bending vibration modes do not show significant differences from the features for unheated (raw) zeolite. These spectral results are consistent with the fact that the three-dimensionally rigid crystal structure of zeolite is more stable than the layer structure of phyllosilicates [40]. Above the temperature of 1000 °C, the structure of zeolite is definitely destroyed and an amorphous phase is formed [21,33,41].

Our previous study [42] concerned the thermal expansion and Young's modulus of samples made from kaolin and zeolite. The samples were not studied in-situ; instead they were preheated at different temperatures from room temperature up to 1100 °C, and selected properties were measured after cooling at room temperature. The next study [43] focused on comparing the thermal expansion of the kaolin–zeolite and illite–zeolite samples. Our previous paper [39] aimed at conducting a thermogravimetric analysis and differential scanning calorimetry of a kaolin–zeolite sample. The aim of this paper was to estimate the influence of natural zeolite in the kaolin–zeolite samples on their thermophysical properties, such as the thermal expansion, thermal diffusivity and conductivity, and specific heat capacity, during heating up to 1100 °C and to determine the possible usage of natural zeolite in ceramic materials.

2. Materials and Methods

Samples were prepared from Sedlec kaolin (Czech Republic) and natural zeolite (Nižný Hrabovec, Slovakia). The major mineral in Sedlec kaolin is kaolinite (77.8%). In addition, there are impurities such as mica clay (17.4%) and quartz (1.5%) [44]. Natural zeolite mainly contains mineral clinoptilolite (58.2%) from the group of heulandite and has impurities such as cristobalite (12.2%), illite with mica and feldspar (albite) (9.6%), quartz (0.7%), and also amorphous phase (19.3%) [45]. The chemical compositions of the Sedlec kaolin and natural zeolite are given in Table 1.

Table 1. The chemical compositions of the Sedlec kaolin and natural zeolite (in mass%).

Oxides	Kaolin	Zeolite
SiO_2	46.9–47.9	65.0–71.3
Al_2O_3	36.6–37.6	11.5–13.1
Fe_2O_3	1.2	0.7–1.9
CaO	0.7	2.7–5.2
MgO	0.5	0.6–1.2
Na_2O	0.1	0.2–1.3
K_2O	0.8–1.1	2.2–3.4
TiO_2	0.5	0.1–0.3

The samples were prepared as follows. Kaolin was partially replaced with natural zeolite in the amounts of 10, 20, 30, 40, and 50 mass%. Pure kaolin and zeolite samples were also prepared. The studied samples were labeled as KZ10, KZ20, KZ30, KZ40, and KZ50, according to the natural zeolite content, whereas the pure kaolin and zeolite samples were labeled as SLA and ZEO (see Table 2). The kaolin pellets were crushed and milled to pass a 100-μm sieve. Zeolite was used as a powder, passing a 50-μm sieve. After these procedures, the powders were mixed with deionized water to obtain a plastic mass. Cylindrical samples with a diameter of 14 mm were extruded from this mass. Then, the samples were dried in open air until an equilibrium of moisture was reached (from 1.4 to 2.9 mass% of the physically bound water). The dry samples were cut to the lengths needed for the analyses.

Table 2. The compositions of the samples made from Sedlec kaolin and natural zeolite (in mass%).

Sample	SLA	KZ10	KZ20	KZ30	KZ40	KZ50	ZEO
Kaolin	100	90	80	70	60	50	–
Zeolite	–	10	20	30	40	50	100

Differential thermal analysis (DTA) and thermogravimetry (TG) of the compact samples (∅14 × 16 mm) with a mass of 3.5 g were performed by means of a Derivatograph 1000 analyzer (MOM Budapest, Budapest, Hungary) [46], in which a pressed alumina reference sample with similar dimensions to the studied sample was used. Thermodilatometry (TDA) was carried out using a horizontal pushrod alumina dilatometer [47] on samples with dimensions of ∅14 × 35 mm. All measurements were performed in the temperature interval from 30 to 1100 °C in static air atmosphere at a heating rate of 5 $°C \cdot min^{-1}$.

Differential scanning calorimetry (DSC) was performed on a Netzsch DSC 404 F3 Pegasus apparatus (NETZSCH Holding, Selb, Germany) in dynamic argon atmosphere with a flow rate of 40 ml/min. Al_2O_3 crucibles with a lid and powder samples with mass of ~30 mg were used. The temperature increased linearly with the heating rate of 5 °C/min and in the temperature interval from 30 up to 1100 °C. In order to determine the enthalpy of reactions, a baseline of the tangential type was selected.

The bulk density was calculated from thermogravimetric and thermodilatometric results to obtain its actual values during firing, according to the following equation:

$$\rho = \rho_0 \frac{\Delta m / m_0}{(1 + \Delta l / l_0)^3}, \tag{3}$$

where ρ_0 is the bulk density of green samples at room temperature.

The open porosity was calculated with the help of the experimentally determined bulk density and matrix density. The bulk density was obtained from the volume and mass of the cylindrical samples. The matrix density was measured by means of helium pycnometry (Pycnomatic ATC, Thermo Fisher Scientific, Waltham, MA USA).

Measurements of the thermal diffusivity (a), thermal conductivity (λ), and specific heat capacity (c_p) of samples were performed by means of the flash method using a Netzsch LFA 427 LaserFlash apparatus (NETZSCH Holding, Selb, Germany) in the temperature interval from 30 to 1000 °C at a heating rate of 5 °C/min and in nitrogen atmosphere with a flow rate of 100 mL/min. The dimensions of samples were ∅12.5 × 2.5 mm. The samples were covered by graphite on both sides before measurements. Measuring the heat capacity of a sample required an additional measurement of a reference material with a known heat capacity and density. The basis of this method is in the application of a laser pulse with the same parameters as the measured and reference samples. In this way, the same amount of heat is provided to both samples. Next, the limit of adiabatic temperature is calculated for both measurements. The specific heat capacity of measured sample is calculated according to the formula:

$$c_p = \frac{m_R \, c_{pR} \, \Delta T_{\infty R}}{m \, \Delta T_\infty}, \tag{4}$$

where m_R is the mass of the reference sample (in software, the mass of a sample is calculated from its bulk density and dimensions), c_{pR} is the heat capacity of the reference sample, $\Delta T_{\infty R}$ is the adiabatic temperature of the reference sample after the amount of thermal energy is received, m is the mass of the studied sample, and ΔT_∞ is the adiabatic temperature of the studied sample after the same amount of thermal energy as for reference sample is received.

Microstructure observations were carried out by means of a scanning electron microscope (FEI Quanta^{TM} FX200, Thermo Fisher Scientific, Waltham, MA, USA) in low vacuum mode (100 Pa) with an accelerating voltage of 10 kV on the compact polished raw samples, and the samples were heated at 1100 °C.

3. Results and Discussion

3.1. *Differential Thermal Analysis*

The DTA results of the studied samples are shown in Figure 1. Three significant peaks are visible. The first peak (up to 300 °C) corresponds to the process of dehydration, where the liberation of physically bound water from pores and surface of crystals occurs [23]. This peak is endothermic and its magnitude increases with the amount of zeolite. The second peak (from 450 to 700 °C) is also endothermic and belongs to the dehydroxylation of kaolinite [19,48]. During this reaction, the chemically bound water is evaporated, and this causes the structure of kaolinite to transform into metakaolinite. The last peak (from 940 to 1030 °C) is exothermic and has been interpreted as the result of the formation of an Al-Si spinel or of spinel and/or mullite by [28] and by many papers extensively discussed in [31]. The magnitudes of last two peaks decrease with the amount of zeolite in the samples. In natural zeolite, only one reaction is observed and it is endothermic (up to 450 °C). This peak corresponds to dehydration, where the physically bound water evaporates from pores and surface of crystals [21].

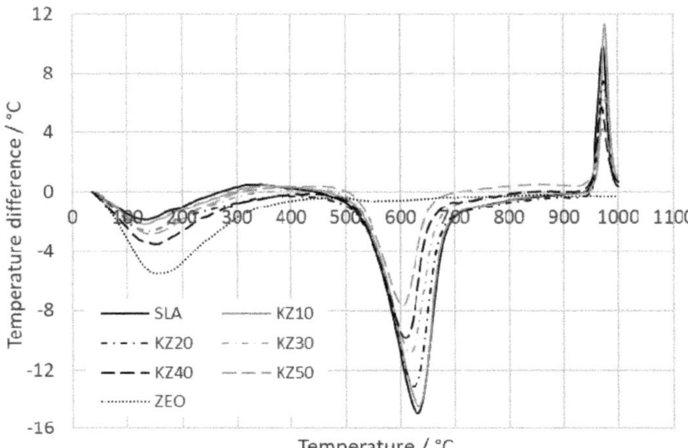

Figure 1. Differential thermal analysis (DTA) of the studied samples.

3.2. Differential Scanning Calorimetry

The DSC results of the kaolin–zeolite samples are shown in Figure 2. Three peaks can be observed, as it was in the DTA results (see Figure 1). The first endothermic peak (from 30 to 200 °C) represents the liberation of physically bound water [23]. The second endothermic peak (from 400 to 600 °C) corresponds to the dehydroxylation of kaolinite [19,48]. The third peak (from 950 to 1000 °C), which is exothermic, is related to the crystallization of high-temperature phases, as indicated before [28,31]. In zeolite and natural clinoptilolite, only two peaks are observed in the curve. The first is the endothermic peak corresponding with dehydration. This means that physically bound water evaporates from the pores and surface of crystals. The second peak is also endothermic, and this reaction begins above 850 °C, when the structure of clinoptilolite is definitely destroyed and an amorphous phase is formed [33,41].

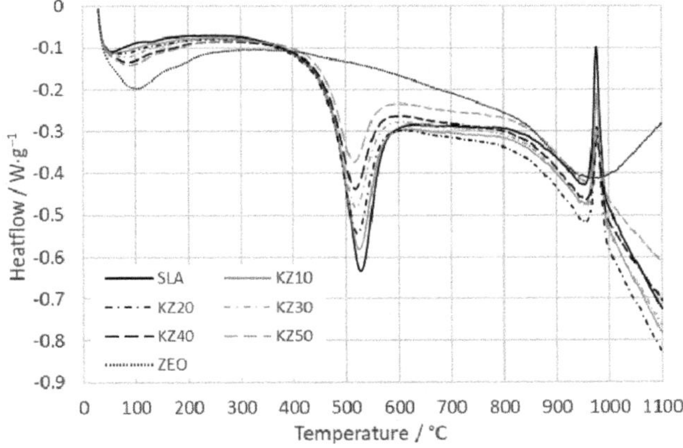

Figure 2. Differential scanning calorimetry (DSC) of the studied samples.

The influence of zeolite in the samples is visible on the enthalpy of reactions (dehydroxylation of kaolinite and Al-Si spinel formation) occurring in the kaolin–zeolite samples during thermal treatment (see Figure 3). The enthalpies were determined from the DSC results (peaks) (see Figure 2). A tangential baseline, which most realistically represents the progress of reactions, was used. The first studied reaction was the dehydroxylation of kaolinite, which is associated with the evaporation of chemically bound water [19,48]. The enthalpy of this reaction decreases linearly with the amount of zeolite in the samples from 248.9 (sample SLA) to 128.1 J/g (sample KZ50). The second reaction was the transformation of metakaolinite into an Al-Si spinel [28]. The values of enthalpy also decrease linearly with the zeolite content, from 58.8 (sample SLA) to 34.4 J/g (sample KZ50).

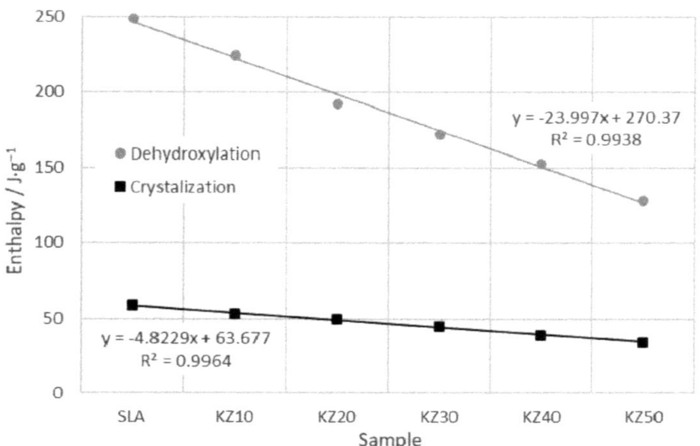

Figure 3. The enthalpy of the studied samples during the dehydroxylation of kaolinite (grey) and crystallization of spinel (black).

As can be seen from the enthalpy results for both reactions, the enthalpies decrease along with the amount of kaolinite. Therefore, it can be concluded that zeolite has no significant influence on both reactions in the studied samples.

3.3. Thermogravimetric Analysis

The relative mass changes of the kaolin–zeolite samples are shown in Figure 4. Two significant mass losses occur. The first loss is in the temperature interval from 30 to 250 °C and corresponds to the process of dehydration, during which physically bound water evaporates [23]. The process of dehydration is the least significant for sample SLA (1.56%) and increases with the amount of zeolite (3.98% for sample KZ50). The second mass loss, in the temperature interval from 450 to 700 °C, corresponds to the dehydroxylation of kaolinite [19,48]. The mass loss for sample SLA is 11.35% and it decreases with the amount of zeolite (6.12% for sample KZ50). The mass loss of the zeolite sample decreases continuously up to 800 °C and reaches 11.27%. Then, the mass loss remains almost constant. During the dehydroxylation, the mass loss is only 0.98%. This is because the dehydroxylation does not occur in zeolite. The reason is that zeolite does not have OH^- ions in the matrix structure. The DTA, DSC, and TG results of kaolin and natural zeolite are very similar to the results presented in [21,26,49]. In these studies, the mass losses of the kaolinite subgroup and clinoptilolite are in good agreement with our results. For kaolinite, it is in the interval from 11.8% to 13.31% (we obtained 11.35%, but the kaolinite content in kaolin was only 77.8%), and for clinoptilolite, it is 9.54% at 1000 °C (we achieved 11.27%). The obtained difference can be explained by a different content of water in the prepared sample from kaolin or natural zeolite. Moreover, the studied zeolite was not pure clinoptilolite as it also contained the amount of cristobalite, illite with mica, and feldspar.

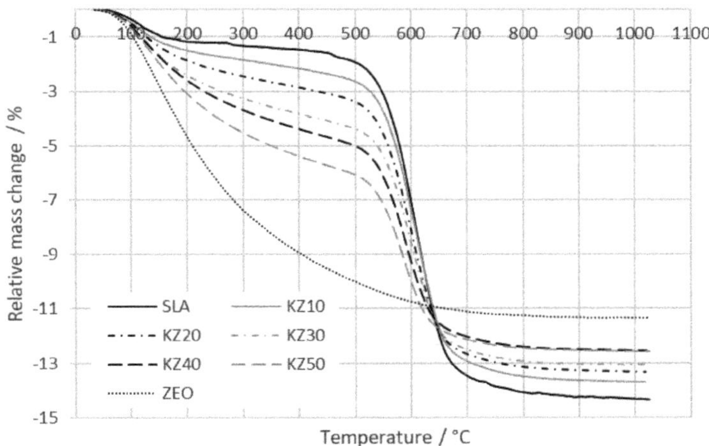

Figure 4. Relative mass change of the studied samples.

The results of the relative mass change of the kaolin–zeolite samples at temperatures of 800, 900, and 1000 °C are plotted in Figure 5. It is visible that the mass loss of the samples increases with temperature and decreases with the amount of zeolite. At a temperature of 1000 °C, it ranges from 14.3% to 12.6%. The trend of decrease is identical for all three selected temperatures and it is almost linear up to 40 mass% of zeolite. Then, the mass loss for samples KZ40 and KZ50 is comparable (the difference is only 0.04%).

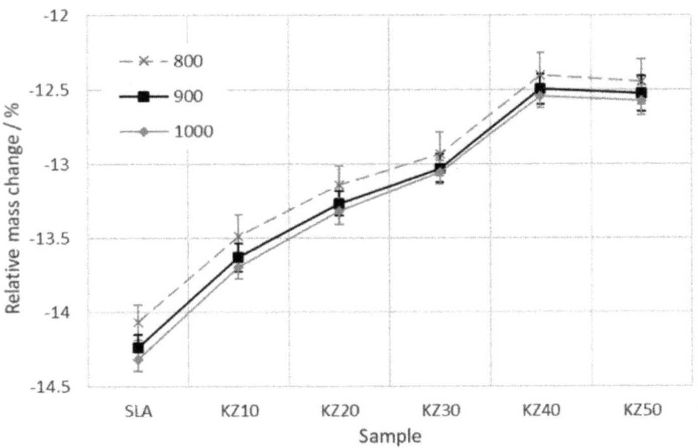

Figure 5. Relative mass change of the studied samples at 800, 900, and 1000 °C.

3.4. Thermodilatometric Analysis

The relative thermal expansion of the kaolin–zeolite samples is shown in Figure 6. In the temperature interval from 30 to 250 °C, the process of dehydration occurs, where physically bound water escapes from the pores and surface of crystals. The length of all samples, except the zeolite sample, increases linearly and reaches about 0.07%. The length of the zeolite sample decreases and its shrinkage is about 0.14%. The expansion of samples decreases with the amount of zeolite. The next process occurring in the studied samples starts at about 500 °C, when the chemically bound water escapes and the structure of kaolinite is transformed into metakaolinite [19,48]. The structure of metakaolinite is similar to the structure of kaolinite. The lattice parameters *a* and *b* remain the same, but the parameter *c* is changed to 0.685 nm. The distance between layers is shorter (about 0.055 nm) than in kaolinite. In addition, it is more defective and less stable than kaolinite [19] Therefore, the shrinkage of samples in the temperature interval from 500 to 700 °C is observed. Nevertheless, after the dehydroxylation is finished, the contraction continues up to 950 °C due to the sintering process [30]. As a result of the reactions above ~950 °C, there is a rapid contraction of the produced bodies. In the temperature interval from 700 to 1100 °C, the sintering process occurs as well. Therefore, this shrinkage is caused by both processes. The total shrinkage of the kaolin–zeolite products increases with the amount of zeolite from 2.59% for sample SLA to 3.91% for sample KZ50. Similar behavior of thermal expansion of kaolin samples was also obtained in [50,51]. The shrinkage of kaolin samples reached 2.5% and 3.2% at 1100 °C, respectively, which is in good agreement with our results.

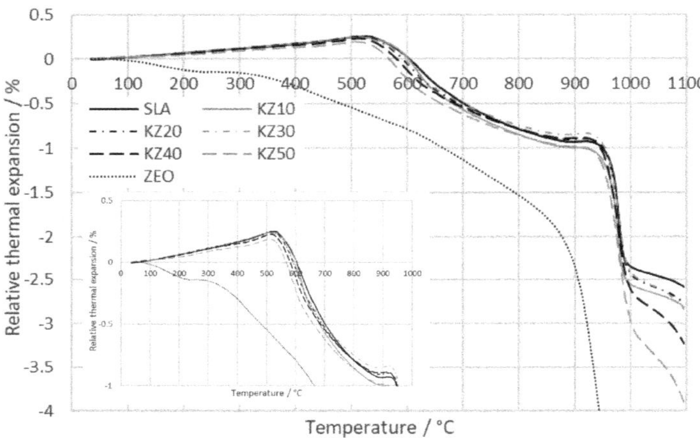

Figure 6. Relative thermal expansion of the studied samples.

Different results were obtained for the zeolite sample because the structure of zeolite does not contain free OH^- ions (zeolite is just a hydrate). The liberation of the physically bound water from the zeolite sample proceeds until up to 850 °C (the shrinkage reaches 1.82%). Above the temperature of 850 °C, very intensive sintering and the formation of a glassy phase occur. The total shrinkage of the zeolite sample at 1100 °C reaches 14.59% Dell'Agli et al. [52] also studied the thermal expansion and mass loss of different types of clinoptilolite, but only up to 700 °C. They found out that the final shrinkage was in the interval from 0.6% to 3%. The shrinkage of our zeolite sample reached 1.2% at 700 °C Nevertheless, the trend of the measured curves is similar.

The linear shrinkage of the kaolin–zeolite samples at different temperatures is shown in Figure 7. The results show that the linear shrinkage at temperatures of 800 and 900 °C with the increasing amount of zeolite is almost constant (the differences are very small, about 0.15%). The curves at the temperatures of 1000 and 1100 °C exhibit a different trend. The shrinkage increases with the amount of zeolite, but the samples with 10, 20, and 30 mass% of zeolite reach almost the same values of shrinkage. Above 30 mass% of zeolite, the linear shrinkage increases. Those results show that the most intensive changes occur when the Al-Si spinel crystallization is finished and the sintering process starts.

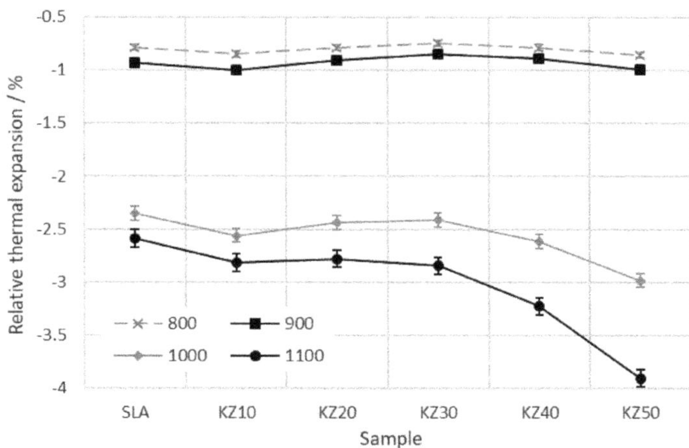

Figure 7. Relative thermal expansion of the studied samples at 800, 900, 1000, and 1100 °C.

3.5. Bulk Density

The results of the bulk density of the kaolin–zeolite samples are shown in Figure 8. The bulk density decreases with the amount of zeolite and also with the temperature increase to 650 °C. The bulk density for the green SLA sample is 1464 kg·m^{-3}, and for the green KZ50 sample, it amounts to 1340 kg·m^{-3}. The first significant decrease is up to 250 °C, which is caused by the liberation of physically bound water [23]. Then, the bulk density decreases almost linearly until the dehydroxylation of kaolinite starts (at about 500 °C) [19,48]. The structural changes (transformation of kaolinite into metakaolinite) cause the decrease in the bulk density. This decrease occurs due to an intensive mass loss (11.35% for sample SLA) (see Figure 4) and contraction (0.68% for sample SLA) (see Figure 6). During dehydroxylation, the bulk density decreases from 135 kg·m^{-3} for sample SLA to 50 kg·m^{-3} for sample KZ50. After dehydroxylation, the bulk density slightly increases for all studied samples up to 950 °C. Then, a sharp increase occurs due to the transformation of metakaolinite into the spinel phase [28]. Above 1000 °C, the bulk density increases only slightly. At 1100 °C, the values of the bulk density are lower than for the green samples. The differences are from 11 kg·m^{-3} for sample KZ50 up to 106 kg·m^{-3} for sample SLA, which means that the difference decreases with the zeolite content.

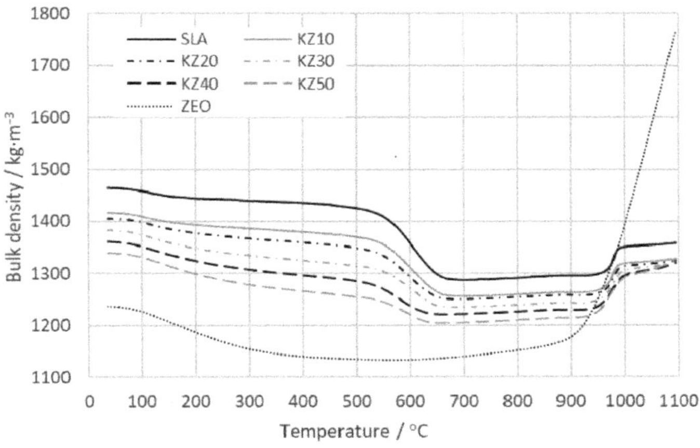

Figure 8. Bulk density of the studied samples.

The bulk density of the zeolite sample decreases up to 550 °C, then a slight increase occurs, and above 900 °C, a sharp increase is visible. Finally, the bulk density reaches 1760 kg·m^{-3} at 1100 °C, which is about 42% higher than for the green sample.

The bulk density of kaolin–zeolite samples at different temperatures is shown in Figure 9. The results show that the bulk density at temperatures of 800 and 900 °C decreases with the amount of zeolite. The differences between the SLA and KZ50 samples are about 40 kg·m^{-3} at 800 °C and about 60 kg·m^{-3} at 900 °C. The values of the bulk density at the temperatures of 1000 and 1100 °C also decrease, almost linearly in this case except for sample KZ10. Those results show that the bulk density decreases with the zeolite content by about 6%.

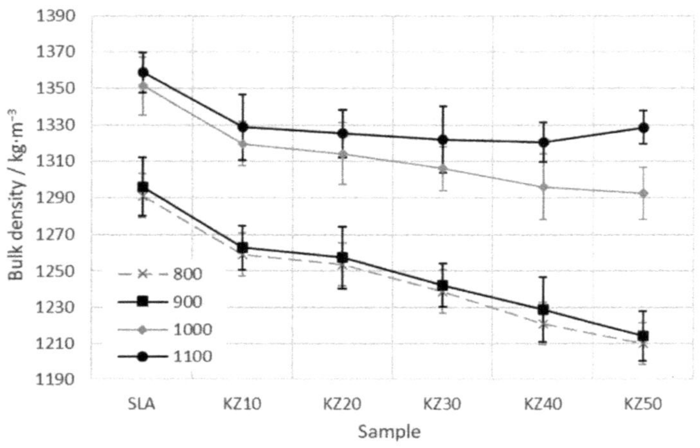

Figure 9. Bulk density of the studied samples at 800, 900, 1000, and 1100 °C.

3.6. Thermal Diffusivity

The results of the thermal diffusivity of the kaolin–zeolite samples in the temperature interval from 30 to 1000 °C are plotted in Figure 10. The thermal diffusivity of the green samples decreases with the amount of zeolite from 0.57 mm^2·s^{-1} for sample SLA to 0.37 mm^2·s^{-1} for sample KZ50. The green zeolite sample has a thermal diffusivity of only 0.21 mm^2·s^{-1}. The thermal diffusivity of all studied samples decreases with temperature except for the zeolite sample. The thermal diffusivity of the zeolite sample is almost the same in the whole temperature range (0.21 mm^2·s^{-1}). The decrease in thermal diffusivity is caused by the escape of physically and chemically bound water in the samples (dehydration and dehydroxylation) [19,23,48]. The differences in thermal diffusivity between samples also decrease with temperature. They are 0.20 mm^2·s^{-1} for the green samples at room temperature and 0.07 mm^2·s^{-1} for the samples measured at 1000 °C. A similar decreasing trend of the thermal diffusivity of clays was obtained in [17,49]. The values were from 0.17 to 0.27 mm^2·s^{-1}. A more interesting study was published by Antal et al. [49], where the thermal diffusivity of textured kaolin samples was assessed. It is visible that the thermal diffusivity of kaolin samples depends significantly on the direction of crystal orientation. The results showed that thermal diffusivity is in the interval from 0.2 to 0.75 mm^2·s^{-1} measured at room temperature and from 0.15 to 0.4 mm^2·s^{-1} at the temperature of 1100 °C.

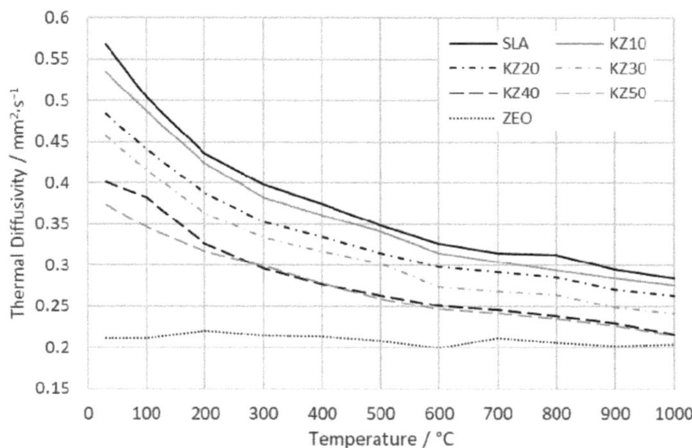

Figure 10. Thermal diffusivity of the studied samples.

The thermal diffusivity of the kaolin–zeolite samples at different temperatures is shown in Figure 11. The thermal diffusivity decreases with the amount of zeolite up to 40 mass%. Sample KZ50 has almost the same value of thermal diffusivity as sample KZ40. All three curves (the results at 800, 900, and 1000 °C) exhibit the same trend. At a temperature of 1000 °C, samples KZ40 and KZ50 have values of thermal diffusivity similar to the zeolite sample. The difference is only 0.01 $mm^2 \cdot s^{-1}$.

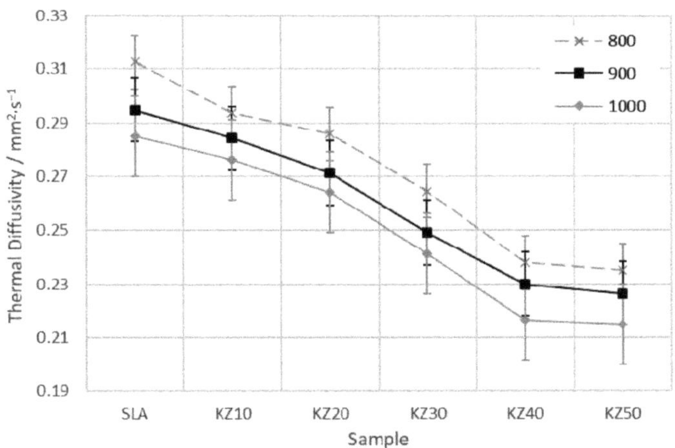

Figure 11. Thermal diffusivity of the studied samples at 800, 900, and 1000 °C.

3.7. Specific Heat Capacity

The results of the specific heat capacity of the kaolin–zeolite samples during thermal treatment up to 1000 °C are shown in Figure 12. The heat capacity of the green samples is in the interval from 1.02 to 1.42 $kJ \cdot kg^{-1} \cdot K^{-1}$. The values of the heat capacity for all samples decrease when the liberation of physically bound water is finished. The heat capacity of water is 4.2 $kJ \cdot kg^{-1} \cdot K^{-1}$ (much higher than for solid materials); therefore, its decrease was expected in the temperature interval up to 300 °C. Then, the specific heat capacity of samples slightly increases up to 800 °C. The values are in the interval from 0.88 to 1.45 $kJ \cdot kg^{-1} \cdot K^{-1}$. The dehydroxylation of kaolinite also influences the heat capacity of samples. The OH^- groups are removed from the structure of kaolinite, and therefore, the heat capacity does not increase, as it is excepted from the Debye model. Above 800 °C a sharp increase in the heat capacity occurs due to crystallization of spinel [28] and a normal characterization of the heat capacity at higher temperatures. At a temperature of 1000 °C, the specific heat capacity of the samples ranges from 1.25 up to 2.24 kJ·kg^{-1}·K^{-1} which is about 50% higher than it was for the green samples. Michot et al. [53] studied the heat capacity and thermal conductivity of kaolinite/metakaolinite in the temperature interval from room temperature up to 1000 °C. Their results showed that the heat capacity of metakaolinite above the temperature of 700 °C is about 1.2 kJ·kg^{-1}·K^{-1}, which is lower than our values. This can be explained by a different content of kaolinite in our kaolin sample and the application of a method which is not accurate for measurement of heat capacity.

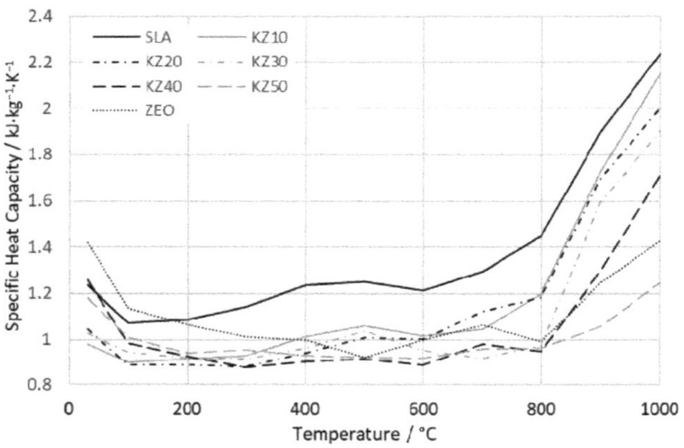

Figure 12. Specific heat capacity of the studied samples.

The specific heat capacity of the kaolin–zeolite samples at temperatures of 800, 900, and 1000 °C is presented in Figure 13. The heat capacities of samples KZ30, KZ40, and KZ50 at 800 °C are almost the same. At temperatures of 900 and 1000 °C, the trend of the heat capacity is similar. Zeolite in the samples causes a decrease iin the specific heat capacity. The difference between samples SLA and KZ50 at 1000 °C is 0.99 kJ·kg^{-1}·K^{-1}, which is about 80%.

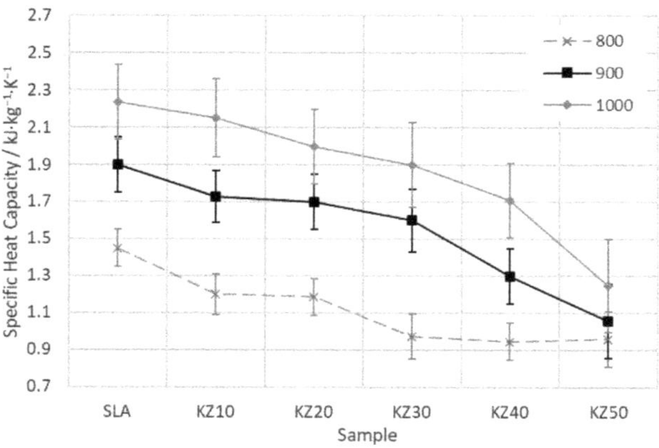

Figure 13. Specific heat capacity of the studied samples at 800, 900, and 1000 °C.

3.8. Thermal Conductivity

The results of the specific heat capacity of the zeolite–kaolin samples during thermal treatment up to 1000 °C are shown in Figure 14. The thermal conductivity of the green samples (at 30 °C) is the highest for sample SLA (1.03 W·m^{-1}·K^{-1}) and the lowest for the zeolite sample (0.37 W·m^{-1}·K^{-1}). It decreases with the amount of zeolite. Then, in the temperature interval from 30 to 300 °C, the thermal conductivity of all studied samples decreases due to the liberation of physically bound water from the pores and crystal surfaces (water has a higher thermal conductivity than ceramic materials) [23]. Significant changes are observed in the temperature interval from 400 to 700 °C. This decrease is caused by dehydroxylation of kaolinite [19,48] and it is smaller with higher zeolite content in the samples. At 700 °C, all samples reach the lowest values of thermal conductivity, which range from 0.52 (sample SLA) to 0.25 W·m^{-1}·K^{-1} (the zeolite sample). Above the temperature of 800 °C, a sharp increase in the thermal conductivity occurs due to the sintering process and crystallization of spinel [28]. The final values at 1000 °C are from 0.86 W·m^{-1}·K^{-1} for sample SLA to 0.35 W·m^{-1}·K^{-1} for sample KZ50. The results obtained by Michot et al. [53] showed a thermal conductivity of the green kaolin body of up to 0.5 W·m^{-1}·K^{-1}, which is in good agreement with the value of 0.52 W·m^{-1}·K^{-1} obtained in this study. On the other hand, our values of thermal conductivity were not measured but calculated from density, heat capacity, and thermal diffusivity. The surface of samples was coated by graphite before the analysis, and above 800 °C, graphite could burn out, although the analysis was performed in nitrogen atmosphere. Therefore, greater uncertainties are expected.

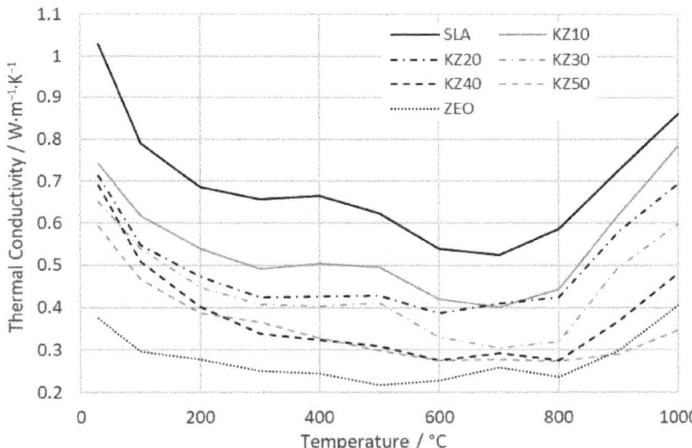

Figure 14. Thermal conductivity of the studied samples.

The thermal conductivity of the kaolin–zeolite samples at temperatures of 800, 900, and 1000 °C is shown in Figure 15. At a temperature of 800 °C, thermal conductivity has a decreasing trend up to 40 mass% of zeolite. Nevertheless, samples KZ40 and KZ50 have similar values of thermal conductivity (the difference is only 0.003 W·m^{-1}·K^{-1}). At temperatures of 900 and 1000 °C, thermal conductivity decreases with the amount of zeolite and this decrease is almost linear. The difference between samples SLA and KZ50 at 1000 °C amounts to 0.45 W·m^{-1}·K^{-1}, which is about 53%. Typically, values of the thermal conductivity of ceramic materials are in the interval from 0.28 to 1.2 W·m^{-1}·K^{-1} [53]. Therefore, the obtained results of the thermal conductivity are in good agreement with the literature.

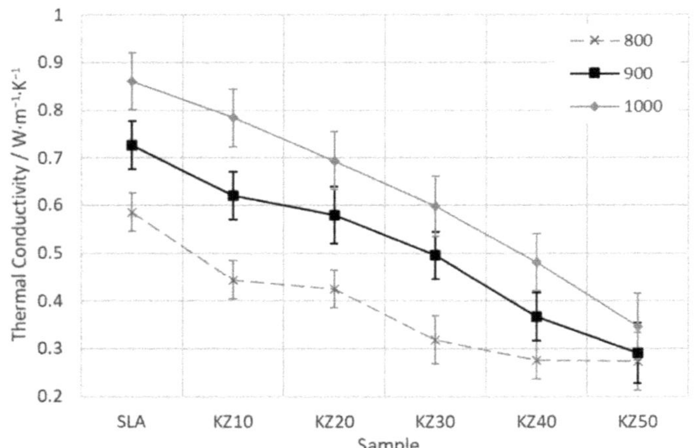

Figure 15. Thermal conductivity of the studied samples at 800, 900, and 1000 °C.

3.9. The Open Porosity and Scanning Electron Microscopy

The results of bulk density were also confirmed with the results of the porosity of kaolin–zeolite samples at temperatures of 800, 900, 1000, and 1100 °C (see Figure 16). The porosity of the studied samples decreases and the bulk density increases with temperature. The porosity at 800 and 900 °C decreases when the sample contains 10 mass% of zeolite. Then, the porosity increases and reaches about 50% for samples KZ30, KZ40, and KZ50. The porosity of sample KZ10 at temperatures of 800, 900, and 1000 °C is almost the same. At a temperature of 1100 °C, porosity decreases from 48.5% (sample SLA) down to 43.0% (sample KZ50). The decrease in the porosity with temperature is caused by the sintering process in the samples [30]. This process causes a decrease in the sample volumes and vanishing of pores. The results show that the porosity decreases with the amount of zeolite in the samples.

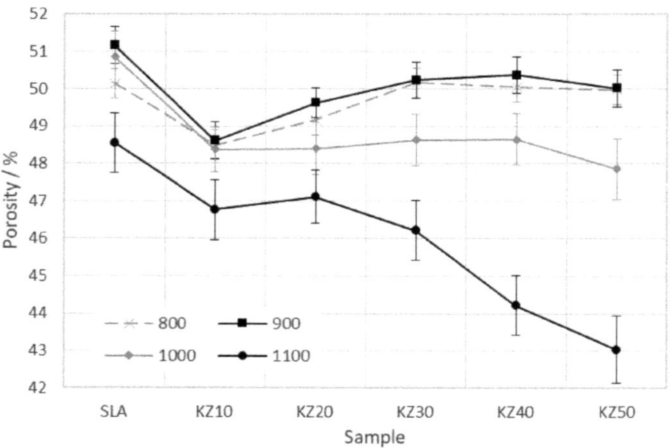

Figure 16. Porosity of the studied samples at 800, 900, 1000, and 1100 °C.

An SEM picture of the green kaolin sample is presented in Figure 17A. Grains of kaolinite with a different size and also agglomerates are visible. Kaolinite crystals have a plate-like shape and they are ordered randomly. The samples contain also many small pores. This is confirmed with high values of the porosity (black or dark points). A picture of the kaolin sample heated at 1100 °C is shown in Figure 17B. The porosity decreased and the kaolin sample stayed more compact. The structure remains in the same condition up to 1100 °C. Moreover, the pseudoamorphous shape of a grain of initial agglomerates is still the same.

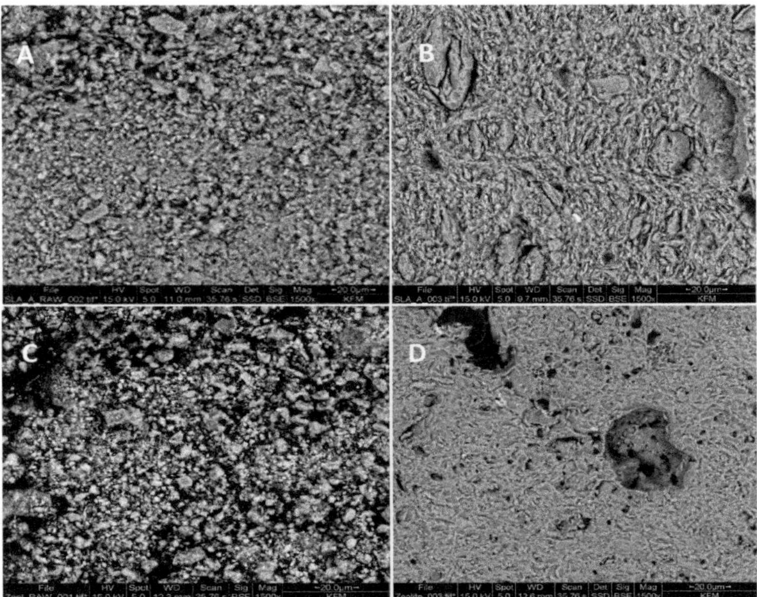

Figure 17. SEM pictures. (**A**) green Sedlec kaolin; (**B**) Sedlec kaolin fired at 1100 °C; (**C**) green zeolite; (**D**) zeolite fired at 1100 °C.

An SEM picture of the green zeolite sample is presented in Figure 17C. Grains as well as agglomerates (clusters of grains) with different sizes and shapes can be seen. A picture of the zeolite sample heated at 1100 °C is shown in Figure 17D, where large pores (black or dark points) are mainly visible. The initial shape of grains is not preserved. The samples are more compact (higher density (see Figure 8) and smaller porosity), and the melting of grains is visible.

4. Conclusions

The thermophysical properties (thermal expansion, thermal diffusivity and conductivity, and specific heat capacity) of kaolin–zeolite samples were investigated during a heating stage of the firing. The samples were prepared from Sedlec kaolin (Czech Republic) and natural zeolite (Nižný Hrabovec, Slovakia). Kaolin was partially replaced with a natural zeolite in the amounts of 10, 20, 30, 40, and 50 mass%. Pure kaolin and zeolite samples were also prepared. From the obtained values for the thermophysical properties (thermal expansion, thermal diffusivity and conductivity, and specific heat capacity) of Sedlec kaolin and blends of that kaolin with natural zeolite, we can conclude that increasing zeolite content of the blends decreases the values of all studied properties (except thermal expansion), which is positive for bulk density, porosity, thermal diffusivity, and conductivity. Zeolite has a negative effect on thermal expansion due to an increased shrinkage. Therefore, the optimal amount of zeolite in the blends (for the studied properties) is 30 mass%.

Author Contributions: Conceptualization, A.T.; methodology, T.H., J.O., Š.C. and A.T.; investigation I.S., T.H., Š.C., J.O., G.Ł., A.S. and A.T.; writing—original draft preparation, A.T.; writing—review and editing, J.O., T.H., Š.C., I.S., G.Ł., A.S. and A.T.; visualization, A.T.; supervision, A.T.; funding acquisition, T.H. and J.O. All authors have read and agreed to the published version of the manuscript.

Funding: This research was funded by Ministry of Education of Slovak Republic, grant numbers VEGA 1/0810/19 and VEGA 1/0425/19, by RVO: 11000.

Acknowledgments: Authors wish to thank the company ZEOCEM, a.s. (Slovakia), for the supply of natural zeolite.

Conflicts of Interest: The authors declare no conflict of interest. The funders had no role in the design of the study; in the collection, analyses, or interpretation of data; in the writing of the manuscript, or in the decision to publish the results.

References

1. Velasco, P.M.; Ortíz, M.M.; Giró, M.M. Fired clay bricks manufactured by adding wastes as sustainable construction material—A review. *Constr. Build. Mater.* **2014**, *63*, 97–107. [CrossRef]
2. Knapek, M.; Húlan, T.; Dobroň, P.; Chmelík, F.; Trnik, A.; Štubňa, I. Acoustic Emission During Firing of the Illite-Based Ceramics with Fly Ash Addition. *Acta Phys. Pol. A* **2015**, *128*, 783–786. [CrossRef]
3. Húlan, T.; Trník, A.; Kaljuvee, T.; Uibu, M.; Štubňa, I.; Kallavus, U.; Traksmaa, R. The study of firing of a ceramic body made from illite and fluidized bed combustion fly ash. *J. Therm. Anal. Calorim.* **2016**, *127*, 79–89. [CrossRef]
4. Lingling, X.; Wei, G.; Tao, W.; Nanru, Y. Study on fired bricks with replacing clay by fly ash in high volume ratio. *Constr. Build. Mater.* **2005**, *19*, 243–247. [CrossRef]
5. Zimmer, A.; Bergmann, C.P. Fly ash of mineral coal as ceramic tiles raw material. *Waste Manag.* **2007**, *27*, 59–68. [CrossRef] [PubMed]
6. Ongwandee, M.; Namepol, K.; Yongprapat, K.; Homwuttiwong, S.; Pattiya, A.; Morris, J.; Chavalparit, O. Coal bottom ash use in traditional ceramic production: Evaluation of engineering properties and indoor air pollution removal ability. *J. Mater. Cycles Waste Manag.* **2020**, *22*, 2118–2129. [CrossRef]
7. Húlan, T.; Štubňa, I.; Ondruška, J.; Trnik, A. The Influence of Fly Ash on Mechanical Properties of Clay-Based Ceramics. *Minerals* **2020**, *10*, 930. [CrossRef]
8. Sokolar, R.; Smetanova, L. Dry pressed ceramic tiles based on fly ash–clay body: Influence of fly ash granulometry and pentasodium triphosphate addition. *Ceram. Int.* **2010**, *36*, 215–221. [CrossRef]
9. Kovac, J.; Trnik, A.; Medved, I.; Štubňa, I.; Vozár, L. Influence of fly ash added to a ceramic body on its thermophysical properties. *Therm. Sci.* **2016**, *20*, 603–612. [CrossRef]
10. Sokolar, R.; Vodova, L. The effect of fluidized fly ash on the properties of dry pressed ceramic tiles based on fly ash–clay body. *Ceram. Int.* **2011**, *37*, 2879–2885. [CrossRef]
11. Chandra, N.; Sharma, P.; Pashkov, G.; Voskresenskaya, E.; Amritphale, S.S.; Baghel, N.S. Coal fly ash utilization: Low temperature sintering of wall tiles. *Waste Manag.* **2008**, *28*, 1993–2002. [CrossRef] [PubMed]
12. Haiying, Z.; Youcai, Z.; Jingyu, Q. Study on use of MSWI fly ash in ceramic tile. *J. Hazard. Mater.* **2007**, *141*, 106–114. [CrossRef]
13. Pontikes, Y.; Esposito, L.; Tucci, A.; Angelopoulos, G. Thermal behaviour of clays for traditional ceramics with soda–lime–silica waste glass admixture. *J. Eur. Ceram. Soc.* **2007**, *27*, 1657–1663. [CrossRef]
14. Hasan, R.; Siddika, A.; Akanda, P.A.; Islam, R. Effects of waste glass addition on the physical and mechanical properties of brick. *Innov. Infrastruct. Solut.* **2021**, *6*, 1–13. [CrossRef]
15. Conte, S.; Zanelli, C.; Molinari, C.; Guarini, G.; Dondi, M. Glassy wastes as feldspar substitutes in porcelain stoneware tiles: Thermal behaviour and effect on sintering process. *Mater. Chem. Phys.* **2020**, *256*, 123613. [CrossRef]
16. Ondruška, J.; Csáki, Š.; Štubňa, I. Influence of waste glass addition on thermal properties of kaolin and illite. *AIP Conf. Proc.* **2019**, *2133*, 020028. [CrossRef]
17. Kovac, J.; Trnik, A.; Medved', I.; Vozár, L. Influence of calcite in a ceramic body on its thermophysical properties. *J. Therm. Anal. Calorim.* **2013**, *114*, 963–970. [CrossRef]
18. Murray, H.H. Traditional and new applications for kaolin, smectite, and palygorskite: A general overview. *Appl. Clay Sci.* **2000**, *17*, 207–221. [CrossRef]
19. Štubňa, I.; Varga, G.; Trnik, A. Investigation of kaolinite dehydroxylations is still interesting. *Építöanyag* **2006**, *58*, 6–9. [CrossRef]
20. Nemecz, E. *Clay Minerals*; Akadémiai Kiadó: Budapest, Hungary, 1981.
21. Foldvári, M. *Hanbook of Thermogarvimetric System of Minerals and Its Use in Geological Practice*; Geological Institute of Hungary: Budapest, Hungary, 2011.
22. Ryan, W. *Properties of Ceramic Raw Materials*; Elsevier BV: Amsterdam, The Netherlands, 1978; pp. 47–50.
23. Štubňa, I.; Sin, P.; Trnik, A.; Veinthal, R. Mechanical Properties of Kaolin during Heating. *Key Eng. Mater.* **2012**, *527*, 14–19. [CrossRef]
24. Norton, F.H. *Fine Ceramics: Technology and Applications*; McGraw-Hill: New York, NY, USA, 1970.

25. Hanykýř, V.; Kutzendorfer, J. *Technology of Ceramics*; Silis Praha: Praha, Czech Republic, 2008.
26. Kurovics, E.; Kotova, O.B.; Ibrahim, J.E.F.M.; Tihtih, M.; Sun, S.; Pala, P.; Gömze, A.L. Characterization of phase transformation and thermal behavior of Sedlecky Kaolin. *Építöanyag* **2020**, *72*, 144–147. [CrossRef]
27. Varga, G. The structure of kaolinite and metakaolinite. *Építöanyag* **2007**, *59*, 6–9. [CrossRef]
28. Ptáček, P.; Šoukal, F.; Opravil, T.; Nosková, M.; Havlica, J.; Brandštetr, J. The kinetics of Al–Si spinel phase crystallization from calcined kaolin. *J. Solid State Chem.* **2010**, *183*, 2565–2569. [CrossRef]
29. Ptáček, P.; Šoukal, F.; Opravil, T.; Nosková, M.; Havlica, J.; Brandštetr, J. Mid-infrared spectroscopic study of crystallization of cubic spinel phase from metakaolin. *J. Solid State Chem.* **2011**, *184*, 2661–2667. [CrossRef]
30. Ondruška, J.; Trnik, A.; Keppert, M.; Medved', I.; Vozár, L. Isothermal Dilatometric Study of Sintering in Kaolin. *Int. J. Thermophys.* **2012**, *35*, 1946–1956. [CrossRef]
31. Chakraborty, A.K. *Phase Transformation of Kaolinite Clay*; Springer Nature: New Delhi, India, 2014.
32. Favvas, E.P.; Tsanaktsidis, C.G.; Sapalidis, A.A.; Tzilantonis, G.T.; Papageorgiou, S.K.; Mitropoulos, A.C. Clinoptilolite, a natural zeolite material: Structural characterization and performance evaluation on its dehydration properties of hydrocarbon-based fuels. *Microporous Mesoporous Mater.* **2016**, *225*, 385–391. [CrossRef]
33. Mansouri, N.; Rikhtegar, N.; Panahi, H.A.; Atabi, F.; Shahraki, B.K. Porosity, characteriza-tion and structural properties of natural zeolite-clinoptilolite—As a sorbent. *Environ. Prot. Eng.* **2013**, *39*, 139–152. [CrossRef]
34. Kocak, Y.; Tascı, E.; Kaya, U. The effect of using natural zeolite on the properties and hydration characteristics of blended cements. *Constr. Build. Mater.* **2013**, *47*, 720–727. [CrossRef]
35. Trnik, A.; Scheinherrova, L.; Medved, I.; Černý, R. Simultaneous DSC and TG analysis of high-performance concrete containing natural zeolite as a supplementary cementitious material. *J. Therm. Anal. Calorim.* **2015**, *121*, 67–73. [CrossRef]
36. Bhattacharyya, T.; Chandran, P.; Ray, S.K.; Pal, D.K.; Mandal, C.; Mandal, D.K. Distribution of Zeolitic Soils in India. *Curr. Sci.* **2015**, *109*, 1305. [CrossRef]
37. Lamprecht, M.; Bogner, S.; Steinbauer, K.; Schuetz, B.; Greilberger, J.F.; Leber, B.; Wagner, B.; Zinser, E.; Petek, T.; Wallner-Liebmann, S.; et al. Effects of zeolite supplementation on parameters of intestinal barrier integrity, inflammation, redoxbiology and performance in aerobically trained subjects. *J. Int. Soc. Sports Nutr.* **2015**, *12*, 40. [CrossRef]
38. Laurino, C.; Palmieri, B. Zeolite: "the magic stone"; main nutritional, environmental, experimental and clinical fields of application. *Nutr. Hosp.* **2015**, *32*, 573–581.
39. Sunitrová, I.; Trnik, A. DSC and TGA of a kaolin-based ceramics with zeolite addition during heating up to 1100 °C. *AIP Conf. Proc.* **2018**, *1988*, 020046. [CrossRef]
40. Che, C.; Glotch, T.D.; Bish, D.L.; Michalski, J.R.; Xu, W. Spectroscopic study of the dehydration and/or dehydroxylation of phyllosilicate and zeolite minerals. *J. Geophys. Res. Space Phys.* **2011**, *116*, 05007. [CrossRef]
41. Földesová, M.; Lukac, P.; Dillinger, P.; Balek, V.; Svetík, Š. Thermochemical Properties of Chemically Modified Zeolite. *J. Therm. Anal. Calorim.* **1999**, *58*, 671–675. [CrossRef]
42. Sunitrová, I.; Trnik, A. Young's modulus and thermal expansion of ceramic samples made from kaolin and zeolite. *AIP Conf. Proc.* **2016**, *1752*, 40026. [CrossRef]
43. Sunitrová, I.; Trnik, A. Thermal expansion of ceramic samples containing natural zeolite. *AIP Conf. Proc.* **2017**, *1866*, 040039. [CrossRef]
44. Ptáček, P.; Křečková, M.; Šoukal, F.; Opravil, T.; Havlica, J.; Brandštetr, J. The kinetics and mechanism of kaolin powder sintering I. The dilatometric CRH study of sinter-crystallization of mullite and cristobalite. *Powder Technol.* **2012**, *232*, 24–30. [CrossRef]
45. Vejmelková, E.; Koňáková, D.; Kulovaná, T.; Keppert, M.; Žumár, J.; Rovnaníková, P.; Keršner, Z.; Sedlmajer, M.; Černý, R. Engineering properties of concrete containing natural zeolite as supplementary cementitious material: Strength, toughness, durability, and hygrothermal performance. *Cem. Concr. Compos.* **2015**, *55*, 259–267. [CrossRef]
46. Podoba, R.; Trník, A.; Podobník, L. Upgrading of TGA/DTA analyzer derivatograph. *Építöanyag* **2012**, *64*, 28–29. [CrossRef]
47. Jankula, M.; Šín, P.; Podoba, R.; Ondruška, J. Typical problems in push-rod dilatometry analysis. *Építöanyag* **2013**, *65*, 11–14. [CrossRef]
48. Ondruška, J.; Trnik, A.; Vozár, L. Degree of Conversion of Dehydroxylation in a Large Electroceramic Body. *Int. J. Thermophys.* **2010**, *32*, 729–735. [CrossRef]
49. Antal, D.; Húlan, T.; Štubňa, I.; Záleská, M.; Trník, A. The influence of texture on elastic and thermophysical properties of kaolin- and illite-based ceramic bodies. *Ceram. Int.* **2017**, *43*, 2730–2736. [CrossRef]
50. Frizzo, R.G.; Zaccaron, A.; Nandi, V.D.S.; Bernardin, A.M. Pyroplasticity on porcelain tiles of the albite-potassium feldspar-kaolin system: A mixture design analysis. *J. Build. Eng.* **2020**, *31*, 101432. [CrossRef]
51. Trnik, A.; Štubňa, I.; Moravčíková, J. Sound Velocity of Kaolin in the Temperature Range from 20 °C to 1100 °C. *Int. J. Thermophys.* **2009**, *30*, 1323–1328. [CrossRef]
52. Dell'Agli, G.; Ferone, C.; Mascolo, G.; Pansini, M. Dilatometry of Na-, K-, Ca- and NH4-clinoptilolite. *Thermochim. Acta* **1999**, *336*, 105–110. [CrossRef]
53. Michot, A.; Smith, D.S.; Degot, S.; Gault, C. Thermal conductivity and specific heat of kaolinite: Evolution with thermal treatment. *J. Eur. Ceram. Soc.* **2008**, *28*, 2639–2644. [CrossRef]

Article

Alkaline Activation of Kaolin Group Minerals

Oliwia Biel, Piotr Rożek *, Paulina Florek, Włodzimierz Mozgawa and Magdalena Król

Faculty of Materials Science and Ceramic, AGH University of Science and Technology, 30 Mickiewicza Av., 30-059 Krakow, Poland; oliwia.biel@gmail.com (O.B.); paulina@agh.edu.pl (P.F.); mozgawa@agh.edu.pl (W.M.); mkrol@agh.edu.pl (M.K.)

* Correspondence: prozek@agh.edu.pl

Received: 2 March 2020; Accepted: 31 March 2020; Published: 2 April 2020

Abstract: Zeolites can be obtained in the process of the alkali-activation of aluminosilicate precursors. Such zeolite–geopolymer hybrid bulk materials merge the advantageous properties of both zeolites and geopolymers. In the present study, the effect of the type and concentration of an activator on the structure and properties of alkali-activated metakaolin, and metahalloysite was assessed. These two different kaolinite clays were obtained by the calcination of kaolin and halloysite, and then activated with sodium hydroxide and water glass. The phase compositions were assessed by X-ray diffraction, the microstructure was observed via scanning electron microscope, and the structural studies were conducted on the basis of the infrared spectra. The structure and properties of the obtained alkali-activated materials depend on both the type of a precursor and the type of an activator. The formation of zeolite phases was observed when the activation was carried out with sodium hydroxide alone, or with a small addition of water glass, regardless of the starting material used. The higher proportion of silicon in the activator solution does not give crystalline phases, but only an amorphous phase. Geopolymers based on metahalloysite have better compressive strength as the result of the better reactivity of metahalloysite compared to metakaolin.

Keywords: zeolite; alkali-activation; geopolymer; metakaolin; metahalloysite

1. Introduction

Zeolites are crystalline aluminosilicates (having the general chemical formula $Me_{2/n}O{\cdot}Al_2O_3{\cdot}xSiO_2{\cdot}yH_2O$ [1], with the possibility of $[SiO_4]$ substitution by another elements, e.g., $[PO_4]$) whose characteristic feature is the microporous, regular structure with a constant channel and pore size. This feature is an advantage of zeolites over other known sorbents, as it enables the selective adsorption of ions and molecules. Zeolites are one of the products of alkali-activated aluminosilicate precursors. The activation is conducted by treating the material with an alkaline solution, and keeping for an appropriate time at an elevated temperature. This method of obtaining zeolites is well known, and has been used since many years; however, an improvement of its parameters is still the subject of research [2]. Depending on the used parameters, geopolymer gel and zeolites can be found among the reaction products. That type of composite is known as a zeolite–geopolymer hybrid bulk material.

Zeolite–geopolymer hybrid bulk materials are composites that combine the advantages of zeolites as a dispersed phase and geopolymer as a matrix. Their structure contains the meso-porosity of the geopolymer and the micro-porosity of the zeolite [3]. The range of porosity in those materials is very wide, and reaches from 5 Å up to 2 μm [4]. It was proven [3] that there is a relationship between the properties of initial kaolinite, its meta phase and the final product: zeolite–geopolymer hybrid bulk material, which gives opportunity to adjust the contents of desirable phases, and in consequence, the range of porosity.

There are known various aluminosilicate precursors, such as synthetic ones (coal fly ash, ash from waste incineration, blast furnace slag, etc.) or natural clays, for instance, kaolinite and halloysite. Using those natural minerals for the synthesis of zeolite and zeolite–geopolymer hybrid bulk materials is a topical issue [3,5,6]. Despite the fact that both kaolinite ($Al_2Si_2O_5(OH)_4$) and halloysite ($Al_2Si_2O_5(OH)_4 \cdot 2H_2O$) have similar compositions [7], their reactivity is different. The source of the differences is the presence of interlayer water in halloysite's structure, which results in its greater porosity, chemical activity and specific surface area [8].

In this work, the results of the structural studies of different composites, obtained using the alkali-activation method and two different kaolinite clays as starting materials, were presented. It is well known that kaolinite and halloysite change into hydroxysodalite by the treatment with sodium hydroxide. It is also known that sodium zeolite type A can be synthesized from both raw materials [9–11]. However, the methodology for preparing composites varies in all available works. Therefore, the aim of this work was to compare the impact of the raw material structure on the course of the synthesis process. As activators, NaOH and water glass were applied. The specific objectives were to compare the properties of materials obtained on the basis of metakaolin and metahalloysite, their structural characterization, and the assessment of the impact of the activator type on the properties and structure of the resulting zeolite–geopolymer hybrid bulk materials. Such composites would possess the synergistic benefits of both zeolites and geopolymers, and could be used as monolithic sorbents, self-supported zeolite sieves or membranes, rather than construction materials.

2. Materials and Methods

2.1. Chemicals

Clays from two different Polish deposits were used in this study: kaolin from the Maria III (KSM Surmin-Kaolin S.A., Nowogrodziec, Poland) and halloysite from the Dunin (Kopalnia Haloizytu DUNINO Sp. z o. o., Krotoszyce, Poland). Sodium hydroxide (analytically pure NaOH as microbeads) and water glass (the content of Na_2O 11.4 wt.% and SiO_2 27.6 wt.%) were used as the activator.

2.2. Pre-Treatment of Natural Clays

The homogenized samples of kaolin and halloysite were thermally treated at 700 °C for 2 h (the temperature was chosen on the basis of the previous work [12]), in order to obtain more reactive materials, metakaolin and metahalloysite.

2.3. Alkali-Activation

The alkali-activated paste samples were synthesized using 4 mL of the activator per 5 g of metakaolin. The compositions of the specimens are summarized in Table 1. The solid and solution were mechanically stirred for several minutes at room temperature. The fresh paste was poured into a silicone mold (20 mm × 20 mm × 20 mm cubic samples) and activated at 80 °C for 24 h. After this time, the samples were demolded and left to mature for another 27 days. With each mixture at least eight samples were prepared.

Table 1. The mixture proportions of geopolymer pastes and their molar ratios of silica, alumina, water and sodium oxide.

Sample Name	MK/MH [g]	8 M NaOH [mL]	Water Glass [mL]	SiO_2/Al_2O_3	Na_2O/Al_2O_3	H_2O/Na_2O
MK1/MH1	1.00	0.80	0.00	2.1/2.1	0.7/0.9	13.2/13.0
MK2/MH2	1.00	0.70	0.10	2.2/2.3	0.7/0.9	13.7/13.5
MK3/MH3	1.00	0.60	0.20	2.4/2.5	0.7/0.9	14.2/13.9
MK4/MH4	1.00	0.50	0.30	2.6/2.7	0.7/0.9	14.7/14.4
MK5/MH5	1.00	0.40	0.40	2.7/2.9	0.7/0.9	15.3/15.0

2.4. Characterization

The solid-state characterization techniques, such as X-Ray Fluorescence (XRF), X-Ray Diffraction (XRD), Infrared Spectroscopy (FT-IR) and Scanning Electron Microscopy (SEM), were carried out on both raw materials, and on the as-synthesized materials.

The chemical compositions of the starting materials were determined using X-ray fluorescence. The spectrum was detected using the wavelength dispersive X-ray fluorescence spectrometer (WD-XRF) Axios mAX 4 kW, PANalytical equipped with Rh source. The PANalytical standardless analysis package Omnian was used for the quantitative analysis of the spectra.

The resulting materials were analyzed in the terms of the phase composition by means of Philips X-ray powder diffraction X'P-ert system (CuK_α radiation). The measurements were carried out in the 2θ angle range of 5–90° for 2 h, with a step of 0.007. Phases were identified with the use of an X'Pert HighScore Plus application, and the International Centre for Diffraction Data.

The existence of zeolite frameworks was also confirmed by the analysis of the spectra in the mid infrared (4000–400 cm^{-1}) that were measured on Bruker VERTEX 70v vacuum FT-IR spectrometer using the standard KBr pellets methods. They were collected in after 64 scans at 4 cm^{-1} resolution.

The microstructure of the resulting samples was observed using a scanning electron microscope FEI Nova NanoSEM 200. The samples were sprayed with graphite.

The bulk density was calculated by dividing the mass of the sample by its volume. The compressive strength of the samples, measured by using the ZwickRoell device machine that works on the principle of a hydraulic press, is defined as the ratio of the sample breaking force to the area on which the force acts. Eight similar samples were tested, and the average of the eight measurements was taken.

3. Results and Discussion

3.1. Characterization of Raw Materials

Both kaolinite and halloysite are considered to be the polytypical forms of aluminum monophyllosilicate with 1:1 packets [13]. They differ in the way the layer packages are arranged in the crystal structure [14]. The kaolinite crystals have a lamellar habit. The shifting of two layers relative to each other, tetrahedral (Si–O) and octahedral (Al–OH), creates a tubular halloysite structure. The characteristic feature of halloysite is also the presence of water molecules in the inter-package space. The presence of water causes an increase in the distance between the layer packs, where for halloysite it is 10.1 Å, while for kaolinite, this is 7.15 Å. Both minerals dehydroxylate at 600–800 °C [15]. As the products, more active amorphous metakaolin and metahalloysite, are obtained.

The chemical compositions of the obtained intermediates, determined on the basis of X-ray fluorescence spectrometry, are summarized in Table 2. By comparing both compositions, it can be stated that the metakaolin sample has a higher silica content, whereas halloysite has higher amounts of iron oxide. In addition, the molar ratio of the main components (Si/Al) in both cases was about 2. The content of CaO in both materials is below 1%, so the resulting products can be called geopolymers (due to the limitation of this name to low-calcium alkali-activated materials).

Table 2. The chemical compositions of the metakaolin and metahalloysite.

Chemical Composition [wt.%]	SiO_2	TiO_2	Al_2O_3	Fe_2O_3	MgO	CaO	Na_2O	K_2O	Residual
MK: metakaolin	53.74	0.47	43.82	0.87	0.27	0.14	0.13	0.36	0.20
MH: metahalloysite	40.39	2.39	32.86	20.13	0.37	0.88	0.43	0.25	2.30

Figure 1 shows the XRD-patterns of both clays (**K** = kaolinite and **H** = halloysite) before and after the calcination. The XRD analysis of kaolin (Figure 1a) showed that it consisted of kaolinite and quartz. After the thermal treatment, only the peaks from quartz, which is not decomposed, are visible. The appearance of an amorphous "halo" in the 20–30° range on the 2-theta scale indicates the distribution of kaolinite to the amorphous metakaolin.

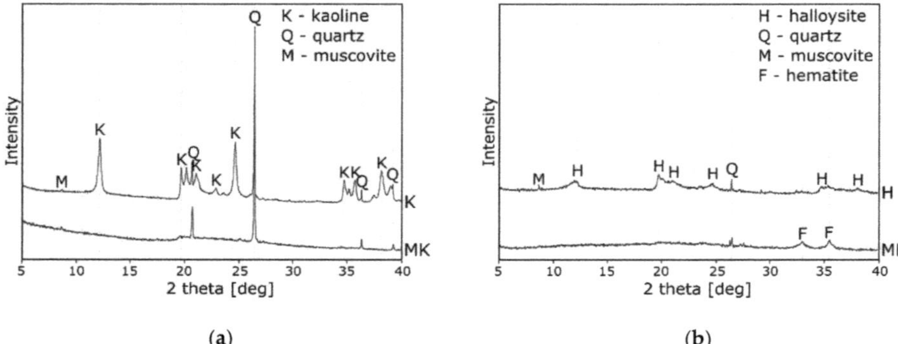

Figure 1. The X-ray diffractometry (XRD) patterns of initial clays and clays after the calcination at 700 °C: (**a**) kaolinite; (**b**) halloysite.

In turn, the analysis of the phase composition of halloysite (Figure 1b) showed that it included, in addition to halloysite, quartz and hematite. For both halloysite and metahalloysite, the reflections are wider, compared to the samples of kaolin and metakaolin, respectively, which indicates a greater defect or the smaller size of crystallites. In addition, the XRD pattern shows reflections from hematite, such that the high iron content in the chemical composition (Table 2) confirms its presence; and it also gave the material a brown color. Attention should also be paid to the quartz content lower than in the case of kaolin, which may result in better reactivity in an alkaline environment, and thus achieving better strength parameters for samples obtained under the analogous conditions.

Figure 2 presents the mid-infrared absorption spectra of kaolin and halloysite (marked as **K** and **H**, respectively) and their calcination products (**MK** and **MH**). The strong similarity between the spectra of the samples of kaolin and halloysite can be observed due to the structural similarity of both materials. The bands in the range of 1100–400 cm^{-1} are associated with the aluminosilicate structure of the material, and come from the vibrations of the Si–O–Si(Al) bridges. The band at 913 cm^{-1} is characteristic of the kaolinite structure, and comes from Al–OH stretching vibrations in the octahedral layer. The band at 695 cm^{-1}, in turn, comes from the vibrations of the Al–O bond for aluminum in the octahedral position. In the spectra of all samples, a characteristic doublet of bands at about 800 and 780 cm^{-1} can also be seen, indicating the presence of quartz.

The spectra of the materials after the calcination are clearly changed. The bands derived from the water and the OH groups disappear, which is caused by the dehydroxylation process. Another difference is the increase in FWHM bands derived from Si–O and Al–O bond vibrations. For the samples of the starting materials, FWHM is small, which means that they had a crystal structure, while in the spectrum of the materials after the heat treatment, they are characterized by a larger FWHM, which indicates a reduced degree of structure ordering. The characteristic band, which indicated the presence of aluminum in the octahedral position (band at 913 cm^{-1}) also disappeared, which indicates that a spatial amorphous aluminosilicate structure was obtained.

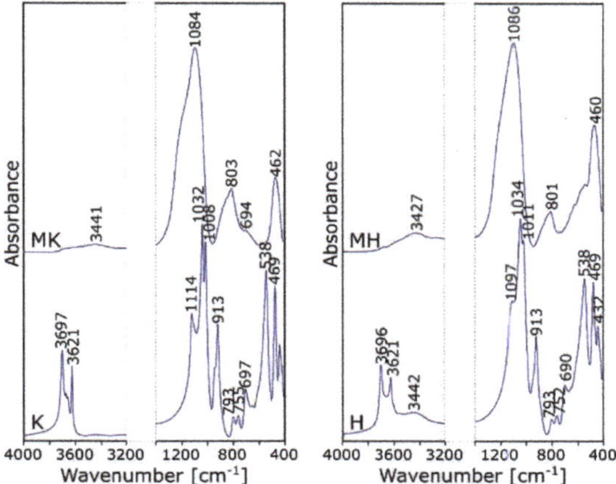

Figure 2. The mid-infrared (MIR) spectra of initial clays (**K** = kaolin (**left**) and **H** = halloysite (**right**)) and clays after the calcination at 700 °C (**MK** = metakaolin (**left**) and **MH** = metahalloysite (**right**)).

3.2. Characterization of Alkali-Activated Clays

Zeolite–geopolymer hybrid bulk materials can be obtained using a thermal activated kaolinitic clay. Due to the chemical composition of metakaolin (Si/Al = 1), the most expected crystalline phase is zeolite A [12], although the appearance of sodalite is not excluded [16].

Figure 3 shows the XRD patterns of the materials obtained as the result of an alkaline activation of both metakaolin (Figure 3a) and metahalloysite (Figure 3b). The analysis of the phase composition showed the presence of quartz (in all samples) and hematite (only in the series based on halloysite). Both mentioned phases were present in the starting material (Figure 1), and did not degrade in the discussed process.

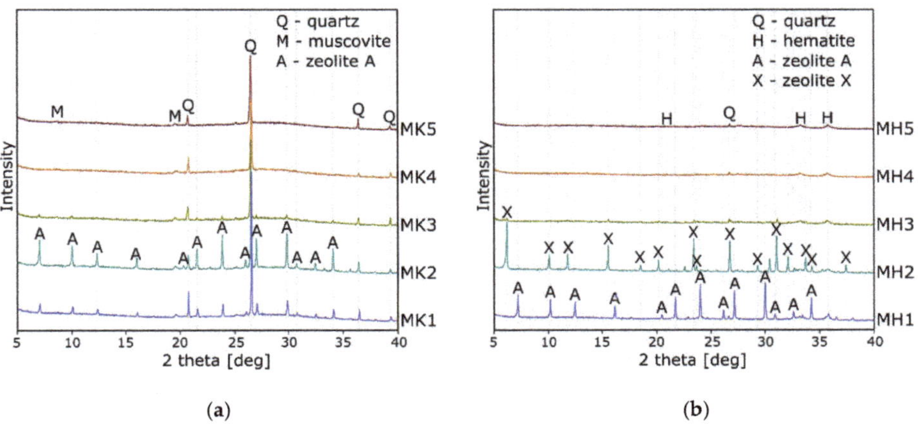

Figure 3. The XRD patterns of alkali-activated: (**a**) metakaolin; (**b**) metahalloysite.

As expected, zeolite A was formed as the result of activation with the highest sodium solution content (**MK1** and **MH1** samples). Interestingly, in the case of metakaolin (Figure 3a), the activation with the solution with a slightly lower Na/Si ratio (comparing **MK1** and **MK2**) gave the greater amounts of zeolite A. The probable reason was the increased proportion of silicon in the solution in the first stages of the geopolymerization process [17,18], which promoted the condensation of tetrahedra into the double-6-ring units, characteristic for LTA structures. In the case of halloysite (**MH2**), zeolite X was formed under the same conditions. The literature data indicate [1] that this zeolite can be obtained using a longer synthesis time, or in reaction systems with a higher silicon content. It can therefore be assumed, that either the metahalloysite dissolved faster than the metakaolin, or the halloysite contained more silicon in the active phase (despite the similar silicon content in both raw materials (Table 1), metakaolin contained inert quartz (Figure 1)). In the systems with higher silicon content (**MK3–MK5** and **MH3–MH5**), zeolite phases were not formed.

The presence of zeolite phase was confirmed by the SEM observations. The selected results are shown in Figures 4 and 5. Especially in the samples containing zeolite A (**MK2**; Figure 4a), the cubic crystallites surrounded and joined with a layer of amorphous phase were visible. Zeolite X obtained in the analogous conditions had less developed morphology (**MH2**, Figure 5a), hence its identification was difficult [16].

(**a**) (**b**)

Figure 4. The microstructure of metakaolin-based composites: (**a**) **MK2**; (**b**) **MK5**.

(**a**) (**b**)

Figure 5. The microstructure of metahalloysite-based composites: (**a**) **MH2**; (**b**) **MH5**.

The higher reactivity of metahalloysite compared to metakaolin was demonstrated by the analysis of samples obtained with the higher proportion of water glass in the activating solution. By comparing the microstructure of **MK5** (Figure 4b) and **MH5** (Figure 5b) samples, it could be easily seen that, while in the case of the sample based on metakaolin (**MK5**), the low ordered amorphous phase was visible, the sample based on metahalloysite (**MH5**) was characterized by a microstructure typical for the N-(A)-S-H phase [19].

In the MIR spectra (Figure 6) the bands due to the characteristic vibrations of bonds observed in both types of oxygen bridges, Si–O–Si and Si–O–Al, were assigned. These bridges constitute basic structural units, forming tetrahedral geopolymer chains. It was found that the slag composition influences the presence of the bands connected with the phases formed during the hydration in the

MIR spectra. Additionally, significant effect of amorphous phases share on the spectra shape was established. Based on IR spectra, it was also possible to determine the influence of the activator type on the products formed.

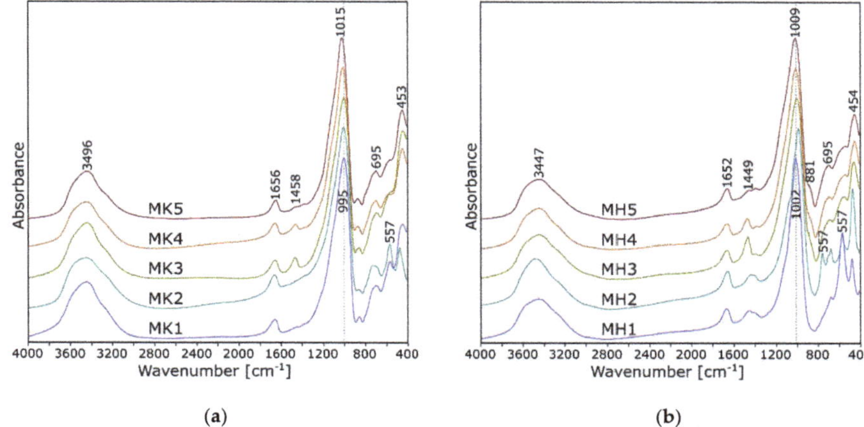

Figure 6. The MIR spectra of alkali-activated: (**a**) metakaolin; (**b**) metahalloysite.

The geopolymerization of alkali-activated metakaolin/metahalloysite (**MK/MH**) was indicated by the bands corresponding to vibrations of Si–O–Si(Al) at 1200–950 cm^{-1} (ν_{as}) and at 750–650 cm^{-1} (ν_s). Their presence in the IR spectra is due to the aluminosilicate character of the structure. These bands do not differ significantly from each other, however, the fact that the envelope in this range is a superposition of several bands should be kept in mind.

First, the slight shift of the most intense band located at about 1000 cm^{-1} with the increasing amount of water glass can be observed. On the one hand, the reason may be the increasing share of silica in the structure. On the other hand, the degree of structure polymerization can be increased, which would explain the systematic increase in the compressive strength of the **MK1–MK5** samples, and which agrees well with the microscopic observations. The analogous conclusions can be drawn by interpreting the spectra of the series of samples based on metahalloysite (**MH1–MH5**). This observation agrees with our previous conclusions [18].

The significant differences in the IR spectra of the **MK1** and **MK2** samples are visible in the so-called pseudolattice range, 800–550 cm^{-1}, in which the vibrations of the over-tetrahedral structural units of the zeolite framework can be visible. Especially, characteristic is the band at 557 cm^{-1}, which is connected with the vibrations realized in the zeolite A structure. In turn, the spectrum of the sample **MH2** shows two bands characteristic for the faujasite-type structure (zeolite X).

The carbonate group bands were identified at about 1450 and 880 cm^{-1}, as they may be the result of the carbonation of N-(A)-S-H or other sodium compounds, or an unreacted alkali-activator. The interesting observation is the fact that the presence of zeolites in the geopolymer structure inhibits the carbonation process. In this case, sodium was probably bound in the zeolite structure, and there was no free OH groups that could be involved in the carbonation. On the other hand, the intensity of this band decreased as the sodium content in the reaction systems decreased (hence the disappearance of this band in the spectra of the MH5 sample, obtained by activation with the solution with the lowest sodium content).

The bulk density and compressive strength of the geopolymer samples after 28 days of maturation were determined. The results are shown in Figure 7. Generally, the strength achieved by geopolymers based on clay minerals is relatively low compared to materials based on fly ash or blast furnace waste [20]. In the case of this work, these values did not exceed 12 MPa. The highest compressive strength values were achieved for the samples **MK5/MH5**. This is in line with the literature data [21], which indicates that the higher the silicon content in the activator, the higher the strength parameters. It is related to the content of the amorphous phase, that is, the higher the glassy content, the more visually homogenous gel structure (Figures 4 and 5), which have possibly contributed to its high strength after the activation.

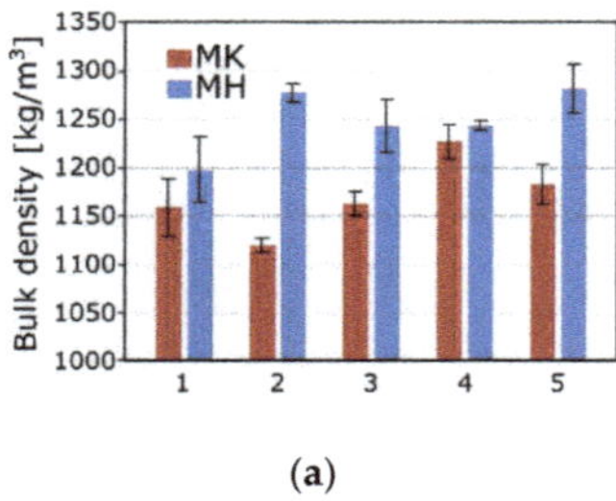

(**a**)

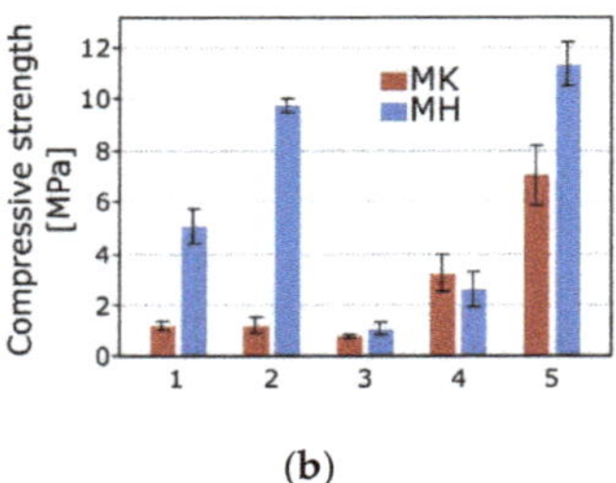

(**b**)

Figure 7. The physical and mechanical properties of the obtained samples: (**a**) bulk density; (**b**) compressive strength.

What is surprising is the high strength of the samples in which zeolite phases were found. Two factors related to the zeolite presence in a geopolymer matrix affect the compressive strength, where one is the amount of zeolites, and the second is the zeolite crystallite size. Their growth causes the decrease in the compressive strength of the composite [2]. The relatively high strength of the obtained samples with zeolites suggest that these factors are at the levels which can be borne by a geopolymer matrix without causing its weakening.

When comparing the two raw materials used, it can be stated that the samples based on metahalloysite were characterized by a higher apparent density, and thus a higher compressive strength. The obtained better strength values may have resulted from the abovementioned absence of quartz in the phase composition of this raw material, as well as from the greater reactivity of metahalloysite compared to metakaolin, which could have affected the formation of phases with the higher structural density. This was confirmed by both the microscopic observations (Figures 4 and 5), and also in the analysis of the infrared spectra (the higher positions of the band originating from Si–O–Si(Al) bridges with similar raw material compositions; (Figure 6)).

A significant difference in the iron content of the compared raw materials (Table 2) can be controversial. The role of iron in the geopolymerization process is widely discussed [22–24]. It is generally accepted that the contribution of iron oxide inhibited the geopolymer formation, and it is necessary to control the content of Fe_2O_3 to enhance the physical characteristics of geopolymer paste [22]. The distribution of iron in geopolymers made with iron-rich precursors has been also investigated [23,24]. The authors agree that iron (Fe^{3+}) occupies primarily octahedral positions. However, information can be found that the coordinated Fe^{3+} replaced Al^{3+} in the aluminosilicate structure of the geopolymer [24]. On the other hand, some observation suggests the replacement of Al^{3+} by Fe^{3+} in octahedral sites most likely within the kaolinite phase, which indicate that iron is not necessarily deleterious to geopolymer formation, as has sometimes been suggested [23].

It was observed that the optimal composition of geopolymer based on fly ash and red mud (Fe_2O_3 content = 33.9 wt.%) was at a weight ratio of 50/50 that gave higher compressive strength than at any other ratio [24]. Geopolymers based on fly ash with a high content of iron oxide (48.8 wt.%) exhibited better heat resistance than Portland cement concrete [25]. Also, a high iron oxide (31.1 wt.%) in fly ash has no negative effect on the strength development of geopolymer [26]. Based on the results described in this paper, it is impossible to clearly assess the role of iron in the formation of the geopolymer structure. Although **MH** contains much higher amounts of iron (Table 2), it is mainly in the form of hematite (Figure 1), and in this form it can also be found in the material after activation (Figure 3). In addition, composites based on **MH** have a higher compressive strength, which is due to the higher degree of structure polymerization observed both in FT-IR and SEM analysis.

4. Conclusions

The aim of this study was to determine the effect of the type and concentration of an activator on the structure and properties of the alkali-activated metakaolin and metahalloysite. Therefore, the first stage of the work was based on the obtaining of the metakaolin and metahalloysite phases as a result of the calcination of the starting materials, kaolin and halloysite, respectively. The obtaining of the reactive phases with a similar structure was confirmed by the analysis of the XRD and IR results. The chemical and phase composition analysis confirmed that both of the starting materials are mainly in the amorphous aluminosilicate phase.

The next stage of the work was an attempt to alkali-activate the previously obtained metakaolin and metahalloysite. A number of activation solutions based on sodium hydroxide and sodium silicate were used. The structure and properties of the resulting geopolymer binders depend on both the type of starting material used, and the type of activator. The activation carried out with sodium hydroxide alone, or with a small addition of water glass, causes the formation of zeolite phases, regardless of the starting material used. The alkaline activation with a solution with a higher proportion of silicon did not give crystalline phases, but only an amorphous N-(A)-S-H phase.

The bulk density and compressive strength were determined. The obtained strength values were in the range of 0.7–7.1 MPa for metakaolin, and in the range of 1.1–11.3 MPa for metahalloysite. The geopolymers based on metahalloysite had better strength results, which was probably due to their higher bulk density. A higher density, and thus better strength parameters, were probably the result of the better reactivity of metahalloysite compared to metakaolin.

Zeolites are known for their ability to immobilize heavy metal ions. Although the resulting materials did not obtain the strength parameters valuable for construction, subsequent tests may prove to be attractive materials for neutralizing hazardous waste and water treatment.

Author Contributions: Conceptualization, M.K.; sample preparation, O.B.; measurement, O.B. and P.R.; data curation, O.B. and M.K; writing—original draft preparation, M.K., P.R. and P.F.; writing—review and editing, P.R. and P.F.; project administration, M.K.; funding acquisition, W.M. All authors have read and agreed to the published version of the manuscript.

Funding: This work was financially supported by The National Science Centre Poland under grant no. 2018/31/B/ST8/03109.

Conflicts of Interest: The authors declare no conflict of interest.

References

1. Breck, D.W. *Zeolite Molecular Sieves: Structure, Chemistry, and Use*; John Wiley and Sons: Hoboken, NJ, USA, 1974.
2. Rożek, P.; Król, M.; Mozgawa, W. Geopolymer-zeolite composites: A review. *J. Clean. Prod.* **2019**, *230*, 557–579. [CrossRef]
3. Takeda, H.; Hashimoto, S.; Yokoyama, H.; Honda, S.; Iwamoto, Y. Characterization of zeolite in zeolite-geopolymer hybrid bulk materials derived from kaolinitic clays. *Materials* **2013**, *6*, 1767–1778. [CrossRef] [PubMed]
4. Papa, E.; Medri, V.; Amari, S.; Benito, P.; Vaccari, A.; Landi, E. Zeolite-geopolymer composite materials: Production and characterization. *J. Clean. Prod.* **2018**, *171*, 76–84. [CrossRef]
5. Krisnandi, Y.K.; Saragi, I.R.; Sihombing, R.; Ekananda, R.; Sari, I.P.; Griffith, B.E.; Hanna, J.V. Synthesis and characterization of crystalline NaY-Zeolite from Belitung Kaolin as catalyst for n-Hexadecane cracking. *Crystals* **2019**, *9*, 404. [CrossRef]
6. Pereira, P.M.; Ferreira, B.F.; Oliveira, N.P.; Nassar, E.J.; Ciuffi, K.J.; Vicente, M.A.; Trujillano, R.; Rives, V.; Gil, A.; Korili, S.; et al. Synthesis of zeolite A from metakaolin and its application in the adsorption of cationic dyes. *Appl. Sci.* **2018**, *8*, 608. [CrossRef]

7. Lima-de-Faria, J. *Structural Classification of Minerals*; Springer: New York, NY, USA, 2003; Volume 2, p. 18.
8. Hillier, S.; Ryan, P.C. Identification of halloysite (7 Å) by ethylene glycol solvation: The 'MacEwan effect. *Clay Miner.* **2002**, *37*, 487–496. [CrossRef]
9. Slaty, F.; Khoury, H.; Wastiels, J.; Rahier, H. Characterization of alkali activated kaolinitic clay. *Appl. Clay Sci.* **2013**, *75–76*, 120–125. [CrossRef]
10. Wang, Q.; Zhang, J.; Wang, A. Alkali activation of halloysite for adsorption and release of ofloxacin. *Appl. Surf. Sci.* **2013**, *287*, 54–61. [CrossRef]
11. San Cristóbal, A.G.; Castelló, R.; Martín Luengo, M.A.; Vizcayno, C. Zeolites prepared from calcined and mechanically modified kaolins. A comparative study. *Appl. Clay Sci.* **2010**, *49*, 239–246. [CrossRef]
12. Król, M.; Rożek, P. The effect of calcination temperature on the metakaolin structure for the synthesis of zeolites. *Clay Miner.* **2019**, *53*, 657–663. [CrossRef]
13. Giese, R.F. Kaolin group minerals. In *Encyclopedia of Sediments and Sedimentary Rocks*; Middleton, G.V., Church, M.J., Coniglio, M., Hardie, L.A., Longstaffe, F.J., Eds.; Springer: Dordrecht, The Netherlands, 2003; pp. 398–400.
14. Velde, B. 1:1 dioctahedral clays (kaolinite, dickite, nacrite halloysite). In *Introduction to Clay Minerals—Chemistry, Origins, Uses and Environmental Significance*; Crowley, S., Ed.; Chapman and Hall: London, UK, 1992; pp. 79–80.
15. Brindley, G.W.; Nakahira, M. Kinetics of dehydroxylation of kaolinite and halloysite. *J. Am. Ceram. Soc.* **1957**, *40*, 346–350. [CrossRef]
16. Król, M.; Minkiewicz, J.; Mozgawa, W. IR spectroscopy studies of zeolites in geopolymeric materials derived from kaolinite. *J. Mol. Struct.* **2016**, *1126*, 200–206. [CrossRef]
17. Duxson, P.; Fernández-Jiménez, A.; Provis, J.L.; Lukey, G.C.; Palomo, A.; Van Deventer, J.S.J. Geopolymer technology: The current state of the art. *J. Mater. Sci.* **2007**, *42*, 2917–2933. [CrossRef]
18. Król, M.; Rożek, P.; Chlebda, D.; Mozgawa, W. ATR/FT-IR studies of zeolite formation during alkali-activation of metakaolin. *Solid State Sci.* **2019**, *94*, 114–119. [CrossRef]
19. Król, M.; Rożek, P.; Mozgawa, W. Preparation and Structure of Geopolymer-Based Alkali-Activated Circulating Fuildized Bed Ash Composite for Removing Ni^{2+} from Wastewater. In *Proceedings of the 12th Pacific Rim Conference on Ceramic and Glass Technology: Ceramic Transactions*; Singh, D., Fukushima, M., Kim, Y., Shimamura, K., Imanaka, N., Ohji, T., Amoroso, J., Lanagan, M., Eds.; John Wiley & Sons: Hoboken, NJ, USA, 2018; Volume 264, pp. 147–154.
20. Abdullah, M.M.A.B.Y.; Liew, M.; Yong, H.C.; Tahir, M.F.M. Clay-Based Materials in Geopolymer Technology. In *Cement Based Materials*; IntechOpen: Rijeka, Croatia, 2018; pp. 239–264.
21. Prochon, P.; Zhao, Z.; Courard, L.; Piotrowski, T.; Michel, F.; Garbacz, A. Influence of Activators on Mechanical Properties of Modified Fly Ash Based Geopolymer Mortars. *Materials* **2020**, *13*, 1033. [CrossRef] [PubMed]
22. Choi, S.C.; Lee, W.K. Effect of Fe_2O_3 on the physical property of geopolymer paste. *Adv. Mater. Res.* **2012**, *586*, 126–129. [CrossRef]
23. Lemougna, P.N.; MacKenzie, K.J.D.; Jameson, G.N.L.; Rahier, H.; Chinje Melo, U.F. The role of iron in the formation of inorganic polymers (geopolymers) from volcanic ash: A ^{57}Fe Mössbauer spectroscopy study. *J. Mater. Sci.* **2013**, *48*, 5280–5286. [CrossRef]
24. Hu, Y.; Liang, S.; Yang, J.; Chen, Y.; Ye, N.; Ke, Y.; Tao, S.; Xiao, K.; Hu, J.; Hou, H.; et al. Role of Fe species in geopolymer synthesized from alkali-thermal pretreated Fe-rich Bayer red mud. *Constr. Build. Mater.* **2019**, *200*, 398–407. [CrossRef]
25. You, S.; Ho, S.W.; Li, T.; Maneerung, T.; Wang, C.H. Techno-economic analysis of geopolymer production from the coal fly ash with high iron oxide and calcium oxide contents. *J. Hazard. Mater.* **2019**, *361*, 237–244. [CrossRef] [PubMed]
26. Kumar, S.; Djobo, J.N.Y.; Kumar, A.; Kumar, S. Geopolymerization behavior of fine iron-rich fraction of brown fly ash. *J. Build. Eng.* **2016**, *8*, 172–178. [CrossRef]

Communication

Size Effects of the Crystallite of ZSM-5 Zeolites on the Direct Catalytic Conversion of L-Lactic Acid to L, L-Lactide

Qintong Huang, Rui Li, Guangying Fu and Jiuxing Jiang *

MOE Key Laboratory of Bioinorganic and Synthetic Chemistry, School of Chemistry, Sun Yat-sen University, Guangzhou 510275, China; huangqt9@mail2.sysu.edu.cn (Q.H.); lirui99@mail2.sysu.edu.cn (R.L.); fugy3@mail.sysu.edu.cn (G.F.)

* Correspondence: jiangjiux@mail.sysu.edu.cn; Tel.: +86-20-84111355

Received: 16 July 2020; Accepted: 31 August 2020; Published: 3 September 2020

Abstract: ZSM-5 zeolites are commonly used as a heterogeneous catalyst for reactions. Four ZSM-5 catalysts (with various crystallite sizes and a similar ratio of Si/Al) and their ball-milling/surface-poisoning derivates were used to convert L-lactic acid to L, L-lactide. The reaction products were analyzed by three independent analytical methods (i.e., Proton nuclear magnetic resonance (^{1}H NMR), high-pressure liquid chromatography (HPLC), and chiral gas chromatography (GC)) for determining the L, L-lactide yield and L-lactic acid conversion. A clear size effect, i.e., smaller catalysts providing better performance, was observed. Further ball-milling/surface-poisoning experiments suggested that the size effect of the ZSM-5 catalysts originated from the diffusion-controlled nature of the reaction under the investigated conditions.

Keywords: ZSM-5 zeolites; L,L-lactide; size effect; diffusion control

---

1. Introduction

Polylactic acid (PLA), produced by the condensation of lactic acid, is widely used in disposable packaging and in vivo biomedical applications [1], among other applications, due to its biodegradability and biocompatibility. Furthermore, its building block, i.e., lactic acid, is a sustainable and renewable biomass-derived platform molecule that is independent of fossil resources.

To avoid the high reaction temperature required for water removal during the condensation process, industrially manufactured PLA is synthesized via the lactide route. The lactide route enables the production of high-quality, monodispersed, higher molecular weight PLA [2]. Hence, pure optical lactides (i.e., L, L-lactide and/or D, D-lactide) are important intermediates for producing high-quality PLA.

Bellis et al. [3] proposed vapor-phase synthesis over a plug flow reactor. Chang and coworkers [4,5] realized a 94% yield with almost 100% enantio-selectivity using a SnO_2–SiO_2 nanocomposite catalyst. However, a high yield was only achieved at a relatively low velocity of 1 or 3 h^{-1}. Sels and co-workers proposed a liquid-phase lactide synthesis process [6]. In their work, zeolite was used as a shape-selective catalyst to produce highly optical pure L-lactide (Scheme 1). The acidic form of beta zeolite was proposed as the best catalyst to achieve an 83% pass yield and a 98% optical purity. However, the work only compared some zeolites and left crystallite size unexplored.

In this work, a size effect and better performance was observed for ZSM-5 in comparison to beta zeolite. Further ball-milling and surface-poisoning experiments were performed to elucidate the size effect.

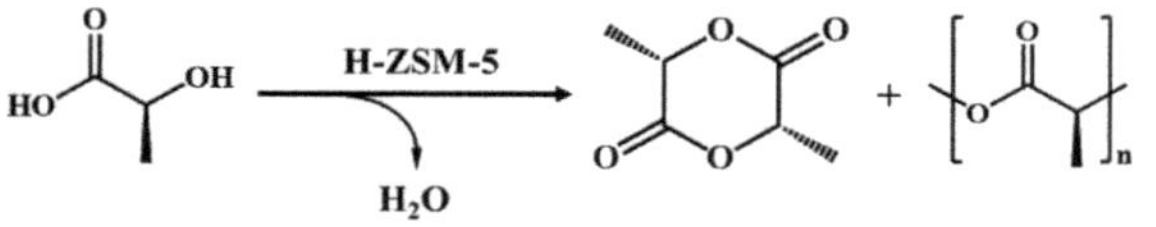

Scheme 1. The current and proposed chemical process for making L, L-lactide and lactyl oligomers from L-lactic acid.

2. Materials and Methods

2.1. Materials

All chemicals were purchased from Aladdin Chemical Co. (Shanghai, China) and Sigma-Aldrich (St. Louis, MO, USA) and used directly after receiving. Commercial nanocrystalline zeolites of NH_4–ZSM-5-I (CBV3020, Si/Al = 15) and NH_4–ZSM-5-II (CBV3024E, Si/Al = 15) were purchased from Zeolyst (Farmsum, The Netherlands). Micrometer-sized ZSM-5-III (Si/Al = 12) and ZSM-5-IV (Si/Al = 10) were provided by Wuhan Zhizhen Zeolite Co. (Wuhan, China). ZSM-5-III-b.m. and ZSM-5-IV-b.m. were obtained by ball milling ZSM-5-III (Si/Al = 12) and ZSM-5-IV (Si/Al = 10) for half an hour at 60 Hz.

All of the ZSM-5 zeolites were calcined before use for 6 h at 550 °C, reaching a rate of 2 °C/min under a smooth flow of air to obtain the acidic form of ZSM-5 as a catalyst for the transformation of L-lactic acid into L, L-lactide. We named the six acidic forms of the ZSM-5 zeolite ZSM-5-I (I), ZSM-5-II (II), ZSM-5-III (III), ZSM-5-I (IV), ZSM-5-III-b.m. (III-b.m.), and ZSM-5-IV-b.m. (IV-b.m.). When ZSM-5-IV (Si/Al = 10) was ball milled for 15 min at 60 Hz, the obtained sample was ZSM-5-IV-b.m.-15 min (IV-b.m.-15 min).

2.2. Methodology

2.2.1. Characterization of the ZSM-5 Zeolites

The X-ray diffraction (XRD) patterns of the ZSM-5 zeolites were obtained using a Rigaku SmartLab diffractometer (Rigaku Co., Tokyo, Japan) with Cu Kα monochromatic radiation (λ = 0.154178 nm) at 40 kV and 30 mA. The crystallite morphology and size of these ZSM-5 catalysts were investigated by ultra-high-resolution scanning electron microscopy (SEM; FE-SEM SU8010, HITACHI, Tokyo, Japan) and transmission electron microscopy (TEM; JEM-1400 Plus, JEOL, Tokyo, Japan). The Si/Al ratios of the ZSM-5 catalysts were determined by inductively coupled plasma optical emission spectrometry (ICP-OES, Perkin ElmerOptima8300, PerkinElmer, Waltham, MA, USA). The N_2 adsorption isotherms of these ZSM-5 catalysts at 77 K were measured using a Micrometrics ASAP 2020 Plus (Micrometrics, Norcross, GA, USA) instrument. To investigate the acidic properties of the ZSM-5 catalysts, temperature-programmed desorption of ammonia (NH_3-TPD) and pyridine adsorption Fourier-transform infrared spectroscopy spectra (py-IR) were used. The NH_3-TPD measurement was conducted using an MFTP-3060 (Xiamen Better Work Intelligent Technology Co., Xiamen, China) instrument with a thermal conductivity detector (TCD), while the py-IR spectra were recorded using a ThermoFisher Scientific RS20 (Thermo Fisher Scientific, Waltham, MA,USA) instrument in situ.

2.2.2. Synthesis of L, L-Lactide

First, 0.83 g of L-lactic acid (9.21 mmol, 98%) dissolved in 0.83 g of water was placed into a 25 mL round-bottom flask, which was equipped with a magnetic stirrer and a Dean–Stark apparatus. Subsequently, 10 mL of *o*-xylene/*p*-xylene (99%) was added, as well as 0.2 g of calcined zeolite. Then, the mixture was heated at the reflux temperature of *o*-xylene (144 °C) for 4 h. After the reaction, the mixture was homogenized by the addition of 10 mL of acetonitrile (99%). The zeolite was removed by centrifugation.

The condition of the surface-poisoning experiment was the same as the above; only 0.031 g of 2,4-dimethylquinoline was added to poison the surface of Sample II to eliminate the influence of the surface reaction on the catalytic results.

2.2.3. Analytical Methods

^{1}H NMR analysis of the production: After accurately measuring the total mass of the homogeneous reaction mixture, a 1 mL sample was taken from the homogeneous reaction mixture, weighed, and then dried in a gentle flowing nitrogen atmosphere to remove the *o*-xylene and acetonitrile solvents. Mild drying in a nitrogen atmosphere can remove the solvent without affecting the less-volatile lactide and lactic acid mixtures. The dry sample was dissolved in 0.5 mL of DMSO-d_6, transferred to NMR tubes, and then measured on a Bruker Avance 400 MHz NMR (Bruker Co., Billerica, MA, USA) spectrometer with an automated sampler.

Chiral GC analysis: GC analysis was performed on an Agilent 7890B system (Agilent Technologies, Santa Clara, CA, USA) with a chiral capillary column, equipped with a flame ionization detector (FID) maintained at 275 °C and Agilent MassHunter Qualitative Analysis software. The injection port temperature was 225 °C, and the initial column temperature was set at 70 °C. This temperature was maintained for 5 min before being increased to 150 °C at 15 °C/min, after which the oven was ramped up to 180 °C at 25 °C/min and held there for 24 min.

HPLC analysis: A 1 mL sample was taken, weighed, and then dried under a flowing nitrogen atmosphere. This dry sample was dissolved in 1 mL of ultra-pure water and analyzed using reverse phase-HPLC on an Agilent 1100 Series Products Spectra System (Agilent Technologies, Santa Clara, CA, USA) equipped with a PL Hi-Plex H Guard 50 × 7.7 mm column, a PL Hi-Plex H 300 × 7.7 mm column, and a 1260 Infinity UV-Vis detector at 210 nm. Lastly, 0.001 M H_2SO_4 solvent was used as a mobile phase, and the flow rate was 0.6 mL/min.

3. Results and Discussion

3.1. Characterization of the ZSM-5 Zeolites with Different Crystallite Sizes

The XRD patterns of the six ZSM-5 zeolites are shown in Figure 1. Characteristic diffraction peaks (2θ = 7.8°, 8.7°, 23.1°, 23.8°, and 24.3°) [7] for the ZSM-5 structure (**MFI**) were found in all the catalysts, and no impurity peaks were surveyed. Across Samples I–IV, the XRD patterns show very sharp reflections and low background signal, indicating their superior crystalline properties. The relative crystallinity of the ZSM-5 zeolites with different crystallite sizes was determined from the peak area between 2θ = 22–25° using Sample I as the standard [7,8], and the results are listed in Table 1. The texture properties of the ZSM-5 zeolites were characterized by nitrogen adsorption/desorption and ICP-OES analysis (Table 1). Examination of the data shows that the Brunauer–Emmett–Teller (BET) surface area and the t-plot micropore volumes of Samples I–IV are similar, but the external surface area increases as the crystallite size decreases. Different shapes and sizes of the catalysts can be observed in the SEM micrographs (Figure 2a,d,g,j) and the TEM images (Figure 2b,e,h,k). Samples I and II are round cuboids, but Samples III and IV are sharp rectangular blocks. All of the ZSM-5 catalysts were agglomerated by microcrystals to form secondary particles. The distribution of the crystallite sizes was obtained from 300 randomly selected microcrystalline samples. Sample I shows the smallest crystallite size (0.05–0.20 μm in length and ~0.08 μm in width), while Sample II (0.08–0.30 μm in length and ~0.1 μm in width) shows a larger crystallite size than Sample I. Sample III displays a length of 0.75–3 μm and a width of 0.5 μm, although Sample IV exhibits the largest crystallite size (a length of 1–4 μm and a width of 1 μm).

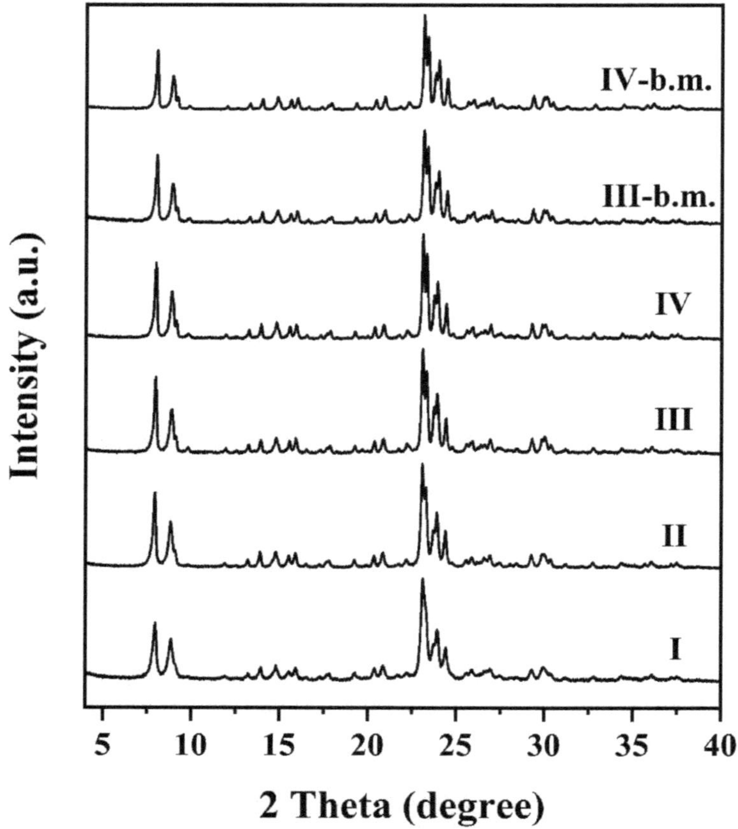

Figure 1. The X-ray diffraction patterns of ZSM-5 zeolites I–IV, III-b.m., and IV-b.m.

Table 1. Structural and textural properties of ZSM-5 zeolites I–V, III-b.m., and IV-b.m.

Zeolite	Relative Crystallinity (%)	Si/Al Molar Ratios [1]	A_{BET} [2] (m^2/g)	V_{micro} [3] (cm^3/g)	External Surface Area [3] (m^2/g)
I	100.0	12.4	397.5	0.15	42.6
II	90.0	12.0	393.4	0.15	36.7
III	89.1	10.7	411.8	0.16	12.0
IV	80.0	9.3	403.7	0.16	6.7
III-b.m.	81.4	10.6	362.9	0.14	27.3
IV-b.m.	70.6	9.5	336.2	0.13	25.1

[1] Measures by ICP-OES; [2] Brunauer–Emmett–Teller (BET) surface area; [3] micropore volume and external surface area determined by t-plot.

When Samples III and IV were milled, we gained III-b.m. (0.1–0.8 μm in length and ~0.2 μm in width) and IV-b.m. (0.2–0.8 μm in length and ~0.3 μm in width), which were irregular pieces (Figure 2m,n,p,q). In Figure 1, for III-b.m. and IV-b.m., the broad XRD peaks around 25° imply the presence of amorphous material impurity because of the ball milling. Meanwhile, the BET surface area and the t-plot micropore volume of III-b.m. and IV-b.m. decreased marginally, but the external surface area increased (Table 1).

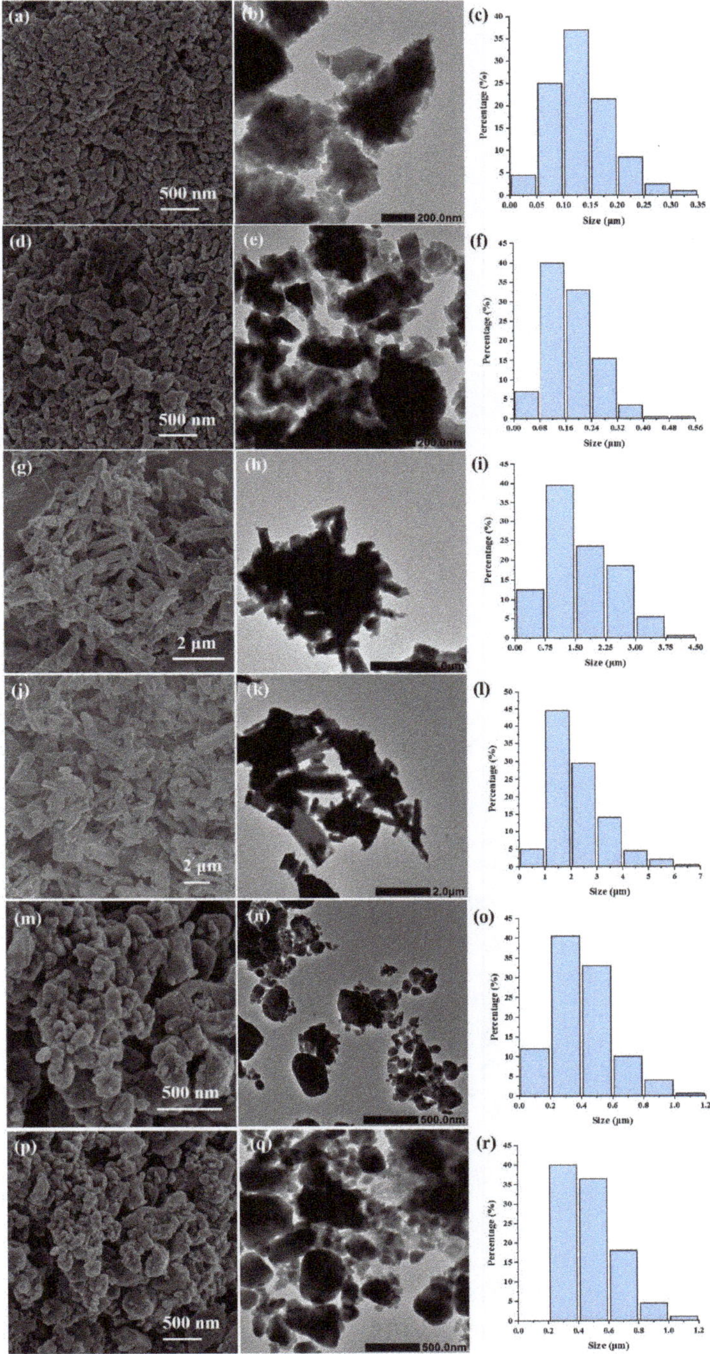

Figure 2. The scanning electron microscopy (SEM) micrographs, transmission electron microscopy (TEM) images, and distribution curves of the crystallite sizes of the ZSM-5 zeolites: (**a**–**c**) I; (**d**–**f**) II; (**g**–**i**) III; (**j**–**l**) IV; (**m**–**o**) III-b.m.; (**p**–**r**) IV-b.m.

The acid strength and the number of the acidic sites on the surfaces of the six kinds of ZSM-5 zeolites were determined by NH_3-TPD measurement (as shown in Figure 3a). All of the ZSM-5 catalysts demonstrated two similar NH_3 desorption peaks—a low-temperature peak at approximately 195 °C and a high-temperature peak at about 405 °C, generally belonging to the NH_3 chemisorbed on weak and strong acidic sites [9–13], respectively. The strength of the acid sites and the number of weak and strong acid sites of these six ZSM-5 samples are basically the same due to the similar Si/Al molar ratio. The py-IR spectra were used to detect the Lewis and Brønsted sites of the ZSM-5 catalysts, and Figure 3b exhibits the py-IR spectra of these acquired ZSM-5 zeolites in the region of 1400–1700 cm^{-1}. The band in the 1450 cm^{-1} region belongs to pyridine adsorbed on the Lewis acid sites, while the band in the 1540 cm^{-1} region is attributed to pyridine adsorbed on the Brønsted acid sites. [14,15] These results indicate that all of the samples have similar amounts of Lewis and Brønsted acid sites. The data in Table 2 show that ZSM-5 zeolites with different crystallite sizes exhibit a similar concentration of both Brønsted and Lewis acidic sites at 150 °C. [16] Thus, the influence of the acidic properties on the catalytic reaction is insignificant.

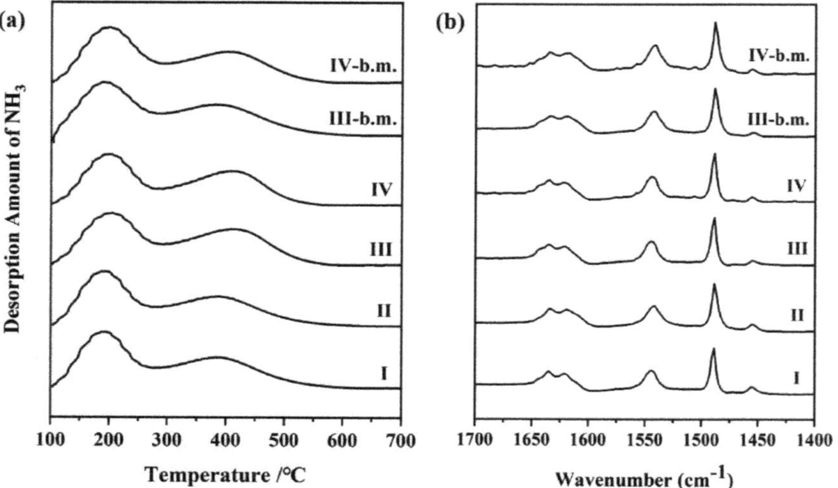

Figure 3. (**a**) Temperature-programmed desorption of ammonia (NH_3-TPD) profiles of various ZSM-5 zeolites; and (**b**) infrared spectra of adsorbed pyridine on various ZSM-5 zeolites at a desorption temperature of 150 °C.

Table 2. Results for the IR spectra of pyridine adsorbed on the different ZSM-5 zeolites after desorption at 150 °C.

Catalyst	I	II	III	IV	III-b.m.	IV-b.m.
B [1] (μmol/g)	148.2	156.5	171.1	187.1	164.5	180.2
L [2] (μmol/g)	48.0	45.3	40.5	39.1	43.8	45.0
Total acid (B + L) (μmol/g)	196.2	201.8	211.6	226.2	208.3	225.2

[1] B: Brønsted acid sites; [2] L: Lewis acid sites.

3.2. *Transformation of L-Lactic Acid into L, L-Lactide*

The catalytic performance of various ZSM-5 zeolites was appraised under the same reaction conditions. In a typical reaction system, the desired amounts of L-lactic acid, water, ZSM-5, and *o*-xylene were mixed in a round-bottom flask, which was then connected to a Dean–Stark apparatus (Figure S1). After that, the mixture was applied under continuous water expunction via the reflux of *o*-xylene. The product yield and selectivity of the desired lactide were determined by ^{1}H NMR, HPLC, and chiral GC. The details of a typical ^{1}H NMR spectrum in DMSO-d_6 in the methine region is given in Figure S2, revealing the conversion of L-lactic acid to L, L-lactide from four reaction mixtures taken at different times. Analyzing the samples in the methine region allows distinguishing between monomeric L-lactic acid; L, L-lactide; the center lactyl units of linear oligomers; and the -COOH and -OH end groups of linear oligomers. [6,17] In general, the L, L-lactide yield; L-lactic acid conversion; and oligomers were measured by ^{1}H NMR integration, relative to the total amount of lactyl-containing products present. HPLC analysis is dependable and allows the admeasurement of L-lactic acid conversion and the yields of all lactyl substances involved. Compared to ^{1}H NMR, which provides the average oligomer length, HPLC analysis can separate all oligomers encountered [18]. This chiral GC enables us to discriminate L, L-lactide; D, D-lactide; and meso-lactide [6]; however, due to the limited volatility of L-lactic acid, GC cannot accurately quantify it. Usually, GC analysis is only used as an auxiliary confirmation tool to confirm the yields of L, L-lactide produced by optimized catalytic reactions and to determine that there is no substantial degree of meso-lactide formation. In this study, there were no other organic species, but condensation products were sought out in the reaction mixtures by HPLC and GC, which corroborated the selectivity of the reaction to the condensation products and the correctness of the ^{1}H NMR analysis. The yields of L, L-lactide obtained by these three independent analytical methods were in good agreement with one another. Figure S3 shows the kinetics of the reaction using Sample I. It can be observed that at the beginning, 2% of oligomers (mainly dimers and trimers) are present. The maximum value of the oligomers was reached after a short contact time. In the third hour, the catalytic reaction was close to completion.

Figure 4 and Table S1 compares the catalytic performance of the H-form of the ZSM-5 catalysts with different crystallite sizes in *o*-xylene at 144 °C for 4 h. The nano-sized I sample shows the highest selectivity for lactide synthesis, yielding nearly 89% at full L-lactic acid conversion, compared to 9% at only 20% conversion in the absence of a catalyst. The L, L-lactide yield, L-lactic acid conversion, and selectivity decreased with the increase in crystallite size.

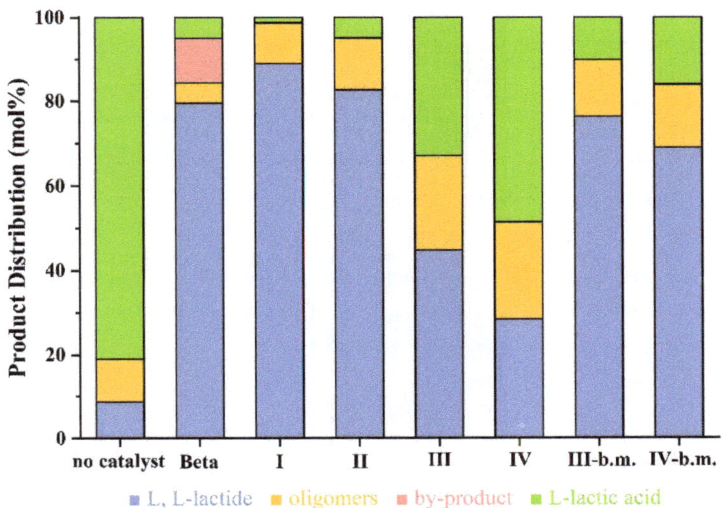

Figure 4. The product distribution of L-lactic acid to L, L-lactide over the ZSM-5 zeolites.

Figure S4 shows the product distributions of the reaction on different solvents with/without surface poisoning. The *p*-xylene provides a subtle difference from *o*-xylene. The surface poisoning experiment reveals that external surface acid sites play important roles, and that the internal acid sites also exhibit activity—however, with limited diffusion.

When we ball milled the micron-sized III and IV samples into the nano-sized III-b.m., IV-b.m.-15 min, and IV-b.m. (30 min) samples, the L, L-lactide conversion and selectivity increased significantly. In particular, IV, IV-b.m.-15 min, and IV-b.m. (30 min) continuously improved the catalytic performance (Figure S5). The ball-milled catalysts not only exposed more internal acid sites but also improved the reactant/product mass transfer, which is also consistent with the results of the surface-poisoning experiments. Furthermore, the change in morphology during the ball-milling process, i.e., break down in the longest dimension, may also have contributed to increasing the exposure of the straight 10-ring channel. As a consequence, the catalytic performance improved due to the shorter diffusion path. In general, besides reducing the particle size of the primary catalyst, the effect of the external (liquid–solid) mass transfer resistance can be eliminated by increasing (a) the amount of catalyst and (b) the reaction temperature [19]. However, the amount of catalyst is limited by the net adsorption of the reactant/product by the catalyst, while the reaction temperature is limited by the side reactions. The internal (pore diffusion) mass transfer is determined by the intrinsic pore size and pore volume of ZSM-5. Based on the results of the experiments, both external and internal mass transfer influence the overall effectiveness of the mass transfer. Further work to evaluate the contributions of both internal and external mass transfer is being conducted in our laboratory.

4. Conclusions

ZSM-5 catalysts with different crystallite sizes were analyzed by XRD, SEM, N_2 adsorption/ desorption, ICP, NH_3-TPD, and pyridine IR. Then, these ZSM-5 catalysts and their derivatives were applied to the reaction of L-lactic acid conversion into L, L-lactide. A clear size effect of the ZSM-5 catalysts, i.e., smaller crystallites providing better performance, was observed. Further ball-milling and surface-poisoning experiments indicated that this size effect mainly originated from the diffusion-controlled nature of the reaction under the investigated conditions. Therefore, this work highlights that any operations to improve mass transfer—e.g., smaller catalyst particle sizes, a shorter b-axis 2D layer, or a hierarchical structure— during synthesis or post-synthesis could enhance the catalytic performance of the reaction.

Supplementary Materials: The following are available online at http://www.mdpi.com/2073-4352/10/9/781/s1. Characterization procedure for ZSM-5 catalysts' acidity; Figure S1: General lab-scale setup; Figure S2: Reaction progress using Sample I as the catalyst according to ^{1}H NMR in DMSO-d_6; Figure S3: Kinetics of the reaction; Table S1: The results of the reaction with the ZSM-5 catalysts; Figure S4. The product distribution of L-lactic acid to L, L-lactide over Sample II in *p*-xylene and the results of poisoning the surface of Sample II with 2,4-dimethylquinoline in *o*-xylene; Figure S5: The SEM images and the distribution curves of the crystallite sizes of Samples IV, IV-b.m.-15 min, and IV-b.m. (30 min), and the product distribution of L-lactic acid to L, L-lactide over Samples IV, IV-b.m.-15 min, and IV-b.m. (30 min).

Author Contributions: Q.H. and J.J. conceived and designed the experiments; Q.H. conducted the experiments; Q.H., R.L., and G.F. performed the characterization and evaluation of the data; Q.H. and J.J. discussed the data and wrote the manuscript. All authors have read and agreed to the published version of the manuscript.

Funding: This research was funded by Natural Science Foundation of China (Grant No. 21971259) and Sinopec.

Conflicts of Interest: The authors declare no conflict of interest.

References

1. Madhavan Nampoothiri, K.; Nair, N.R.; John, R.P. An overview of the recent developments in polylactide (PLA) research. *Bioresour. Technol.* **2010**, *101*, 8493–8501. [CrossRef] [PubMed]
2. Dechy-Cabaret, O.; Martin-Vaca, B.; Bourissou, D. Controlled Ring-Opening Polymerization of Lactide and Glycolide. *Chem. Rev.* **2004**, *104*, 6147–6176. [CrossRef] [PubMed]
3. Harold, E.B.; Kamlesh, K.B. Continuous catalyzed vapor phase dimeric cyclic ester process. U.S. Patent 5138074A, 11 August 1992.
4. Upare, P.P.; Yoon, J.W.; Hwang, D.W.; Lee, U.H.; Hwang, Y.K.; Hong, D.-Y.; Kim, J.C.; Lee, J.H.; Kwak, S.K.; Shin, H.; et al. Design of a heterogeneous catalytic process for the continuous and direct synthesis of lactide from lactic acid. *Green Chem.* **2016**, *18*, 5978–5983. [CrossRef]

5. Park, H.W.; Chang, Y.K. Economically Efficient Synthesis of Lactide Using a Solid Catalyst. *Org. Process Res Dev.* **2017**, *21*, 1980–1984. [CrossRef]
6. Dusselier, M.; Van Wouwe, P.; Dewaele, A.; Jacobs, P.A.; Sels, B.F. Shape-selective zeolite catalysis for bioplastics production. *Science* **2015**, *349*, 78–80. [CrossRef] [PubMed]
7. Gao, Y.; Zheng, B.; Wu, G.; Ma, F.; Liu, C. Effect of the Si/Al ratio on the performance of hierarchical ZSM-5 zeolites for methanol aromatization. *RSC Adv.* **2016**, *6*, 83581–83588. [CrossRef]
8. Palcic, A.; Ordomsky, V.V.; Qin, Z.; Georgieva, V.; Valtchev, V. Tuning Zeolite Properties for a Highly Efficient Synthesis of Propylene from Methanol. *Chemistry* **2018**, *24*, 13136–13149. [CrossRef] [PubMed]
9. Xia, W.; Chen, K.; Takahashi, A.; Li, X.; Mu, X.; Han, C.; Liu, L.; Nakamura, I.; Fujitani, T. Effects of particle size on catalytic conversion of ethanol to propylene over H-ZSM-5 catalysts—Smaller is better. *Catal. Commun.* **2016**, *73*, 27–33. [CrossRef]
10. Jabbari, A.; Abbasi, A.; Zargarnezhad, H.; Riazifar, M. A study on the effect of SiO2/Al2O3 ratio on the structure and performance of nano-sized ZSM-5 in methanol to propylene conversion. *React. Kinet. Mech. Catal.* **2017**, *121*, 763–772. [CrossRef]
11. Lónyi, F.; Valyon, J. On the interpretation of the NH3-TPD patterns of H-ZSM-5 and H-mordenite. *Microporous Mesoporous Mater.* **2001**, *47*, 293–301. [CrossRef]
12. Bian, C.; Wang, X.; Yu, L.; Zhang, F.; Zhang, J.; Fei, Z.; Qiu, J.; Zhu, L. Generalized Methodology for Inserting Metal Heteroatoms into the Layered Zeolite Precursor RUB-36 by Interlayer Expansion. *Crystals* **2020**, *10*, 530. [CrossRef]
13. Ma, F.; Li, H.L.; Jiang, J.X. Furfural Reduction via Hydrogen Transfer from Supercritical Methanol. *Chem. Res. Chin. Univ.* **2019**, *35*, 498–503. [CrossRef]
14. Martin, A.; Wolf, U.; Nowak, S.; Lucke, B. Pyridine-i.r. investigations of hydrothermally treated dealuminated HZSM-5 zeolites. *Zeolites* **1991**, *11*, 85–87. [CrossRef]
15. Yang, L.; Liu, Z.; Liu, Z.; Peng, W.; Liu, Y.; Liu, C. Correlation between H-ZSM-5 crystal size and catalytic performance in the methanol-to-aromatics reaction. *Chin. J. Catal.* **2017**, *38*, 683–690. [CrossRef]
16. Emeis, C.A. Determination of Integrated Molar Extinction Coefficients for Infrare Absorption Bands of Pyridine Adsorbed on Solid Acid Catalysts. *J. Catal.* **1993**, *141*, 347–354. [CrossRef]
17. Espartero, J.L.; Rashkov, I.; Li, S.M.; Manolova, N.; Vert, M. NMR Analysis of Low Molecular Weight Poly(lactic acid)s. *Macromolecules* **1996**, *29*, 3535–3539. [CrossRef]
18. Vu, D.T.; Kolah, A.K.; Asthana, N.S.; Peereboom, L.; Lira, C.T.; Miller, D.J. Oligomer distribution in concentrated lactic acid solutions. *Fluid Phase Equilib.* **2005**, *236*, 125–135. [CrossRef]
19. Ercan, C.; Dautzenberg, F.M.; Yeh, C.Y.; Barner, H.E. Mass-Transfer Effects in Liquid-Phase Alkylation of Benzene with Zeolite Catalysts. *Ind. Eng. Chem. Res.* **1998**, *37*, 1724–1728. [CrossRef]

Article

Formation Process of Columnar Grown (101)-Oriented Silicalite-1 Membrane and Its Separation Property for Xylene Isomer

Motomu Sakai [1,*], Takuya Kaneko [2], Yukichi Sasaki [3], Miyuki Sekigawa [3] and Masahiko Matsukata [1,2,4]

1. Research Organization for Nano and Life Innovation, Waseda University, 513 Waseda-Tsurumaki-cho, Shinjuku-ku, Tokyo 162-0041, Japan; mmatsu@waseda.jp
2. Department of Applied Chemistry, Waseda University, 513 Waseda-Tsurumaki-cho, Shinjuku-ku, Tokyo 162-0041, Japan; t.kaneko.waseda@gmail.com
3. Nanostructures Research Laboratory, Japan Fine Ceramics Center, 2-4-1 Atsuta-ku, Nagoya-shi, Aichi 456-8587, Japan; sasaki@jfcc.or.jp (Y.S.); m.sekigawa.waseda@gmail.com (M.S.)
4. Advanced Research Institute for Science and Engineering, Waseda University, 3-4-1 Okubo, Shinjuku-ku, Tokyo 169-8555, Japan

* Correspondence: saka.moto@aoni.waseda.jp; Tel.: +81-3-5286-3850

Received: 16 September 2020; Accepted: 16 October 2020; Published: 17 October 2020 

Abstract: Silicalite-1 membrane was prepared on an outer surface of a tubular α-alumina support by a secondary growth method. Changes of defect amount and crystallinity during secondary growth were carefully observed. The defect-less well-crystallized silicalite-1 membrane was obtained after 7-days crystallization at 373 K. The silicalite-1 membrane became (*h0l*)-orientation with increasing membrane thickness, possibly because the orientation was attributable to "evolutionally selection". The (*h0l*)-oriented silicalite-1 membrane showed high *p*-xylene separation performance for a xylene isomer mixture (*o*-/*m*-/*p*-xylene = 0.4/0.4/0.4 kPa). The *p*-xylene permeance through the membrane was 6.52×10^{-8} mol m^{-2} s^{-1} Pa^{-1} with separation factors $\alpha_{p/o}$, $\alpha_{p/m}$ of more than 100 at 373 K. As a result of microscopic analysis, it was suggested that a very thin part in the vicinity of surface played as effective separation layer and could contribute to high permeation performance.

Keywords: membrane; separation; zeolite; silicalite-1; xylene; orientation; evolutionally selection

1. Introduction

The separation and purification process requires a huge amount of energy in petroleum and petrochemical fields. To reduce energy consumption, an energy-saving process for hydrocarbon separation is desired. Membrane separation and distillation-membrane hybrid separation techniques are expected as promised novel energy-efficient processes [1]. Although many kinds of polymeric membranes have been widely used for gas separation and water treatment, polymeric membranes are hardly used for hydrocarbon separation due to their scarce thermal, and chemical resistance required for membrane materials used in petroleum and petrochemical industries. Therefore, inorganic membranes such as silica [2,3], carbon molecular sieve [4,5], and zeolite [6–11] for hydrocarbon separation have been extensively studied for recent decades due to their high stability. In particular, zeolite is one of the most promised membrane materials since zeolite has unique molecular sieving properties based on its uniformly sized micropore.

MFI-type zeolite is one of the most popular synthetic zeolites with straight channels (0.56 × 0.53 nm) along *b*-axis and sinusoidal channels (0.51 × 0.55 nm) along *a*-axis [12,13]. As these pore sizes are close to sizes of middle hydrocarbons, an MFI-type zeolite membrane is anticipated to separate these mixtures by the molecular sieving effect. Therefore, much research on the separation of ethanol/water [14], butane isomer [6], hexane isomer [6,7], and xylene isomer [6,8–11] with MFI-type zeolite membranes have been reported. Xylene isomer separation is particularly important, however a difficult target of the MFI-type zeolite. Whereas the demand for *p*-xylene as feedstocks of polyesters is larger than those of the other isomers, *p*-xylene selectivity in isomerization is limited to 20% by thermodynamic equilibria under common reaction conditions. For this reason, *p*-xylene is separated from large amounts of other xylene isomers.

The principle of *p*-xylene separation from xylene isomer mixture by using the MFI-type zeolite membrane is the molecular sieving effect that enables us to separate them by size. In other words, *p*-xylene preferentially penetrates through the MFI-type zeolite membrane because a diffusivity of *p*-xylene is a magnitude larger than those of the other isomers owing to its small molecular size. To exhibit such size-exclusion ability, the MFI-type zeolite membrane has to be synthesized with very few defects.

Preparation methods of defect-less membranes have been widely studied for the last decades. In particular, the calcination step is quite an important problem for the preparation of membranes with high separation ability. Calcination is necessary to remove the organic structure-directing agent (OSDA) to activate the micropore of the MFI-type zeolite membrane prior to use. However, crack formation often occurs in the calcination step due to the large differences in thermal expansion coefficients between support and membrane, or by the volume change in the zeolite crystal followed by removal of OSDA [15,16]. It had been reported that membrane thickness strongly influenced crack formation in the calcination step [17,18]. Thicker membranes are easily broken by calcination and spoiled their size exclusion ability. Hedlund et al. pointed out that a thin membrane layer contributes to obtaining a defect-free membrane because these membranes hardly collapsed in calcination steps [17].

Because *p*-xylene permeance through membranes were still insufficient for practical use, improvement of the permeance is an important issue in development as well. To enlarge *p*-xylene permeance many researchers have focused on crystal orientation and membrane thickness. Lai et al. reported a preparation method of oriented silicalite-1, a pure silica MFI-type zeolite, membranes [9,10]. In their study, a *b*-oriented silicalite-1 membrane where through-pores existed vertically to the support surface showed high *p*-xylene permeance for xylene isomer separation. Hedlund et al. prepared an ultra-thin MFI-type zeolite membrane by using the masking technique [6,19]. In their masking technique, the pores of porous support were plugged with polymethyl methacrylate (PMMA) prior to crystallization to avoid the formation of resistance layer, amorphous, and/or crystals, inside of support. The PMMA was removed and an ultra-thin, ca. 500 nm, membrane layer remained on the surface of support after crystallization. The ultra-thin membrane exhibited high permeance for C_4–C_8 hydrocarbon separation. Recently, Tsapatsis et al. reported the *b*-oriented MFI nanosheet membrane with the thickness of ca. 200 nm [20,21]. Their nanosheet membrane exhibited the *p*-xylene permeance of 2.9×10^{-8} mol m^{-2} s^{-1} Pa^{-1} from *o*- and *p*-xylene mixture with separation factor of ca 8000 at 423 K.

These reports described above clearly showed that controlling crystal orientation and reducing membrane thickness were effective approaches for improvement of permeation ability. However, such preparation methods are complicated to apply in practical use and lead to escalating a membrane cost. Here, we suppose a simple synthesis method of a silicalite-1 membrane with a thin effective separation layer by a secondary growth method along with two strategies. The seed crystal is only loaded on an outer surface of the support to avoid the formation of a resistance layer inside the support and to obtain a highly permeable membrane. In addition, the seed crystal is grown by hydrothermal treatment in relatively mild conditions in which nucleation hardly occurs and crystals slowly grow. As soon as defects among crystals are occluded, hydrothermal treatment is stopped, resulting in that a membrane having a thin compact layer will be obtained.

In this study, the formation process of the silicalite-1 membrane was carefully observed. Changes in defect amount and separation property were specifically investigated. In addition, permeation and separation properties for xylene isomer mixture were evaluated.

2. Experimental

2.1. Preparations of Seed Crystal and Seeded Support

Silicalite-1 membranes were synthesized by a secondary growth method. Seed crystal and seed slurry for preparation of seeded support were prepared as follows.

Synthesis solution of $25SiO_2$:$2(TPA)_2O$:$1100H_2O$:100EtOH:$0.1Na_2O$ [22] was prepared by mixing of tetraethylorthosilicate (TEOS, 98 wt.%, Merck Co., Gernsheim, Germany), tetrapropylammonium hydroxide solution (TPAOH, 1.0 M, Sigma-Aldrich Co., Missouri, US), distilled water and sodium hydroxide (97%, Kanto Chemical, Tokyo, Japan). TEOS was slowly added into a mixture of the other chemicals, and the obtained synthesis solution was aged at room temperature for 24 h under stirring condition before a hydrothermal treatment. The hydrothermal treatment was carried out at 373 K for 24 h. After crystallization, white precipitation was procured by a filtration. The filtered powder was washed by boiling water and dried at 383 K overnight. Finally, silicalite-1 seed crystal was obtained. Organic structure-directing agents (OSDA) were removed in the seed crystal by calcination at 773 K for 8 h prior to preparing seed slurry. The calcined crystal was dispersed in an appropriate amount of distilled water to form the slurry. A solid content in the slurry was adjusted at 5.0 g L^{-1}.

Seeded support was prepared by a dip-coating method using the seed slurry. Seed crystal was loaded on an outer surface of tubular α-alumina support (i.d. = 7.0 mm, o.d. = 10 mm, average pore size = 150 nm, NORITAKE Ltd., Nagoya, Japan). Both ends of tubular support were plugged with polytetrafluoroethylene caps to avoid penetration of the slurry. Support was dipped in the slurry for 1 min, withdrawn vertically at ca. 3 cm s^{-1}, and then dried at 343 K over 40 min. This process was run twice. The dip-coated support was calcined at 773 K for 2 h to bind seed crystals on a support surface during secondary growth. Finally, seeded support was obtained.

2.2. Membrane Preparation

Silicalite-1 membranes were prepared on an outer surface of tubular support by a secondary growth method as follows.

The seeded tubular support calcined was placed vertically in a PTFE-lined stainless-steel autoclave with a synthesis solution. The synthesis solution had the molar composition of $25SiO_2$:$1.5(TPA)_2O$:$1650H_2O$:200EtOH and was prepared by mixing TEOS, TPAOH, distilled water, and ethanol (99.5 wt.%, Kanto Chemical Co.). TEOS was slowly added into a mixture of the other chemicals as in the case of seed crystal preparation. This solution was aged under the stirring condition at 333 K for 4 h prior to use. The autoclave was closed and placed in a preheated oven for hydrothermal crystallization at 373 K. Crystallization periods were changed from 6 to 336 h (6, 24, 72, 120, 168, 252, and 336 h) to investigate a membrane formation process. After crystallization, the membranes were washed with boiling water and dried at 383 K overnight. The silicalite-1 membrane was calcined at 773 K for 8 h to remove OSDA before use in separation tests and characterizations.

2.3. Microscopic Observations

Membrane shape, thickness, and orientation were observed by using field emission scanning electron microscopy (FE-SEM, S-4800, Hitachi, Japan) and X-ray diffraction (XRD) measurement (Ultima IV, Rigaku, Japan).

Zeolite membranes were also observed using transmission electron microscopes (TEM). TEM samples were prepared by the ion milling method with Ar ions. The observations were performed using a JEOL JEM-2010 microscope at an accelerating voltage of 200 kV. Low dose irradiation conditions were necessary to avoid damage to the zeolite structure.

2.4. N_2 Adsorption Measurement

Micropore volume was evaluated by non-destructive N_2 adsorption measurement. The measurement was performed by using BELSORP-max (MicrotracBEL Corp., Osak, Japan) with a special sample holder which enables us to insert a whole membrane without destruction [23]. The samples were outgassed at 623 K for 8 h under vacuum conditions prior to the adsorption test. Adsorption measurements were carried out at 77 K.

2.5. Nano-Permporometry Test

Nano-permporometry was performed by using Porometer nano-6 (MicrotracBEL Corp., Osak, Japan). Permeate flow rate of inert gas through the membrane was measured while the relative pressure of vapor was increased in a step-wise manner. In this process, pores in the membrane were plugged by condensation of vapor in the order of pore size from smallest to largest, and then permeate flow rate of inert gas decreased. Pore size distribution could be calculated by a relationship between the relative pressure of vapor and flow rate.

Thermal treatment at 573 K for 2 h was taken place to remove any adsorbed molecules before the nano-permporometry test. The nano-permporometry test was carried out at 333 K with relative pressure of vapor from 0 to 0.2. In the measurement, helium and *n*-hexane were used for inert gas and condensable vapor, respectively.

2.6. Permeation Test

Permeation test was carried out in vapor permeation mode. Hydrocarbon mixture was pumped into pre-heater to vaporize and the vapor was fed to the outer surface of tubular support. The membrane temperature was controlled by using a heating jacket. Permeate side was swept with amounts of flowing gas, argon. Both the feed and permeate sides were kept at atmospheric pressure. Permeate flow rate was determined by gas chromatography (GC-8A, Shimadzu, Tokyo, Japan) using internal standard gas, methane.

Flux, J and permeance, Π was calculated using the following Equations (1) and (2):

$$J\ (\mathrm{mol\ m^{-2}\ s^{-1}}) = u\ A^{-1} \tag{1}$$

$$\Pi\ (\mathrm{mol\ m^{-2}\ s^{-1}\ Pa^{-1}}) = J\ \Delta p^{-1} \tag{2}$$

where u is the flow rate (mol s^{-1}), A is the membrane area (m^2) and Δp is the partial pressure difference between the feed and permeate sides (Pa). The separation factor α was calculated from the following Equation (3):

$$\alpha_{A/B} = (Y_A / Y_B) / (X_A / X_B) \tag{3}$$

where X_A and X_B are mol fraction of component A and B in the feed. Y_A and Y_B are mol fraction of component A and B in permeate, respectively.

3. Results and Discussion

3.1. Microscopic Observation of Membrane Formation

Microscopic analysis by using FE-SEM was adopted to observe the change in the appearance of silicalite-1 membranes during secondary growth. Figures 1 and 2 show typical FE-SEM images of seeded support and silicalite-1 membranes with different crystallization periods. Solid and dashed lines in Figure 2 are guides for the eye, and they show the seed crystal layer and columnar crystal layer appeared in the secondary growth step, respectively.

Spherical seeds with a diameter of ca. 300 nm orderly covered an outer surface of porous support with several layers as shown in Figure 1a. Judging from the cross-sectional view in Figure 2a, the thickness of the seed layer was about 2 μm. On the other hand, seed crystals were hardly observed inside of the support. Distribution of particle size in the slurry used for dip-coating was measured by a dynamic light scattering (DLS) method (ELSZ-1000ZS, Otsuka Electronics, Kyoto, Japan). The particle size had a narrow distribution from 200 to 500 nm (see Figure S1 in Supplementary materials). It is noted that the particle size in the slurry was larger than the average pore size of 150 nm in the support, resulting in that seed crystals hardly entered through the pore of support. In addition, the particle size in a slurry measured by the DLS was almost the same as the crystal size observed in FE-SEM images, suggesting that seed crystal was well dispersed in the seed slurry. For these reasons, seed crystals would be uniformly supported only on an outer surface by a dip-coating method.

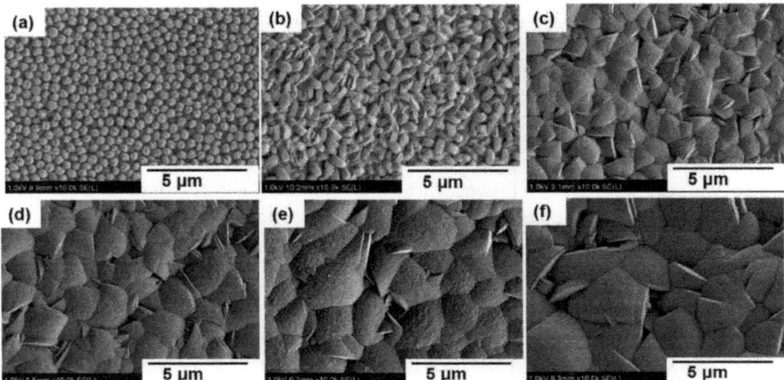

Figure 1. Typical FE-SEM images of surface of (**a**) a seeded support and silicalite-1 membranes crystallized for (**b**) 72, (**c**) 120, (**d**) 168, (**e**) 252, and (**f**) 336 h.

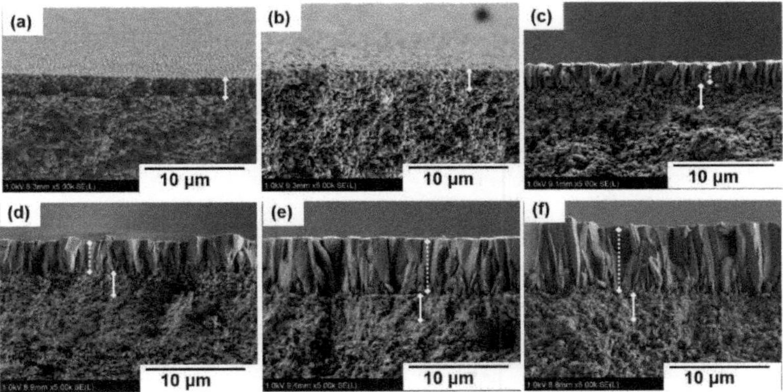

Figure 2. Typical FE-SEM images of cross-section of (**a**) a seeded support and silicalite-1 membranes crystallized for (**b**) 72, (**c**) 120, (**d**) 168, (**e**) 252, and (**f**) 336 h.

The growth of outermost crystals from the seed layer was followed from the top-view images shown in Figure 1b–f. In the early stage of a secondary growth step up to 72 h, the shape of the crystal was little increased around 500 nm and the facet of crystal became to be observed. The size of surface crystal increased and the number of the visible crystals decreased with prolonging the secondary growth period. After 336 h crystallization, the size of the surface crystal finally reached around 2 µm which is an order of magnitude larger than that of the seed crystal. In addition, defects among crystals were hardly observed judging from FE-SEM images after 120 h crystallization.

The formation of the columnar growth layer was clearly observed from cross-sectional views shown in Figure 2b–f. Although crystal size and shape changed in the top-view image of the membrane crystallized for 72 h, the growth layer was yet to be observed from a cross-sectional view. After 120 h crystallization, crystal growth progressed from the outermost crystals on the seed layer. For membranes crystallized for 120 h and more, the thickness of the columnar growth layer increased with an increasing crystallization period. The thickness of the columnar growth layer increased almost linearly, and then these values of membranes crystallized for 120, 168, 252, and 336 h were approximately 2.5, 4.0, 6.0, and 7.5 μm, respectively. In contrast, nucleation and crystal growth hardly observed toward the inside of a support. Loading of seed crystal only an outer surface and relatively mild conditions of secondary growth could contribute to inhibition of nucleation inside a support.

The silicalite-1 membranes shown in Figures 1 and 2 were prepared by using calcined seeded support. Here, we discuss the effect of calcination after dip-coating on the secondary growth process. Seeded support without calcination after dip-coating was immersed into a synthesis solution for 10 min, and then the surface was observed by FE-SEM (see Figure S2 in Supplementary materials). As a result, the seed crystal layer was partially peeled and a bare support surface was observed. In contrast, a bare surface was not observed on calcined seeded support as shown in Figure 1a, indicating that the calcination for seeded support would contribute to inhibit peeling the seed layer in hydrothermal treatment as a previous report by Pan and Lin [24].

3.2. Changes of Crystallinity and Orientation

Change of zeolite pore volume which was an index of crystallinity was measured by non-destructive N_2 adsorption. Zeolite pore volumes in silicalite-1 membranes were calculated as follows. The Amount of N_2 adsorbed at $p\ ps^{-1} = 1 \times 10^{-4}$ was adopted as a saturated adsorbed amount for zeolite pore because this amount meant saturated adsorbed amount for a cylindrical pore with diameter 0.55 nm which was almost the same as the pore size of MFI-type zeolite in Saito-Foley model [25,26].

Figure 3a shows adsorption isotherms on silicalite-1 seed crystal and membranes. Figure 3b shows the calculated zeolite pore volumes from the isotherms in each sample. In Figure 3b, a plot at the crystallization period of 0 and dashed line show the zeolite pore volumes of the seed crystal. At first, seed crystals had 0.118 $cm^3\ g^{-1}$ zeolite pore volume before crystallization. With the onset of crystallization, zeolite pore volumes in a membrane declined about 30% in the early stage. For example, the volume was 0.081 $cm^3\ g^{-1}$ in 72 h crystallized membrane. This would be caused by the partial dissolution of seed crystals and the generation of amorphous and crystals having low crystallinity. In contrast, zeolite pore volumes swelled in the later stage. The volume in the membrane crystallized for 168 and 336 h were 0.107 and 0.124 $cm^3\ g^{-1}$. As a result, the volume overtook that of seed crystal with prolonging the crystallization period. It indicated that most of the generated materials in the later stage had high crystallinity, in other words, the whole of a membrane consisted of well-crystallized silicalite-1.

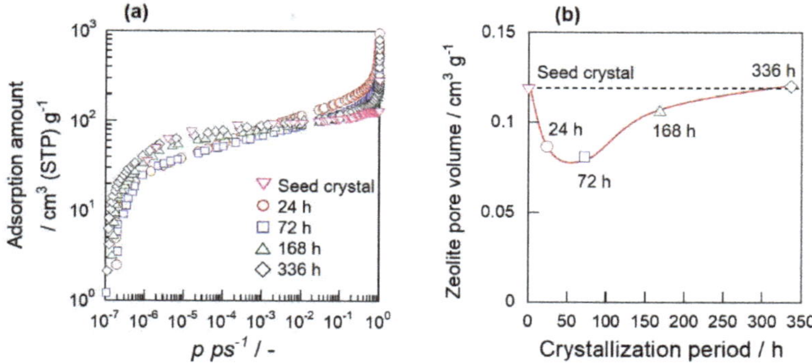

Figure 3. Results of non-destructive N_2 adsorption test: (**a**) isotherm and (**b**) zeolite pore volume change. ▽, Seed crystal; membranes crystallized for O, 24; □, 72; △, 168; ◇, 336 h.

Figure 4 shows XRD patterns of seeded support and silicalite-1 membranes and Figure 5 shows membrane weight and thickness of the columnar crystal layer as a function of the crystallization period. The intensity of each peak was almost constant until 72 h crystallization as shown in Figure 4. At a later stage in the secondary growth step, the intensity increased with the prolonging crystallization period. This result is consistent with the pore volume change in the N_2 adsorption test and the observation by FE-SEM. On the other hand, weight gain was observed from the early phase in the crystallization period as shown in Figure 5. These results strongly suggest that amorphous and crystal having low crystallinity generated up to 72 h crystallization, resulting in that pore volume per unit weight decreased in early times.

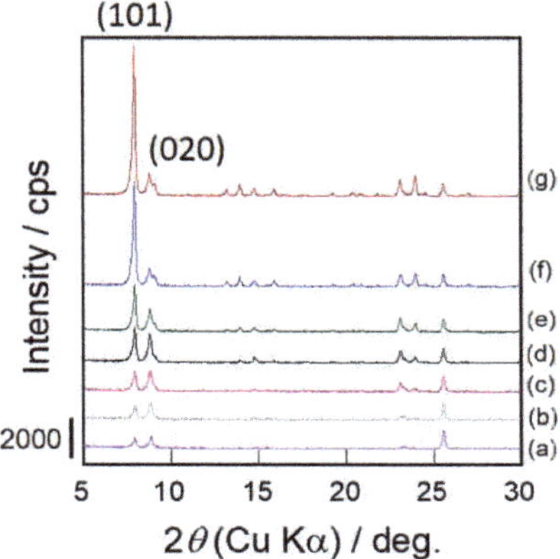

Figure 4. XRD patterns of (**a**) a seeded support and silicalite-1 membranes crystallized for (**b**) 24, (**c**) 72, (**d**) 120, (**e**) 168, (**f**) 252, and (**g**) 336 h.

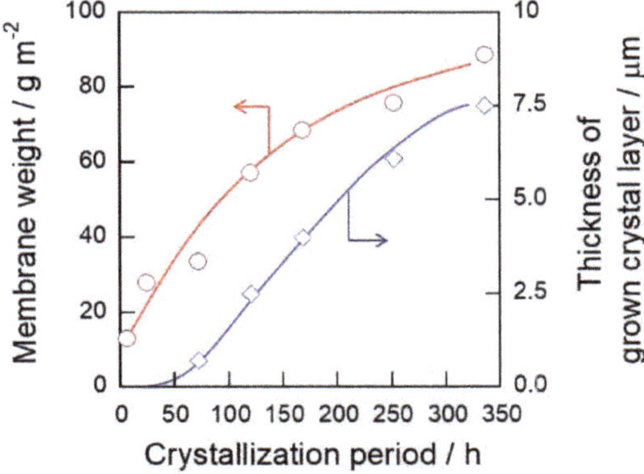

Figure 5. Membrane weight and thickness as a function of the crystallization period. O, membrane weight; ◇, the thickness of the grown crystal layer.

Here, we discuss the orientation of the silicalite-1 membrane obtained from XRD results and selected area diffraction patterns by using TEM. Judging from the XRD results, the intensity ratio of (101) to (020) increased as the growth of the columnar crystal layer. The ratio of (101) to (020) in membranes crystallized for 72, 168 and 336 h were 1.0, 2.0, and 6.5, respectively. In addition, electron diffraction patterns derived from (202), (002), and (200) were observed as shown in Figure 6, meaning that the surface of the layer consisted of (*h0l*) plane. We deduce that the appearance of (*h0l*)-orientation is attributable to "evolutionally selection" as in the case of (001)-oriented MOR-type zeolite membrane preparation by a secondary growth method [27].

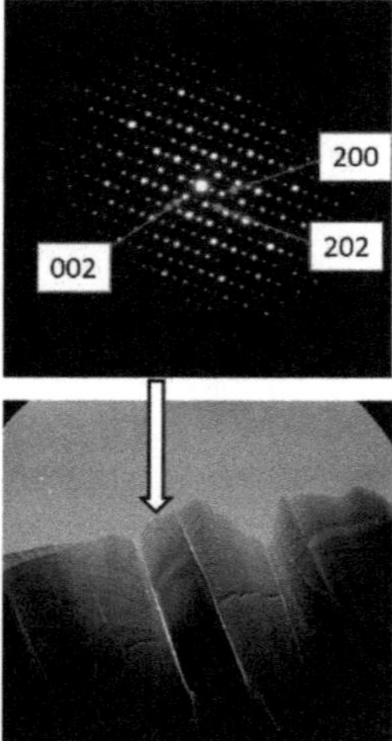

Figure 6. TEM image and electron diffraction pattern of silicalite-1 membranes crystallized for 336 h.

An orientation by "evolutionally selection" appears as follows. Anisotropic crystal generally has a different growth rate along each axis. At the early stage of secondary growth, seeds can freely grow in all directions because the seed layer had no distinct orientation. When the axis of the highest growth rate in a crystal lies on a support, the crystal collides against a lateral side of another crystal and its crystal growth stops. Finally, crystals which have the axis of highest growth rate perpendicular to support are selected and continue to grow in "evolutionally selection" mode. In that case, many crystals are embedded under the growing layer and the number of outermost crystals decreased by the selection as shown in Figures 1 and 2. A schematic diagram of evolutional selection was drawn in Figure S3 in Supplementary materials.

3.3. Molecular Sieving Properties and Defect Amount

Separation test for an equimolar mixture (50:50 kPa) of *n*-hexane (*n*-Hex) and 2,2-dimethylbutane (DMB) was carried out at 573 K in vapor permeation mode for silicalite-1 membranes with different crystallization period. The separation factor of *n*-Hex/DMB ($\alpha_{n\text{-Hex/DBM}}$) was a kind of index for the molecular sieving property of the silicalite-1 membrane. DMB has a kinetic diameter of 0.63 nm which is close to the pore size of silicalite-1 [28]. Its diffusivity in micropore of silicalite-1 is three orders of magnitude smaller than that of *n*-Hex with a kinetic diameter of 0.43 nm [29]. In other words, the defect-less silicalite-1 membrane exhibits superior *n*-Hex selectivity based on the molecular sieving effect.

Figure 7 shows the results of separation tests for *n*-Hex and DMB mixture. Permeation and separation performance drastically changed by the crystallization period. Although *n*-Hex permeance was almost constant, DMB permeance changed by orders of magnitude. DMB permeance decreased with an increasing crystallization period up to 168 h. The permeance were 45.8, 0.466, and 0.169 × 10^{-9} mol m^{-2} s^{-1} Pa^{-1} through membranes crystallized for 72, 120, and 168 h, respectively. On the other hand, DMB permeance increased through membrane crystallized for 252 h and more. As a result, DMB permeance had a minimum value at a crystallization period of 168 h, and then the separation factor had the maximum value of 781 at that time.

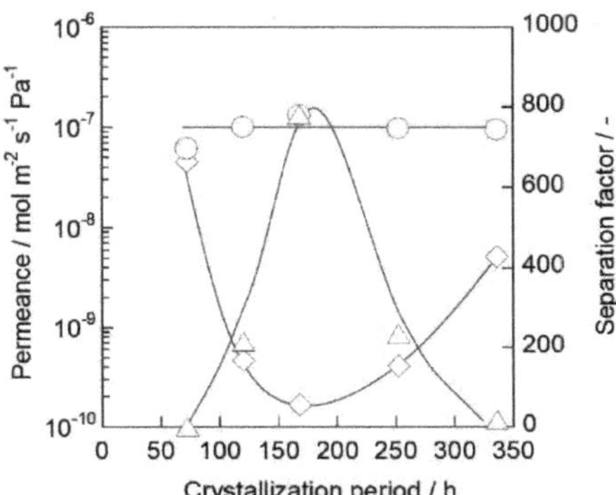

Figure 7. Permeances of C_6 hydrocarbons through silicalite-1 membranes with different crystallization period. O, *n*-hexane; ◇, 2,2-dimethylbutane; △, separation factor.

To investigate the reason for the difference in separation ability, the number of defects in each silicalite-1 membrane was evaluated by using nano-permporometry. Kelvin equation shown in Equation (4) was used to estimate pore size in prepared membranes from the results of the nano-permporometry test.

$$D_K = -4\ \nu\ \sigma\ \cos\theta\ (R\ T\ \ln(p\ ps^{-1}))^{-1} \tag{4}$$

where, D_K, ν, σ, θ, R, T and $p\ ps^{-1}$ represent Kelvin diameter [m], molar volume [$m^3\ mol^{-1}$], surface tension [$kg\ s^{-2}$], contact angle [°], gas constant [$m^2\ kg\ s^{-2}\ K^{-1}\ mol^{-1}$], temperature [K] and relative pressure of condensable vapor [-], respectively. Here we assumed that the contact angle θ is 0°. Although the Kelvin equation cannot be adapted to micropore theoretically, good agreement of the equation in micropore was previously reported [30]. When $p\ ps^{-1}$ is 0.7×10^{-3}, kelvin diameter derived from Equation (4) shows 0.55 nm which is almost the same as the pore size of silicalite-1. Therefore, helium permeation through defect-free membrane should be completely blocked by condense of vapor in micropore above $p\ ps^{-1}$ of 0.7×10^{-3}.

Figure 8 shows the results of nano-permporometry tests for membranes with different crystallization period. The dashed line shows a quantification limit of helium permeance at 1.0×10^{-9} mol $m^{-2}\ s^{-1}\ Pa^{-1}$. Helium permeance was almost constant through a membrane crystallized for 24 h despite the concentration of condensable vapor, suggesting that many defects among crystals remained after the secondary growth. Similarly, many defects still remained even after 72 h crystallization. Because of such insufficient crystal growth, the membrane prepared with a shorter crystallization period showed quite a low separation performance in the separation test.

On the other hand, helium permeance decreased above $p\ ps^{-1}$ of 4.0×10^{-3} through membranes crystallized over 120 h which exhibited good molecular sieving properties as shown in Figure 7. The decrease of helium permeance means that very few defects among crystals remained in these membranes.

Here we defined a ratio of the non-zeolitic pathway to all pathways for quantitative discussion about the effect of the non-zeolitic pathway on molecular sieving ability. The ratio of the non-zeolitic pathway was calculated as dividing helium permeance at $p\ ps^{-1}$ of 0.2 by the permeance at $p\ ps^{-1}$ of 0 according to a previous report by Hedlund et al. [6]. Helium permeance at $p\ ps^{-1}$ of 0 means the permeance through all pathways. Besides, helium permeance at $p\ ps^{-1}$ of 0.2 means through non-zeolitic pathway having more than ca. 2 nm of diameter determined based on the Kelvin equation. Therefore, the ratio of the non-zeolitic pathway was larger than 2 nm to all pathways can be evaluated.

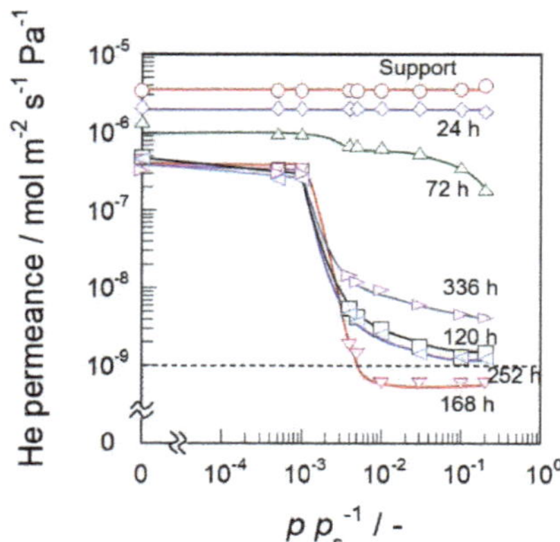

Figure 8. Results of nano-permporometry tests for silicalite-1 membranes with different crystallization period. O, Support; membrane crystallized for ◇, 24; △, 72; □, 120; ▽, 168; ◁, 252; ▷, 336 h.

A relationship between the calculated ratio of non-zeolitic pathway and separation factor, $\alpha_{n\text{-Hex/DMB}}$ was shown in Figure 9. The ratio drastically decreased from 0.27 to 2.7×10^{-3} between the crystallization period of 72 and 120 h, suggesting that most of the defects among crystals were occluded in this term. This ratio had a minimum ($<2.0 \times 10^{-3}$) in the membrane crystallized for 168 h. After that, the ratio increased with increasing crystallization period up to 336 h; the ratios are 2.8×10^{-3} and 1.4×10^{-2} in membranes crystallized for 252 and 336 h, respectively. It suggests that defects are newly generated in these membranes for any reason which is discussed in the following section. At any rate, a relationship between the ratio of non-zeolitic pathway and separation factor, $\alpha_{n\text{-Hex/DMB}}$ clearly showed that separation ability of silicalite-1 membrane by molecular sieving effect was dominated by the amount of non-zeolitic pathway in a membrane. It is noteworthy that only 1% of the non-zeolitic pathway well spoiled a molecular sieving property of the silicalite-1 membrane.

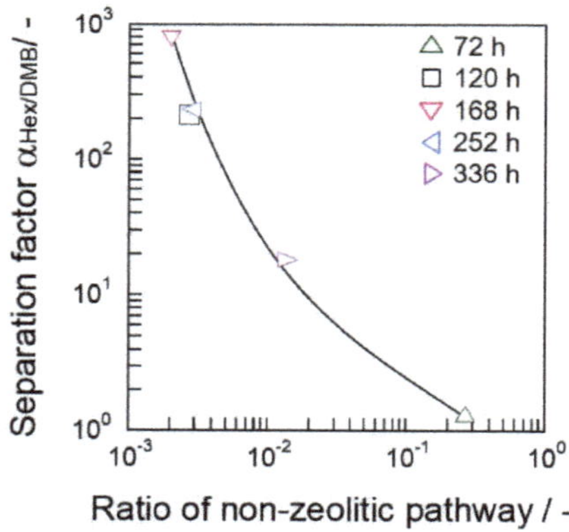

Figure 9. Relationships between the ratio of non-zeolitic pathway and selectivity of hydrocarbon. Membranes were crystallized for △, 72; □, 120; ▽, 168; ◁, 252; ▷, 336 h.

We shed light on the reason for defects appearance in membranes crystallized over 252 h. Nano-permporometry tests were also performed for these membranes before calcination, and then amounts of non-zeolitic pathways were compared between before and after calcination. As a result, helium hardly penetrated through both 252 and 336 h crystallized membranes before calcination, suggesting that these membranes had very few non-zeolitic pathways. The result showed that crack must be formed in these membranes in the calcination step. There are many reports about crack formation in zeolite membranes in the calcination step to remove OSDA described in the introduction section [15–18]. The thicker membrane would have larger stress by thermal expansion of support and volume change of crystals, resulting in that crack formed in membranes prepared with longer crystallization period.

3.4. Xylene Isomer Separation

Finally, the separation test of xylene isomer mixture was performed for the silicalite-1 membrane crystallized for 168 h. *o*-, *m*-, *p*-Xylene equimolar mixture (0.4:0.4:0.4 kPa) was fed to the membrane with dilution gas, Argon. The membrane temperature was adjusted at 373 K.

Table 1 listed the results of xylene isomer separation tests. Our silicalite-1 membrane showed *p*-xylene permeance of 6.52×10^{-8} mol m^{-2} s^{-1} Pa^{-1} with *o*- and *m*-xylene permeances below quantification limit of 6.0×10^{-10} mol m^{-2} s^{-1} Pa^{-1}, resulting in that separation factors, $\alpha_{p/o}$ and $\alpha_{p/m}$ of above 100. The (*h0l*)-oriented silicalite-1 membrane prepared by a simple secondary growth method exhibited good permeation and separation performance compared with MFI-type zeolite membranes previously reported as shown in Table 1 [6,8–11,20,21].

Prepared (*h0l*)-oriented membrane exhibited relatively high permeation performance despite the large membrane thickness of ca. 6 μm including the seed layer as shown in Figure 2b. The reason for high permeation performance was discussed from the microstructure of the obtained membrane as shown in Figure 10. As a result of microscopic analysis, many defects among crystals remained in the seed layer after 168 h crystallization. In addition, many grain boundaries existed inside of columnar crystal layer along the direction as molecular permeation. However, the membrane crystallized for 168 h exhibited superior separation performance based on a molecular sieving effect, suggesting that the grain boundaries would not penetrate through the membrane. From the enlarged view of grain boundary near-surface, as shown in Figure 10b, the grain boundary between two crystals was plugged with a thin amorphous layer. From these results, we conclude that only a very thin part in the vicinity of the surface is compact without defects and plays as an effective separation layer. The thin effective layer less than 1 μm can contribute to high permeation performance. The deduced model of a formation process of the silicalite-1 membrane is shown in Figure 11.

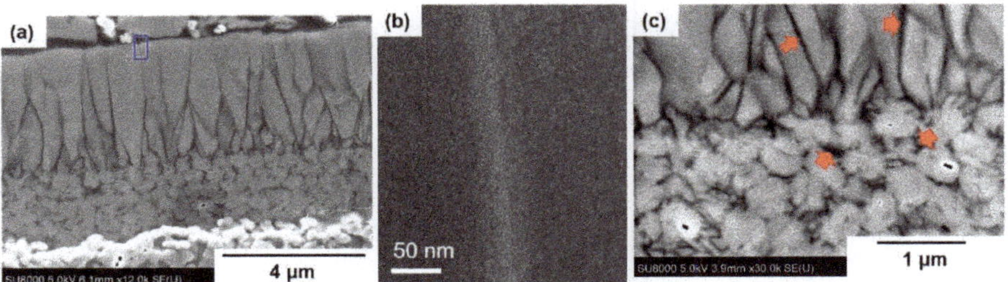

Figure 10. TEM images of a cross-section of a silicalite-1 membrane crystallized for 168 h. (**a**) Wide view, (**b**) grain boundary near-surface (an area divided blue square), and (**c**) interface between seed and columnar crystal layer. Arrowheads in (**c**) show defects among seed crystals and grain boundaries between columnar crystals.

Table 1. Comparison of permeation and separation performance of the MFI-type zeolite membrane for xylene isomers.

Support	**Membrane**	**Feed Pressure/kPa**			**Temperature/K**	***p*-Xylene Permeance**			**Ref.**
		***p*-Xylene**	***o*-Xylene**	***m*-Xylene**		**$/10^{-8}$ mol m^{-2} s^{-1} Pa^{-1}**	$\alpha_{p/o}$	$\alpha_{p/m}$	
α-Alumina Disc	**Ultra-Thin MFI**	0.27	0.59	–	373	60	3.2	–	[6]
Stainless steel tube	Ba-ZSM-5	0.23	0.26	0.83	423	7.0	13.5	8.2	[8]
Stainless steel tube	Ba-ZSM-5	0.23	0.26	0.83	673	5.4	8.4	6.2	[8]
α-Alumina disc	*b*-Oriented siliceous ZSM-5	0.45	0.35	–	373	20	600	–	[9]
α-Alumina disc	*b*-Oriented siliceous ZSM-5	0.50	0.45	–	373	25	60	–	[10]
α-Alumina tube	Composite MFI-alumina tube	0.63	0.32	0.27	473	1.1	>400	–	[11]
Sintered silica fibre	*b*-Oriented MFI nanosheet	0.30	0.31	–	323	0.85	32.3	–	[19]
Sintered silica fibre	*b*-Oriented MFI nanosheet	0.37	0.35	–	423	55.5	1989	–	[19]
Sintered silica fibre	*b*-Oriented MFI nanosheet	0.50	0.50	–	373	2.5	ca. 3500	–	[20]
Sintered silica fibre	*b*-Oriented MFI nanosheet	0.50	0.50	–	423	29	ca. 8000	–	[20]
α-Alumina tube	(*h0l*)-Oriented silcialite-1	0.40	0.40	0.40	373	6.5	>100	>100	This work

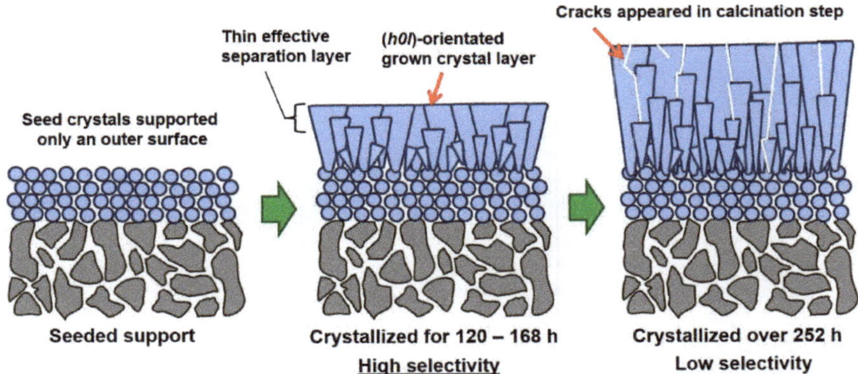

Figure 11. Model of the silicalite-1 membrane formation process.

4. Conclusions

The silicalite-1 tubular membrane was prepared by a simple secondary growth method and its formation process was discussed.

Uniformly ordered seed crystals and relatively mild conditions of secondary growth could contribute to forming a columnar growth layer on the outermost crystal of the seed layer. In addition, the columnar growth layer showed (*h0l*)-orientation, possibly because the orientation is attributable to "evolutionally selection".

Change of defect amount and separation performance during secondary growth were investigated by using nano-permporometry and hydrocarbon separation tests. Most of the defects among crystals were occluded between crystallization periods of 72 to 168 h. Separation performance was spoiled in membranes crystallized over 252 h by the crack formation in the calcination step to remove OSDA, suggesting that thicker membranes would have larger stress by thermal treatment. As a result, the membrane crystallized for 168 h showed maximum separation factor $\alpha_{n\text{-Hex/DMB}}$ of 781 for the equimolar mixture of *n*-hexane and 2,2-dimethylbutane by molecular sieving effect. From a microscopic analysis, it was suggested that a very thin part in the vicinity of the surface was an effective separation layer.

Separation performance from xylene isomer mixture (*o*-/*m*-/*p*-xylene = 0.4/0.4/0.4 kPa) through the silicalite-1 membrane was evaluated. *p*-Xylene permeance through the membrane was 6.52 × 10^{-8} mol m^{-2} s^{-1} Pa^{-1} with separation factors $\alpha_{p/o}$, $\alpha_{p/m}$ of more than 100 at 373 K. Our (*h0l*)-oriented silicalite-1 membrane prepared by a simple secondary growth method exhibited good permeation and separation performance for xylene isomer mixture compared with MFI-type zeolite membranes previously reported.

Supplementary Materials: The following are available online at http://www.mdpi.com/2073-4352/10/10/949/s1, Figure S1: Particle size distribution in seed slurry for dip-coating. Figure S2: Typical FE-SEM images of the surface of seeded supports immersed into synthesis solution without calcination after dip-coating: (a) wide and (b) enlarged view. Figure S3: A schematic diagram of evolutional selection. Figure S4: Results of nano-permporometry tests for silicalite-1 membranes crystallized for O 252; □, 336 h before calcination.

Author Contributions: Conceptualization, M.S. (Motomu Sakai); methodology, M.S. (Motomu Sakai), T.K., M.S., (Miyuki Sekigawa) and Y.S.; validation, M.S. (Motomu Sakai), Y.S. and M.M.; investigation, M.S. (Motomu Sakai), Y.S. and M.M.; resources, Y.S. and M.M.; data curation, M.S. (Motomu Sakai) and Y.S.; writing—original draft preparation, M.S. (Motomu Sakai); supervision, M.M.; project administration, M.M.; funding acquisition, M.M. All authors have read and agreed to the published version of the manuscript.

Funding: This research was partially supported by JST-CREST (Japan Science and Technology Agency, Create REvolutionary technological seeds for Science and Technology innovation program), Grant Number JPMJCR1324, Japan.

Conflicts of Interest: The authors declare no conflict of interest.

References

1. Sholl, D.S.; Lively, R.P. Seven chemical separations to change the world. *Nature* **2016**, *532*, 435–437. [CrossRef]
2. Kanezashi, M.; Shazwani, W.N.; Yoshioka, T.; Tsuru, T. Separation of propylene/propane binary mixtures by bis(triethoxysilyl) methane (BTESM)-dericed silica membranes fabricated at different calcination temperatures. *J. Membr. Sci.* **2012**, *415–416*, 478–485. [CrossRef]
3. Ikeda, A.; Nomura, M. Preparation of Amorphous Silica Based Membranes for Separation of Hydrocarbons. *J. Jpn. Petrol. Inst.* **2016**, *59*, 259–265. [CrossRef]
4. Xu, L.; Rungta, M.; Koros, W.J. Matrimid® derived carbon molecular sieve hollow fiber membranes for ethylene/ethane separation. *J. Membr. Sci.* **2011**, *380*, 138–147. [CrossRef]
5. Hayashi, J.; Mizuta, H.; Yamamoto, M.; Kusakabe, K.; Morooka, S.; Suh, S.-H. Separation of Ethane/Ethylene and Propane/Propylene Systems with a Carbonized BPDA-pp'ODA Polyimide Membrane. *Ind. Eng. Chem. Res.* **1996**, *35*, 4176–4181. [CrossRef]
6. Hedlund, J.; Sterte, J.; Anthonis, M.; Bons, A.-J.; Carstensen, B.; Corcoran, N.; Cox, D.; Deckman, H.; Gijnst, W.D.; Moor, P.-P.; et al. High-flux MFI membranes. *Microporous Mesoporous Mater.* **2002**, *52*, 179–189. [CrossRef]
7. Funke, H.H.; Argo, A.M.; Falconer, J.L.; Noble, R.D. Separations of Cyclic, Branched, and Linear Hydrocarbon Mixtures through Silicalite Membranes. *Ind. Eng. Chem. Res.* **1997**, *334*, 137–143. [CrossRef]
8. Tarditi, A.M.; Lombardo, E.A. Influence of exchanged cations (Na^+, Cs^+, Sr^{2+} and Ba^{2+}) on xylene permeation through ZSM-5/SS tubular membranes. *Sep. Purif. Technol.* **2008**, *61*, 136–147. [CrossRef]
9. Lai, Z.; Bonilla, G.; Diaz, I.; Nery, J.G.; Sujaoti, K.; Amat, M.A.; Kokkoli, E.; Terasaki, O.; Thompson, R.W.; Tsapatsis, M.; et al. Microstructural Optimization of a Zeolite Membrane for Organic Vapor Separation. *Science* **2003**, *300*, 456–460.
10. Lai, Z.; Tsapatsis, M.; Nicolich, J.P. Siliceous ZSM-5 Membranes by Secondary Growth of *b*-Oriented Seed Layers. *Adv. Funct. Mater.* **2004**, *14*, 716–729. [CrossRef]
11. Daramola, M.O.; Burger, A.J.; Giroir-Fendler, A.; Miachon, S.; Lorenzen, L. Extractor-type catalytic membrane reactor with nanocomposite MFI-alumina membrane tube as separation unit: Prospect for ultra-pure para-Xylene production from *m*-Xylene isomerization over Pt-HZSM-5 catalyst. *Appl. Catal., A* **2010**, *386*, 109–115. [CrossRef]
12. Kokotailo, G.T.; Lawton, S.L.; Olson, D.H.; Meier, W.M. Structure of synthetic zeolite ZSM-5. *Nature* **1978**, *272*, 437–438. [CrossRef]
13. Olson, D.H.; Kokotailo, G.T.; Lawton, S.L.; Meler, W.M. Crystal Structure and Structural-Related Properties of ZSM-5. *J. Phys. Chem.* **1981**, *85*, 2238–2243. [CrossRef]
14. Ueno, K.; Negishi, H.; Okano, T.; Tawarayama, H.; Ishikawa, S.; Miyamoto, M.; Uemiya, S.; Ouni, Y. Effects of Silica-Particle Coating on Silica Support for the Fabrication of High-Performance Silicalite-1 Membranes by Gel-Free Steam-Assisted Conversion. *Membranes* **2019**, *9*, 46. [CrossRef] [PubMed]
15. Dong, J.; Lin, Y.S.; Hu, M.Z.-C.; Peascoe, R.A.; Payzant, E.A. Template-removal-associated microstructural development of porous-ceramics-supported MFI zeolite membranes. *Microporous Mesoporous Mater.* **2000**, *34*, 241–253. [CrossRef]
16. Choi, J.; Jeong, H.-K.; Snyder, M.A.; Stoeger, J.A.; Masel, R.I.; Tsapatsis, M. Grain Boundary Defect Elimination in a Zeolite Membrane by Rapid Thermal Processing. *Science* **2009**, *325*, 590–593. [CrossRef]
17. Akhtar, F.; Ojuva, A.; Wirawan, S.K.; Hedlund, J.; Bergstrom, L. Hierarchically porous binder-free silicalite-1 discs: A novel support for all-zeolite membranes. *J. Mater. Chem.* **2011**, *21*, 8822–8828. [CrossRef]
18. Akhtar, F.; Sjoberg, E.; Korelskiy, D.; Rayson, M.; Hedlund, J.; Bergstrom, L. Preparation of graded silicalite-1 substrates for all-zeolite membranes with excellent CO_2/H_2 separation performance. *J. Membr. Sci.* **2015**, *493*, 206–211. [CrossRef]
19. Hedlund, J.; Jareman, F.; Bons, A.-J.; Anthonis, M. A masking technique for high quality MFI membranes. *J. Membr. Sci.* **2003**, *222*, 163–179. [CrossRef]
20. Jeon, M.Y.; Kim, D.; Kumar, P.; Lee, P.S.; Rangnekar, N.; Bai, P.; Shete, M.; Elyassi, B.; Lee, H.S.; Narasimharao, K.; et al. Ultra-selective high-flux membranes from directly synthesized zeolite nanosheets. *Nature* **2017**, *543*, 690–694. [CrossRef]

21. Kim, D.; Jeon, M.Y.; Stottrup, B.L.; Tsapatsis, M. *para*-Xylene Ultra-selective Zeolite MFI Membranes Fabricated from Nanosheet Monolayers at the Air–Water Interface. *Angew. Chem. Int. Ed.* **2018**, *57*, 480–485. [CrossRef] [PubMed]
22. Ren, N.; Bronic, J.; Subotic, B.; Lv, X.-C.; Yang, Z.-J.; Tang, Y. Controllable and SDA-free synthesis of sub-micrometer sized zeolite ZSM-5. Part 1: Influence of alkalinity on the structural, particulate and chemical properties of the products. *Microporous Mesoporous Mater.* **2011**, *139*, 197–206. [CrossRef]
23. Sakai, M.; Fujimaki, N.; Kobayashi, G.; Yasuda, N.; Oshima, Y.; Seshimo, M.; Matsukata, M. Formation process of *BEA-type zeolite membrane under OSDA-free conditions and its separation property. *Microporous Mesoporous Mater.* **2019**, *284*, 360–365. [CrossRef]
24. Pan, M.; Lin, Y.S. Template-free secondary growth synthesis of MFI type zeolite membrane. *Microporous Mesoporous Mater.* **2001**, *43*, 319–327. [CrossRef]
25. Saito, A.; Foley, H.C. Curvature and Parametric Sensitivity in Models for Adsorption in Micropores. *AIChE J.* **1991**, *37*, 429–436. [CrossRef]
26. Saito, A.; Foley, H.C. Argon porosimetry of selected molecular sieves: Experiments and examination of the adapted Horvath-Kawazoe model. *Microporous Mater.* **1995**, *3*, 531–542. [CrossRef]
27. Li, G.; Kikuchi, E.; Matsukata, M. The control of phase and orientation in zeolite membranes by the secondary growth method. *Microporous Mesoporous Mater.* **2003**, *62*, 211–220. [CrossRef]
28. Sommer, S.; Melin, T.; Falconer, J.L.; Noble, R.D. Transport of C_6 isomers through ZSM-5 zeolite membranes. *J. Membr. Sci.* **2003**, *224*, 51–67. [CrossRef]
29. Cavalcante, L., Jr.; Ruthven, D.M. Adsorption of Branched and Cyclic Paraffins in Silicalite. 2. Kinetics. *Ind. Eng. Chem. Res.* **1995**, *34*, 185–191. [CrossRef]
30. Tsuru, T.; Hino, T.; Yoshioka, T.; Asaeda, M. Permporometry characterization of microporous ceramic membranes. *J. Membr. Sci.* **2001**, *186*, 257–265. [CrossRef]

Article

Quasi Natural Approach for Crystallization of Zeolites from Different Fly Ashes and Their Application as Adsorbent Media for Malachite Green Removal from Polluted Waters

Denitza Zgureva [1,*], Valeria Stoyanova [2], Annie Shoumkova [2], Silviya Boycheva [3] and Georgi Avdeev [2]

1 College of Energy and Electronics, Technical University of Sofia, 8 Kl. Ohridsky Blvd, 1000 Sofia, Bulgaria
2 Bulgarian Academy of Sciences, Institute of Physical Chemistry "R. Kaischew", 1113 Sofia, Bulgaria; valeriastoyanova@gmail.com (V.S.); anishu@abv.bg (A.S.); g_avdeev@ipc.bas.bg (G.A.)
3 Department of Thermal and Nuclear Power Engineering, Technical University of Sofia, 8 Kl. Ohridsky Blvd, 1000 Sofia, Bulgaria; sboycheva@tu-sofia.bg
* Correspondence: dzgureva@tu-sofia.bg

Received: 1 November 2020; Accepted: 20 November 2020; Published: 23 November 2020

Abstract: Worldwide disposal of multi-tonnage solid waste from coal-burning thermal power plants (TPPs) creates serious environmental and economic problems, which necessitate the recovery of industrial waste in large quantities and at commercial prices. Fly ashes (FAs) and slag from seven Bulgarian TPPs have been successfully converted into valuable zeolite-like composites with various applications, including as adsorbents for capturing CO_2 from gases and for removal of contaminants from water. The starting materials generated from different types of coal are characterized by a wide range of SiO_2/Al_2O_3 ratio, heterogeneous structure and a complex chemical composition. The applied synthesis procedure resembles the formation of natural zeolites, as the raw FAs undergo long-term self-crystallization in an alkaline aqueous solution at ambient temperature. The phase and chemical composition, morphology and N_2 adsorption of the final zeolite products were studied by scanning electron microscopy (SEM), energy-dispersive X-ray (EDX), X-Ray Diffraction (XRD) and Brunauer–Emmett–Teller (BET) analyses. The growth of faujasite (FAU) crystals as the main zeolite phase was established in all samples after 7 and 14 months of alkaline treatment. Phillipsite (PHI) crystals were also observed in several samples as an accompanying phase. The final products possess specific surface area over 400 m^2/g. The relationships between the surface properties of the investigated samples and the characteristics of the raw FAs were discussed. All of the obtained zeolite-like composites were able to remove the highly toxic dye (malachite green, MG) from water solutions with efficiency over 96%. The experimental data were fitted with high correlation to the second-order kinetics.

Keywords: waste management; fly ash zeolites; faujasite; wastewater remediation

1. Introduction

In the recent years, one of the most current topics is related to the deteriorating quality of the environment, especially the purity of the atmospheric air and water basins. Their pollution is mainly due to anthropogenic factors and the intensive development of various industries. An illustrative example for this impact is the continuous increase in electricity consumption by both industry and households, and the necessity for greater energy production from power plants. According to the annual report of the International Energy Agency in 2018, the world gross electricity production was

3.9% higher than in 2017 [1]. Coal-fired thermal power plants (TPPs) still possess the largest share of the electricity market, 38% in 2018. At the same time, coal combustion is the main contributor to the green house emission into the atmosphere and to the production of huge amounts of solid waste classified as hazardous in some countries [2,3]. The main challenge facing the industry today is to meet energy needs without having a negative impact on the environment. It could be expected that the future of the energy sector will be based on the combination of variety of energy sources with a key role of the traditional technologies, such as fossil fuel burning or nuclear fission, but at stricter environmental requirements [4]. In recent decades, the capture of carbon dioxide (CO_2) in TPPs has been intensively studied. At the beginning, the concept for liquefaction of CO_2 and its further deposition in deep geological formation have been widely explored, referred to as carbon capture and storage (CCS). Lately, this approach was displaced by carbon capture and utilization (CCU), in which the captured CO_2 is utilized in commercial products [5,6]. The CCS concept was gradually displaced by CCUS, which is carbon capture utilization and storage [7]. It could be expected that CCUS technologies will be commercialized soon and the requirements of the Paris Agreement to reduce CO_2 emissions will be met in time for TPPs [8]. However, this does not apply to solid waste from the combustion of coal, which is produced as a by-product of the burning process. This residual consists of the mineral part from the fuel, which is incombustible. It could be obtained as slag or fly ash (FA) depending on the boiler design. Coal ash consists mainly of SiO_2, Al_2O_3, Fe_2O_3, CaO, Na_2O and traces of other oxides. According to the World Wide Coal Combustion Products Network (WWCCCPN), the global annual ash production for 2016 is 1221.9 Mt and almost a half is further utilized [9].

The most common approach for utilization of coal ash on an industrial scale is for production of constructional materials and cement. Based on their composition, FAs are classified in two classes C and F (50 wt. % < SiO_2 + Al_2O_3 + Fe_2O_3 < 70 wt. % for class C and SiO_2 + Al_2O_3 + Fe_2O_3 > 70 wt. % for class F) according to the ASTM C618. Class C fly ash can be used as a porland cement replacement due to its self-cementing and pozzolanic properties. Class F fly ash contains less CaO and higher amounts of glassy aluminosilicates and quartz, and it is used as a pozzolanic additive with less or without cementing values [10]. The unused part of the coal combustion wastes is stored in landfills, which leads to the destruction of agricultural land and creates a potential risk of ground water pollution. There are much published investigations on the opportunities for FA utilization but with limited applications on industrial scales mostly because of the cost ineffectiveness [11–13]. One promising approach for FA processing, first proposed in 1985, is the conversion of coal FA into zeolites due to their compositional similarities [14]. The most researched approach for synthesis of zeolites is the hydrothermal activation of FA in alkaline solution and a pilot installation for scale production is already in operation [15,16]. The biggest challenge facing this technique is the controlled production of a determined zeolite structure with well-defined properties, which predetermines the practical affordability of the material. Different zeolite structures or mixtures of them can be obtained depending on the chemical composition of the raw FA and mostly on the SiO_2/Al_2O_3 ratio. For FA with a SiO_2/Al_2O_3 ratio < 1.7 at low molarity of NaOH (1–2 M) and an activation temperature in the region of 100–140 °C for a wide hydrothermal period from 2 to 96 h, the dominant structure in the synthesized material is gismondine (GIS) [17–19]. At the same molarity of the NaOH activator, but with an increase in the temperature of 150–200 °C, analacime (ANA) is obtained [17]. The lower synthesis temperature (90 °C) leads to the crystallization of a mixture of linde (LTA) and faujasite (FAU) zeolite structures in the final product [20]. For raw FA with 1.7 < SiO_2/Al_2O_3 < 2.4, the tendency in the crystallization of zeolite structures by hydrothermal alkaline activation is the same as in the previous case. For these FAs, increasing the NaOH molarity from 1 M to 2 M, the main GIS phase is replaced by phillipsite (PHI). A further increase in the molarity to 3.1–12.5 always leads to the crystallization of sodalite (SOD) [21,22]. Low quantities of LTA have been observed for these FAs at temperatures of 80–90 °C [22,23]. For raw FA with SiO_2/Al_2O_3 > 2.4, the main zeolite phases that crystallize at high activation temperatures (140–150 °C) and 3–4 M of NaOH are PHI, SOD, and chabazite (CHA) [24]. At temperature of 60 °C and 168 h of hydrothermal treatment, a negligible yield of FAU < 5 wt. % is obtained [25]. The two-stage synthesis has been developed by Shigemoto et al. in 1992. They introduced initial high temperature treatment of the mixtures of the raw FA and NaOH prior to the hydrothermal treatment [26–28]. In this step, soluble silica and alumina salts are produced, which facilitates the formation of hydrogel in water solution. The hydrogel then crystallizes onto unreacted parts of the FA particles, which act as crystallization centers. In this approach the raw FA composition does not affect the final zeolite structure, but the synthesis conditions are of key meaning. At high temperatures above 100 °C and NaOH molarity above 3, SOD is crystallized [29]. At activation temperatures below 100 °C and 2.0–2.5 M NaOH, the resultant structure is LTA, while at molarity higher than 2.5 M NaOH, the only crystalized zeolite is FAU [30,31].

FAU is a typical large pore zeolite with well-defined textural characteristics and easily accessible framework. These peculiarities make FAU suitable material for a wide range of applications mainly as an adsorbent, catalyst and ion exchanger. The fly ash zeolites (FAZ) of FAU type have been broadly studied for gas purification, soil recovery, heavy metal and radioactive species uptake, water softening, desalination and purification, including removal of dyes, microorganisms, and various organic and inorganic compounds [32,33]. In our previous studies, zeolite FAU obtained from lignite FA by two-stage synthesis was successfully applied as a CO_2 adsorbent, catalyst for VOCs degradation, matrix for optical sensor, Fenton-like catalyst for removal of organic contaminants from water, adsorbent for retention of heavy metals and dyes from waste effluents [34]. The obtained results are promising but for applications on an industrial scale the two-stage synthesis would increase drastically the waste processing costs. The water remediation systems are intended for treatment of large water streams, which requires large amounts of the filtration materials.

In this study, we investigate the opportunity for a controlled obtaining of FAU-structured zeolite from raw fly ashes with a wide range of SiO_2/Al_2O_3 ratios. The proposed synthesis approach resembles the formation of a natural zeolite as the raw FA undergoes self-crystallization in an alkaline water solution at ambient temperatures. Until recently, this technique was considered inappropriate due to the long crystallization time. However, the strategy of modern production is the development of a circular economy, which imposes the need to recover industrial waste in large quantities and at commercial prices. An advantage of recovering coal ash by zeolitization is that long-deposited ash can also be processed to repair already-caused environmental damages. The obtained FA zeolites are investigated in regard to their ability to remove dyes from contaminated water. The adsorption of malachite green (MG) is studied because of its high toxicity and presence in water used for fish farming. It has been established that MG remained in food could cause carcinogenesis, mutagenesis, chromosomal fractures, teratogenicity and respiratory toxicity [35].

2. Materials and Methods

2.1. Materials

The crystallization of zeolites was investigated for raw FAs sampled from seven different sources obtained by combustion of lignite, brown, and anthracite coal in Bulgarian thermal power plants (TPPs). Five of the raw samples are fine powders collected from the electrostatic precipitators of TPPs, one is a solid slag sampled from bottom ash, and another one is a mixture of FA with slag. Prior to the synthesis, the slag was screened to obtain a fraction with particle sizes smaller than 0.2 mm. Detailed analyses of the raw materials were reported in our previous studies, where mainly amorphous aluminosilicates and crystalline phases of quartz, gypsum, mullite, and magnetite were detected [36,37]. The chemical composition of FAs, SiO_2/Al_2O_3 ratios, the FA classification and the weight part of the amorphous ingredients calculated by integral deconvolution of XRD spectra of analyzed samples (uncertainty of ±2.5%) are summarized in Table 1.

Table 1. Chemical composition and amorphous part of fly ashes (FAs) subjected to zeolitization.

Raw FA [1]	SiO_2 wt. %	Al_2O_3 wt. %	Fe_2O_3 wt. %	Na_2O wt. %	CaO wt. %	MgO wt. %	SO_3 wt. %	K_2O wt. %	TiO_2 wt. %	SiO_2/Al_2O_3 wt. ratio	FA Class	AP [2] wt. %
FAB	40.29	20.49	13.59	0.79	13.11	2.14	3.21	4.09	2.30	1.97	F	58
FAG	12.50	8.36	45.87	0.65	6.42	0.49	25.72	0.00	0.00	1.50	C	36
FAM	39.51	22.89	16.09	1.78	9.88	2.58	5.34	1.34	0.59	1.73	F	57
FAR	54.63	28.67	5.94	0.00	1.53	1.94	1.08	4.87	1.35	1.91	F	54
FARE	49.30	27.02	14.40	0.00	3.23	2.08	0.00	3.98	0.00	1.82	F	94
FAV	46.75	27.13	10.31	1.08	3.73	1.55	2.99	6.46	0.00	1.72	F	65
SS	39.47	17.17	20.09	1.11	10.72	1.55	5.94	3.95	0.00	2.30	F	100

[1] FAB (FA from TPP "Bobovdol", 630 MW, brown coals with high ash content); FAG (from TPP "AES Galabovo", 670 MW, lignite coals); FAM (FA from TPP "Maritza 3", 900 MW, lignite coals); FAR (from TPP "Republika", 180 MW, brown coals); FARE (FA mixed with grinded in a planetary ball mill slag from TPP "Rousse East", 400 MW); FAV (FA from TPP "Varna", 1260 MW, highly calorific and low sulphur anthracite coals); SS (slag from TPP "Sviloza", Svishtov, 120 MW, bituminous coals). [2] AP is the amorphous part in the samples calculated by deconvolution of the XRD spectra.

2.2. Aging Method

The self-crystallization is performed as 10 g FA was added to 150 mL 5 M NaOH (purity > 98%, Valerus, Sofia, Bulgaria), the slurries were filled in closed reaction vessels and stayed for aging at ambient temperature. The mixtures were stirred for homogenization over long time intervals. A small portion of the samples was taken out from the vessels after 7 and 14 months, washed with distilled water until pH = 8 and dried at room temperature. The synthesized samples were named with prefix Z-zeolite before FA acronyms, followed by dash and time of the conversion. The powdered products were subjected to various analyses in order to evaluate their structure, morphology, and surface properties.

2.3. XRD Analysis

X-ray diffractograms of the all samples were measured on a Philips PW 1050 powder diffractometer with CuKα filtered radiation and Bragg–Brentano geometry in the diffraction angle range 2θ from 6 to 84.95° at a step width of 0.05° (2θ) for scan time of 1 s. An equal mass of powdered samples and identical sample holders was used in all XRD measurements. The identification of all crystalline phases was performed by the X'Pert HighScore 3.0d software (2011 PANalytical B.V., The Netherlands) and the database of X-ray patterns collection of the International Zeolite Association.

2.4. SEM and EDX Analysis

The morphology and the integral chemical composition of the samples were studied by scanning electron microscopy (SEM) on a JSM6390 apparatus, equipped with an Oxford Instruments EDX analyzer (Highcombe, London, UK). The sample images were investigated in both secondary electron imaging (SEI) and backscattered electron imaging (BEI) modes. SEI images provided by secondary electrons originated from the surface or the near-surface regions of the samples give information for the morphology, size and surface topography of the objects. Micrographs in BEI mode are derived by the backscattered electrons detected out of the specimen interaction volume, which are proportional to the atomic number of the attended chemical elements. As a result, the objects rich in heavy metals look brightly shining on the dark background of the other particles (alumosilicates, calcites, etc.). All energy-dispersive X-ray analyses (EDX) were performed at 20 keV. Elements with an atomic number less than that of nitrogen are not included in the determination of the elements weight percentage (wt. %). The average composition of the samples was estimated on the base of so-called "integral spectra" taken at low magnification (~30×) from an area comprising several hundred to thousand particles, scanned for at least 300 s. The elemental spectrum of single particles was determined using "in point" analysis at substantially higher magnifications (~1000–300,000×), recorded for about 60 s.

2.5. Nitrogen Adsorption/Desorption Tests

The studies on the surface properties of all samples were performed with N_2-adsorption/desorption analysis by a volumetric gas analyzer Micromeritics TriStar 3020 (Micromeritics Instrument Corp.,

Norcross, GA, USA), according to the ISO 9277:2010 standard [38]. The samples were pretreated at 260 °C for 4 h under continuous Helium flow to release moisture and atmospheric gases. The samples were further degassed by evacuation with a rate of 0.67 kPa/s for 1 h provided by an Edwards vacuum pump. The isotherms were built using N_2 gas of analytical purity (5N) at cryogenic temperature of 77 K provided by liquid nitrogen. The adsorption/desorption tests were performed at 92 experimental points in the pressure range p/p_0 = 0.001–0.995, where p_0 is the saturation pressure depending on the ambient temperature (approximately 104 kPa). To the experimental isotherms, standardized Brunauer–Emmett–Teller (BET), t-plot and Barrett–Joyner–Helenda (BJH) models were applied for calculation of specific surface area (SSA, m^2/g), micropore area (S_{micro}, m^2/g), external surface area ($S_{external}$, m^2/g), micropore volume (V_{micro}, cm^3/g), mesopore volume (V_{meso}, cm^3/g), and average pore width (Å). The relative pressure range of the monolayer adsorption for application of the BET model was calculated from the extreme value of the function $n_a.(1 - p/p_0) = f(p/p_0)$ (n_a is quantity of N_2 adsorbed, mmol/g).

2.6. Dye Adsorption

The adsorption tests were performed in 100 mL water solution of MG (96% of purity, Valerus, Sofia, Bulgaria) with stock concentration of 10 mL/l by immersing of 5 g fly ash zeolites obtained after 14 months of crystallization (FAZs-14). The adsorption kinetics was studied by measuring MG concentration by the help of UV spectrophotometer, 1200 Series, Cole Palmer Instruments Company, East Bunker Court, IL, USA. Various kinetic models were applied to the experimental data to evaluate the parameters of the water purification process.

3. Results and Discussions

3.1. Phase Composition of the Coal Fly Ash Zeolites

The results from XRD analyses are plotted in Figure 1, presenting the patterns of the raw FAs, synthesized products after 7 and 14 months, and referent zeolite FAU for comparison. The most intensive reflexes of FAU could be found in all FAZs, and at $2\theta \approx 26.6°$ the zeolite peak overlaps with that of the quartz transferred into the products from the raw FA. Thus, the presence of unreacted quartz could not be confirmed unambiguously by XRD analysis, but it could be assumed when expansion or splitting of the peak characteristic of both phases is observed. Only at XRD patterns of FAZs obtained from completely amorphous raw fly ashes, such as SS and FARE, it can be considered undoubtedly that the intensive peak at this angle belongs to FAU. For samples of the all series, the longer synthesis time leads to a slight increase in the intensity of the reflexes. At the zeolite obtained from FAB after 14 months of alkaline aging, a mixture of zeolite structures is observed, as, in addition to FAU reflexes, two intensive peaks of PHI are identified on its patterns. A single PHI reflex is also appeared at the XRD spectra of the samples synthesized from FAR and FAV. On their XRD patterns, a shoulder at the base of the peak assigned to FAU could be also seen, which is indicative for incomplete assimilation of the quartz in the zeolitization process. For the samples aged 14 months, the shoulder is more pronounced as compared to FAZs obtained under shorter time of incubation. This indicates that, due to its thermodynamic instability, FAU begins to recrystallize after prolonged incubation in the reaction mixture. Considering that the same amounts of all samples are subjected of XRD analysis, it could be supposed that maximum yield of zeolite FAU is obtained at FAZs from the series FAR, FARE, and FAV because their patterns are characterized by the largest total areas described by FAU peaks.

3.2. Morphological Study of Coal Fly Ash Zeolites

More detailed studies on the crystallization of zeolites from FA were achieved by performing the morphological SEM analysis of the obtained samples. The results for the samples originated from FAB, FAG, and FAM are shown in Figure 2, while for those from FAR, FARE, FAV, and SS are presented in Figure 3. The SEM images of ZFAB-7 (zeolite from FAB obtained after 7 months of exposure) show the presence of single FAU crystals with average diameter of 3.5 μm and agglomerates of smaller FAU particles (1 μm). On the second image of the same sample, crystals with prismatic morphology typical for zeolite PHI are also observed. The longer synthesis time leads to intensive growth of PHI crystals which corresponds to an increase in the intensity of the peaks of this zeolite phase on the XRD pattern of the sample ZFAB-14.

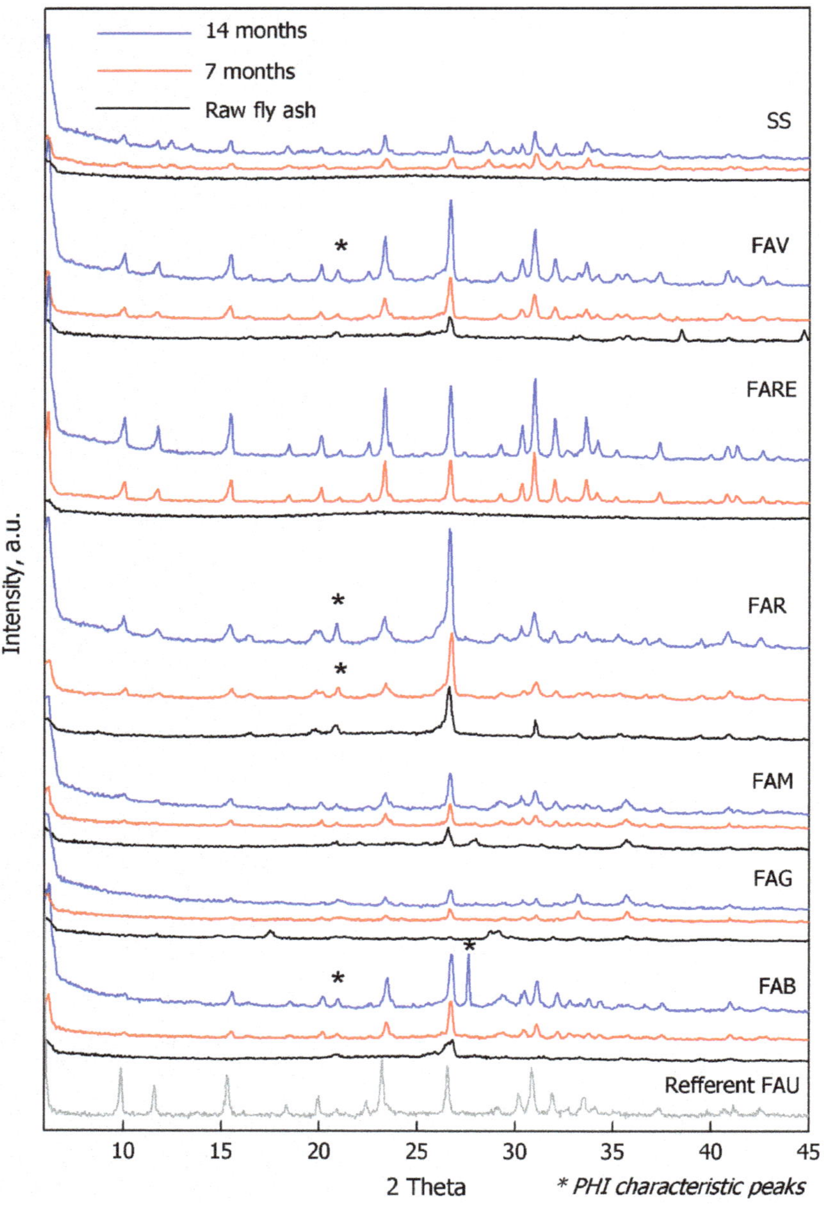

Figure 1. X-ray diffraction patterns of raw fly ashes and synthesized samples after 7 months (red spectra) and 14 months (blue spectra) of quasi-natural crystallization.

Figure 2. Scanning electron microscopy (SEM) images of fly ash zeolites obtained from FAB, FAG, and FAM after 7 and 14 months of crystallization.

On the SEM images of the materials synthesized from FAG, typical FAU crystals could also be identified but in a strong morphological disorder. The first image of ZFAG-7 shows the crystallization of new structures onto particles of irregular shape probably remnants of unreacted fly ash. The images with higher magnification indicate the presence of disordered FAU crystals with sizes ranging from 0.1 to 1.5 μm. Prolongation of the reaction time does not lead to a substantial change in the morphology of the sample (ZFAG-14). The morphology analyses of samples derived from FAM show intensive crystallization of FAU, as submicron individual crystals were obtained after 7 months of reaction, as it is revealed by the SEM image of ZFAM-7. The doubling of the reaction time leads to an increase in the diameters of the individual crystals, and to their agglomeration, as it is observed in the images of ZFAM-14.

The SEM images of the ZFAR-7 sample correspond very well to the XRD data, as the simultaneous presence of FAU and PHI crystals is observed at a shorter synthesis time. The crystallization of zeolite structures occurs over unreacted FA particles, which act as crystallization centers and this could be observed in the SEM micrograph of ZFAR-14. Unlike previous samples, SEM analyses of the FARE series indicate FAU crystallization from FA with homogeneous distribution of the individual crystals by their size. Similar morphology with respect to the crystalline shape is observed for the sample obtained from FAV after 7 months of synthesis, but the size of the individual particles is almost twice lower. The SEM image of ZFAV-14 confirms the observation on the XRD pattern of the same sample, as prismatic crystals of PHI grow among the FAU structure. At the sample ZFAB-7, the crystallization of PHI is related to the XRD peak at $2\theta \approx 20.9$, while at ZFAB-14 the growth of PHI crystals is associated to the intensive peak at $2\theta \approx 27.6$, which is the most pronounced reflex for the pure phase. At the series of samples obtained from SS, the presence mainly of the FAU structure is also confirmed, as for a shorter time of 7 months the crystals are well-shaped with sizes of 5–7 μm. However, the prolongation of the reaction leads to agglomeration and to obtaining of dense aggregates of particles.

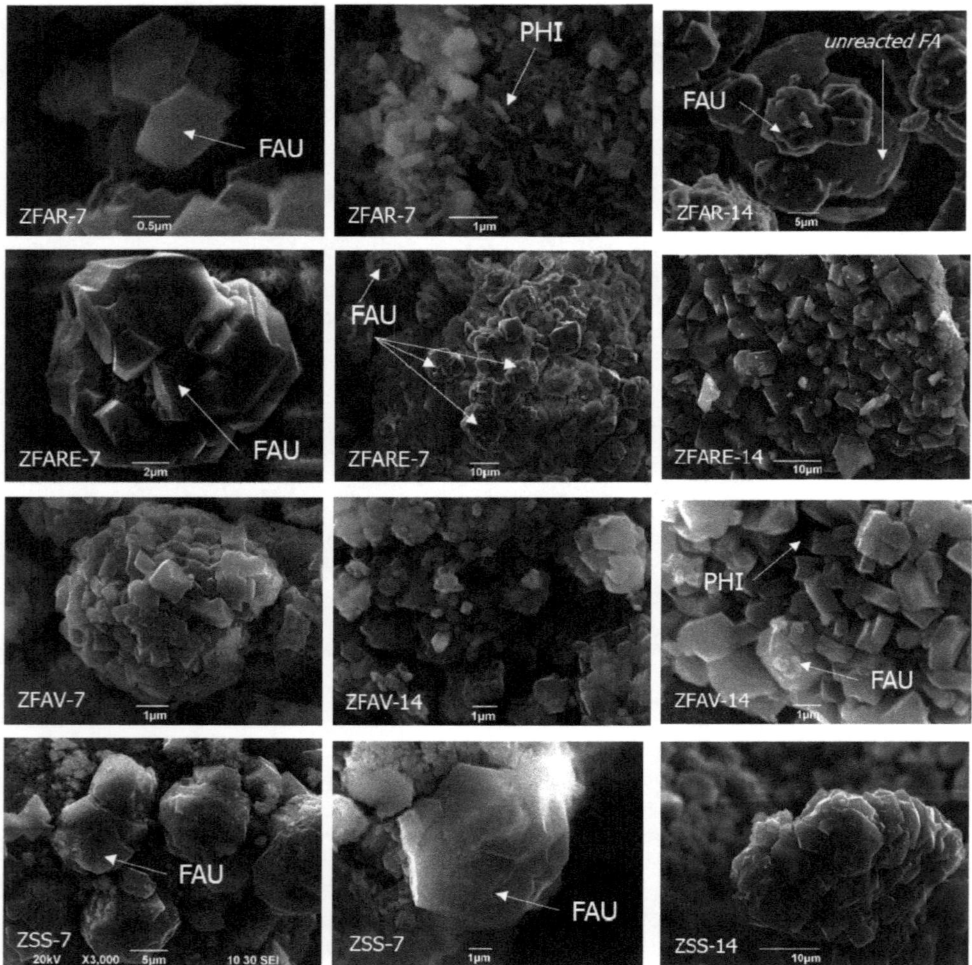

Figure 3. SEM images of fly ash zeolites obtained by FAR, FARE, FAV and SS after 7 and 14 months of crystallization.

3.3. Surface Properties of Coal Fly Ash Zeolites

The isotherms of N_2-adsorption/desorption for the raw FAs and FAZs of different raw materials converted for 7 and 14 months are plotted in Figure 4. The raw FAs are characterized by weak N_2 adsorption and isotherms of Type III corresponding to non-porous materials, as it is demonstrated for FAR and FAM (Figure 4). The isotherms of all FAZs refer to Type IV according to the International Union of Pure and Applied Chemistry (IUPAC) classification, and their shape is typical for porous solids with mixed micro-mesoporosity. At all plots, fast adsorption at low pressure range $p/p_0 = 0.001–0.05$ is observed due to the micropores filling and monolayer formation. The part of the isotherms at $p/p_0 = 0.05–0.85$ corresponds to the multilayer adsorption of molecules and the filling of mesopores.

The different slope of this region determines the share of mesopores into the structure. The region of the adsorption isotherms at p/p_0 = 0.85–0.99 represents the filling in macropores or gas adsorption in the inter-particles free volume of the powdered material. The desorption branches go above the adsorption isotherms, describing hysteresis loops which are closed at p/p_0 = 0.45 for all samples. The type of all hysteresis loops is H3, which corresponds to parallel plate-shaped pores with different widths, depending on the hysteresis area [39]. The most intensive monolayer filling is observed for samples from series FAV and FARE, as the quantity of the adsorbed gas reaches 4.3 mmol/g at p/p_0 = 0.025, and their specific surface area is expected to be the highest. The sample ZFAV-14 possesses higher adsorption in the low pressure region as compared to ZFAV-7, which could be explained by the higher yield of zeolite FAU after a longer time of synthesis. The other sample sets follow the same trend with some exceptions, mostly for the ZSS series. ZSS-14 obtained after 14 months of crystallization has weak adsorption, which is not typical for zeolite FAU. This could be explained by the degradation of the crystallized FAU after its longer incubation into the alkaline solution. Zeolite FAU is obtained as a metastable phase and, upon prolonged alkaline treatment, it recrystallizes to a more stable zeolite phase, but with a lower surface area. Similar observations have been made in our previous studies on self-crystallization of FA investigating the influence of the reaction mixture alkalinity on the final product [11]. The raw fly ash SS is highly amorphous, which predetermines faster kinetics of its alkaline conversion. Most likely, the crystallization of FAU from SS has reached the maximum rate at a shorter reaction time and further incubation of the product in the alkaline solution is led to the destruction of FAU.

The data from N_2 isotherms of FAZs are processed by different mathematical models to calculate their surface characteristics. The determined surface properties of FAZs are listed in Table 2.

Table 2. Surface properties of FAZs calculated by the N_2-adsorption/desorption experimental isotherms.

Sample	SSA m^2/g	S_{micro} m^2/g	$S_{external}$ m^2/g	V_{micro} cm^3/g	V_{meso} cm^3/g	V_{total} cm^3/g	W_{pores} Å	M_{share} %
ZFAB-7	296	210	86	0.0834	0.1755	0.2589	65.55	32
ZFAB-14	282	201	81	0.0801	0.1607	0.2407	64.60	33
ZFAG-7	123	37	86	0.0154	0.1275	0.1429	50.47	11
ZFAG-14	139	43	96	0.0176	0.1241	0.1417	46.25	12
ZFAM-7	224	140	84	0.0568	0.1862	0.2430	63.65	23
ZFAM-14	210	133	77	0.0539	0.1669	0.2209	61.41	24
ZFAR-7	238	174	64	0.0684	0.1338	0.2022	68.46	34
ZFAR-14	248	182	66	0.0717	0.1264	0.1981	63.48	36
ZFARE-7	404	357	47	0.1344	0.0742	0.2086	48.51	64
ZFARE-14	389	339	50	0.1279	0.0867	0.2147	51.97	60
ZFAV-7	341	275	66	0.1050	0.1196	0.2247	65.14	47
ZFAV-14	387	319	68	0.1222	0.1190	0.2412	62.22	51
ZSS-7	169	131	38	0.0502	0.0687	0.1189	67.68	42
ZSS-14	43	13	30	0.0057	0.0588	0.0646	67.00	9

Note SSA—specific surface area, S_{micro}—micropore area, $S_{external}$—external surface area, V_{micro}—micropore volume, V_{meso}—mesopore volume, V_{total}—total pore volume, W_{pores}—average pore width, M_{share}—volume share of micropores.

Although, the zeolite FAU was identified by XRD and SEM at the all synthesized samples, FAZs differ in their textural properties. The specific surface area of the samples varies in a wide range as the highest obtained value is 404 m^2/g for the sample ZFARE-7. Slight decrease in SSA and in the share of micropores is obtained for longer time of synthesis at ZFARE-14, which could be indicative for the beginning of recrystallization process. The SSAs of the samples of ZFAV series are comparable, as the yield of zeolite FAU remains high after 14 months of reaction. The volume fraction of micropores for ZFAVs is relatively lower than that for ZFAREs, because of the different amorphous part in the structure of the raw materials. Similar textural properties are calculated for the samples from the series ZFAB, ZFAM, and ZFAR with SSAs in the range of 210–296 m^2/g. Despite the lower surface area compared to the previous series of samples, the total pore volume of these FAZs is higher, reaching values of 0.2589 cm^3/g due to the wider pores with sizes up to 68.46 Å. The well-developed mesopore structure of these materials is beneficial for their application in systems for adsorption of gases or for removal of pollutants from waters. The lowest values for all surface characteristics are calculated for the samples from the ZFAG and ZSS series, which is in accordance to the observation from XRD, SEM, and N_2-adsorption analyses.

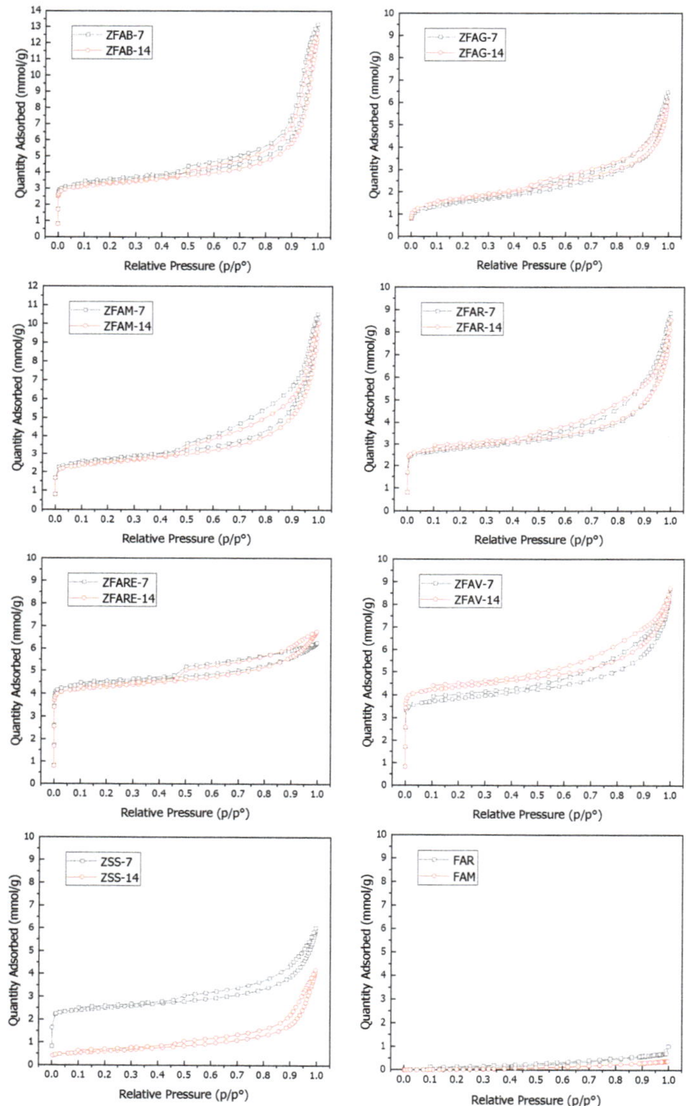

Figure 4. Isotherms of N_2-adsorption/desorption of FAs and FAZs obtained from different raw materials for 7 and 14 months.

There is no unified approach to estimate the zeolitic yield in the products of FA conversion. In the literature, data from thermogravimetric analyses, water vapor adsorption and release, XRD deconvolution, and SSA comparison have been applied for the estimation of the zeolitization extent [40]. In this study, the SSA of the synthesized samples was used to evaluate the influence of the raw fly ashes properties onto the crystallization of zeolite FAU. The dependences of SSAs and M_{share} of the investigated samples (excluding SS because of the FAU destruction observed) on the SiO_2/Al_2O_3 ratio and the amorphous constituent of FA are represented in Figure 5. The increase in SSA and the yield of zeolite FAU are directly related to the enhancement of the amorphous content in the raw fly ash. The same dependence is observed for the micropores share in the synthesized samples (index μ on the plot) predetermined by the better solubility of alumina and silica in alkaline media and further crystallization of the obtained hydrogel in a determined zeolite phase. The presence of PHI in some of the samples does not influence the porosity of the FAZs. It was established that the chemical composition of the raw FA does not affect significantly the zeolitization process in the SiO_2/Al_2O_3 range of 1.5–2.3, and the FAU structure is obtained by all of the investigated raw materials.

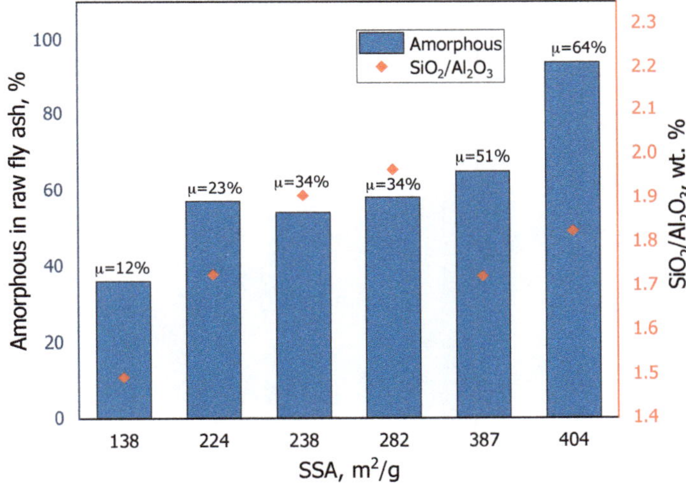

Figure 5. Dependences between surface properties of investigated samples and raw fly ash properties.

3.4. Adsorption of Malachite Green from Coal Ash Zeolites

The adsorption of MG onto the surface of the synthetic zeolites is investigated at 10 mg/L concentration of the stock solution by adding 3 g/L of the zeolite products. The removal efficiency RE (%) of coal fly ash zeolites toward MG at different times of analyses is calculated by the following equation:

$$RE(\%) = ((C_0 - C_e)/C_0) \cdot 100 \tag{1}$$

where C_0 is the initial concentration of MG in the test solutions; C_e is the measured equilibrium concentration of MG in the liquid phase after the adsorption, mg/L.

The results from the MG kinetic study on FAZs are plotted in Figure 6.

The experimental results were described applying the pseudo first- (Equation (2)) and second-order (Equation (3)) kinetic models. The linear forms of the model equations used are as follows:

$$-\ln(C_{\mathbf{MG},\tau}/C_{MG,0}) = k_1\tau \tag{2}$$

$$\tau/C_{\mathbf{MG},\tau} = 1/(k_2 \cdot C_{MG,0}) + \tau/C_{MG,0} \tag{3}$$

where k_1 and k_2 are the reaction rate constants for the kinetic models of the pseudo first- and second-order, min^{-1} and $M^{-1}{\cdot}min^{-1}$, correspondingly; τ is the reaction time, min.

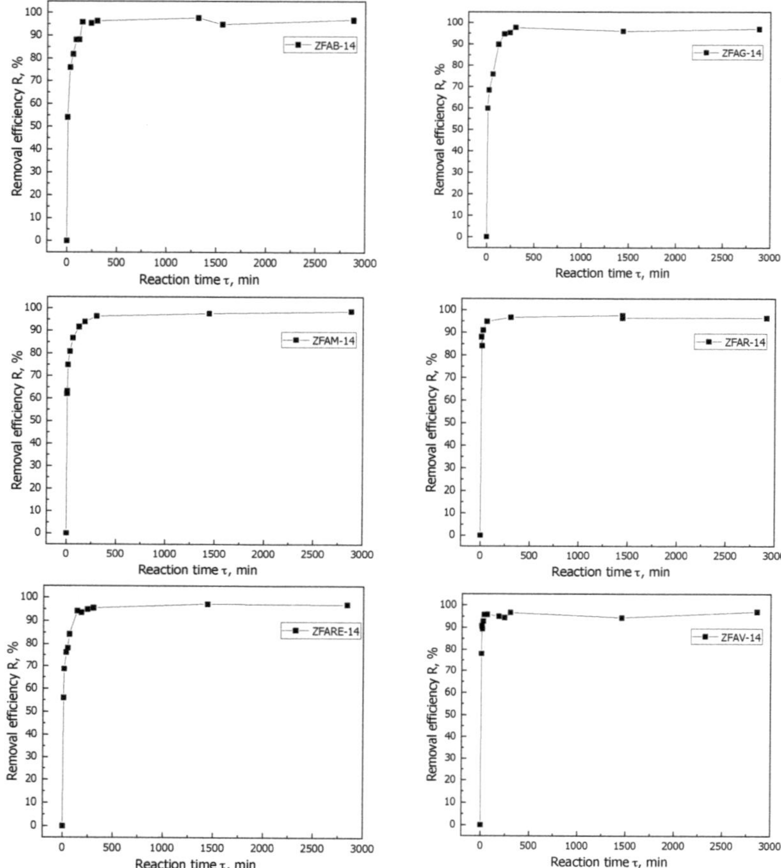

Figure 6. Adsorption kinetics of Malachite green onto the surface of synthetic fly ash zeolites obtained after 14 months of self-crystallization.

Linear plots of pseudo first- (PFO) and second- (PSO) order kinetic models are shown in Figure 7 together with the results from regression analyses. The PFO does not correlate to the experimental data, as R-Square varies in the range of 0.22–0.63, while the PSO effectively describes the kinetics of the adsorption with R-Square of 0.999 for all samples.

The calculated rate constant k_2 ($M^{-1}{\cdot}min^{-1}$) and half-life time $\tau_{1/2}$ (min^{-1}) from the PSO kinetics modelling and the obtained maximum removal efficiencies of FAZ toward MG are listed in Table 3. The fastest adsorption kinetics is calculated for the samples ZFAR-14 and ZFAV-14 but the highest removal efficiency is achieved at ZFAR-14. Although the investigated samples are characterized with surface areas varying in a wide range, all of them reach almost full remediation of the water solutions from MG. At these concentration levels of MG, the adsorption process probably takes place on the external surface of the ZFAs and the MG molecules do not penetrate into the pore volume composed of micro and mesopores. This is a prerequisite for the FAZs to have the capacity to remove completely even higher levels of this contaminant.

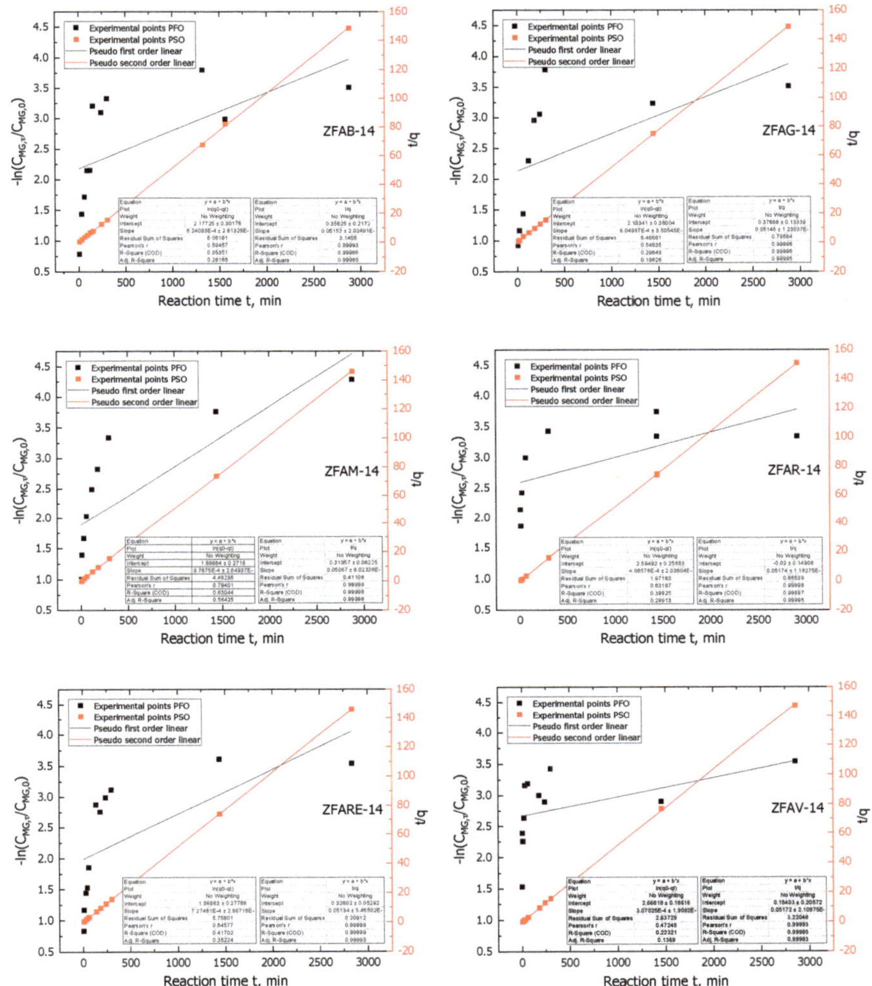

Figure 7. Regression analyses of the experimental data by pseudo first- (PFO) and second- (PSO) order kinetic models (linear plots).

Table 3. Parameters of pseudo-second-order kinetic model for adsorption of Malachite green onto fly ash zeolite surface and the obtained removal efficiencies.

PSO		ZFAB-14	ZFAG-14	ZFAM-14	ZFAR-14	ZFARE-14	ZFAV-14
k_2	$M^{-1}\cdot min^{-1}$	0.0075	0.0070	0.0080	0.0180	0.0081	0.0174
$\tau_{1/2}$	min	13.42	14.22	12.45	5.57	12.37	5.76
RE_{max}	%	96.6	97.0	98.6	97.5	97.0	97.0

The study of the sorption kinetics of wastewater treatment is important to clarify the process mechanism, the rate of retention of contaminants, and the time for exhaustion of the sorbent capacity. These characteristics are important in the design of efficient sorption treatment plants for water treatment of specific pollutants. It was found that the pseudo-second-order model provides the best correlation to the adsorption kinetics of MG onto FA, which means that the rate-limiting stage is chemisorption involving valence forces by electronic sharing or exchange between sorbent and sorbate [41]. This is confirmed by the fact that no unambiguous tendency of increase in the adsorption rate constant with increase in the specific surface of coal ash zeolites is established. In a case that the adsorption process proceeds with diffusion control, then the surface characteristics of the adsorbent, such as particle size, specific surface area and pore diameter will be essential for the process kinetics, which is not observed for the present adsorption studies. It could be assumed that the chemisorption mechanism of MG onto FAZ is obeyed by the cationic behavior of this dye upon its dissolution into water. Cationic dyes carry positive charges in their molecules and dissociate into cations in the aqueous solutions [42]. In this case, an ion exchange process can occur between the mobile cations in the zeolite structure and the dye cations from the solution. Finally, it could be supposed that the rate of adsorption of MG on FAZ will initially be dominated by surface factors, and as the process progresses, the concentration and the distribution of the ion exchange sites in the zeolite framework will play a key role. Some studies reveal that the raw fly ash can also behave a high retention capacity for some pollutants, for example, toward anionic surfactants [43]. However, adsorbents based on zeolitized ash definitely have a number of advantages in water purification. Coal ash zeolites have a higher specific retention capacity per unit mass, which means less need for adsorbent to treat a unit volume of water. Another advantage is that zeolitized coal ash releases fewer impurities into the remediated water as compared to the parent ash, which is characterized by substantial leaching. The ability of coal zeolites to act as ion exchangers and adsorbents makes them applicable in the cleaning of industrial sludges, acid mine drainages, industrial, agricultural and domestic waters by a variety of techniques, such as filtration, coagulation and catalytic degradation of contaminants. Moreover, they can serve as permeable reactive barriers for remediation and barriers against infiltrations of toxic components into ground water.

4. Conclusions

The focus of this study is on the crystallization of zeolites from different coal fly ashes at ambient conditions. The obtained materials were investigated for two times of aging in regard to identify the phase composition and morphology of the final products. It was established that the self-crystallization process at low temperature ensures long-term controllable synthesis of zeolite FAU from FA of different coal types, at a wide range of SiO_2/Al_2O_3 ratios, and varying extent of amorphization. It was also observed that the yield of the zeolite FAU is mainly favored by the higher amounts of the amorphous aluminosilicates in the raw material. The values of the experimentally determined specific surface area over 400 m^2/g and the textural properties of the synthesized zeolites are beneficial for a wide range of possible applications. This synthesis approach is promising for large-scale waste utilization because of its simplicity, zero energy demands and low operational costs. It could be easily applied in the places of fly ash landfills and long-accumulated waste can be recovered. The quasi natural synthesis procedure would meet the environmental challenges of waste storage management, and in the same time would produce a valuable commercial product. In this study, one of the possible approaches for useful utilization of the coal fly zeolites obtained by different sources was investigated. The successful adsorption of malachite green was achieved with removal efficiency over 96%. The experimental data were fitted to the second-order kinetic reaction model with high correlation.

Author Contributions: Conceptualization, D.Z.; methodology, D.Z., A.S. and V.S.; software, G.A., D.Z.; validation, V.S., A.S. and D.Z.; formal analysis, D.Z. and A.S.; investigation, A.S., V.S., D.Z. and S.B.; resources, S.B.; data curation, G.A., D.Z.; writing—original draft preparation, D.Z.; writing—review and editing, S.B, D.Z. and V.S.; visualization, D.Z., V.S. and A.S.; supervision, S.B. and A.S.; project administration, S.B.; funding acquisition, S.B. All authors have read and agreed to the published version of the manuscript.

Funding: This research and article processing charge were funded by National Science Fund, Ministry of Education and Science of R. Bulgaria under the Grant DN 17/18 (2017).

Acknowledgments: The authors express their gratitude to all colleagues from the Electron Microscopy Lab of Institute of Physical Chemistry at Bulgarian Academy of Sciences (IPC-BAS) for their technical assistance in the SEM and EDX analysis.

Conflicts of Interest: The authors declare no conflict of interest.

References

1. IEA Statistics 2018 CO_2 Emissions from Fuel Combustion: Overview. Available online: https://www.iea.org/subscribe-to-data-services/CO2-emissions-statistics (accessed on 21 November 2020).
2. Yang, H.; Xu, Z.; Fan, M.; Gupta, R.; Slimane, R.B.; Bland, A.E.; Wright, I. Progress in carbon dioxide separation and capture: A review. *J. Environ. Sci.* **2008**, *20*, 14–27. [CrossRef]
3. Król, A. The role of the silica fly ash. *Sustain. Waste Manag.* **2016**, *10*. [CrossRef]
4. Brook, B.W.; Alonso, A.; Meneley, D.A.; Misak, J.; Blees, T.; van Erp, J.B. Why nuclear energy is sustainable and has to be part of the energy mix. *Sustain. Mater. Technol.* **2014**, *1*, 8–16. [CrossRef]
5. Shukla, A.K.; Ahmad, Z.; Sharma, M.; Dwivedi, G.; Verma, T.N.; Jain, S.; Verma, P.; Zare, A. Advances of carbon capture and storage in coal-based power generating units in an Indian context. *Energies* **2020**, *13*, 4124. [CrossRef]
6. Zhang, Z.; Pan, S.-Y.; Li, H.; Cai, J.; Olabi, A.G.; Anthony, E.J.; Manovic, V. Recent advances in carbon dioxide utilization. *Renew. Sustain. Energy Rev.* **2020**, *125*, 109799. [CrossRef]
7. Stuardi, F.M.; MacPherson, F.; Leclaire, J. Integrated CO_2 capture and utilization: A priority research direction. *Curr. Opin. Green Sustain. Chem.* **2019**, *16*, 71–76. [CrossRef]
8. Lee, B.J.; Lee, J.I.; Yun, S.Y.; Lim, C.; Park, Y. Economic evaluation of carbon capture and utilization applying the technology of mineral carbonation at coal-fired power plant. *Sustainability* **2020**, *12*, 6175. [CrossRef]
9. World Wide Coal Combustion Products Network (WWCCPN): Member information 2018/2019. Available online: https://wwccpn.com/ (accessed on 21 November 2020).
10. *ASTM C618-17, Standard Specification for Coal Fly Ash and Raw or Calcined Natural Pozzolan for Use in Concrete*; ASTM International: West Conshohocken, PA, USA, 2017.
11. Zgureva, D.; Boycheva, S.; Behunová, D.; Václavíková, M. Smart- And zero-energy utilization of coal ash from thermal power plants in the context of circular economy and related to soil recovery. *J. Environ. Eng.* **2020**, *146*, 04020081. [CrossRef]
12. Baek, C.; Seo, J.; Choi, M.; Cho, J.; Ahn, J.; Cho, K. Utilization of CFBC fly ash as a binder to produce in-furnace desulfurization sorbent. *Sustainability* **2018**, *10*, 4854. [CrossRef]
13. Podgorodetskii, G.S.; Gorbunov, V.B.; Agapov, E.A.; Erokhov, T.V.; Kozlova, O.N. Challenges and opportunities of utilization of ash and slag waste of TPP (thermal power plant). Part 1. *Izv. Ferr. Metall.* **2018**, *61*, 439–446. [CrossRef]
14. Bergk, K.H.; Porsch, M.; Wolf, F. Transformation of primary- and secondary raw materials in zeolite containing products. part I: Production of zeolite-A containing products from power plant fly ash. *Chem. Tech.* **1985**, *37*, 253–256.
15. Querol, X.; Moreno, N.; Alastuey, A.; Juan, R.; Andrés, J.M.; López-Soler, A.; Ayora, C.; Medinaceli, A.; Valero, A. Synthesis of high ion exchange zeolites from coal fly ash. *Geol. Acta* **2007**, *5*, 49–57.
16. Bukhari, S.; Rohani, S. Continuous flow synthesis of zeolite—A from coal fly ash utilizing microwave irradiation with recycled liquid stream. *Mater. Eng. Sci. Div. 2016—Core Program. Area 2016 AIChE Annu. Meet.* **2016**, *2*, 558–565.
17. Adamczyk, Z.; Białecka, B. Hydrothermal synthesis of zeolites from polish coal fly ash. *Pol. J. Environ. Stud.* **2005**, *14*, 713–719.
18. Fukui, K.; Nishimoto, T.; Takiguchi, M.; Yoshida, H. Effects of NaOH concentration on zeolite synthesis from fly ash with a hydrothermal treatment method. *KONA Powder Part. J.* **2006**, *24*, 183–191. [CrossRef]
19. Inada, M.; Tsujimoto, H.; Eguchi, Y.; Enomoto, N.; Hojo, J. Microwave-assisted zeolite synthesis from coal fly ash in hydrothermal process. *Fuel* **2005**, *84*, 1482–1486. [CrossRef]
20. Molina, A.; Poole, C. A comparative study using two methods to produce zeolites from fly ash. *Miner. Eng.* **2004**, *17*, 167–173. [CrossRef]
21. Fukui, K.; Arai, K.; Kanayama, K.; Yoshida, H. Phillipsite synthesis from fly ash prepared by hydrothermal treatment with microwave heating. *Adv. Powder Technol.* **2006**, *17*, 369–382. [CrossRef]
22. Shoumkova, A.; Stoyanova, V. Zeolites formation by hydrothermal alkali activation of coal fly ash from thermal power station "Maritsa 3", Bulgaria. *Fuel* **2013**, *103*, 533–541. [CrossRef]

23. Querol, X.; Moreno, N.; Umaa, J.C.; Alastuey, A.; Hernández, E.; López-Soler, A.; Plana, F. Synthesis of zeolites from coal fly ash: An overview. *Int. J. Coal Geol.* **2002**, *50*, 413–423. [CrossRef]
24. Murayama, N.; Yamamoto, H.; Shibata, J. Mechanism of zeolite synthesis from coal fly ash by alkali hydrothermal reaction. *Int. J. Miner. Process.* **2002**, *64*, 1–17. [CrossRef]
25. Ferret, L.; Fernandes, I.D.; Khahl, C.A.; Endres, J.C.T.; Maegawa, A. Zeolitification of ashes obtained from the combustion of southern's Brazil candiota coal. In Proceedings of the International ash Utilization Symposium, Lexington, KY, USA, 18–20 October 1999.
26. Singer, A.; Berkgaut, V. Cation exchange properties of hydrothermally treated coal fly ash. *Environ. Sci. Technol.* **1995**, *29*, 1748–1753. [CrossRef] [PubMed]
27. Shigemoto, N.; Hayashi, H.; Miyaura, K. Selective formation of Na-X zeolite from coal fly ash by fusion with sodium hydroxide prior to hydrothermal reaction. *J. Mater. Sci.* **1993**, *28*, 4781–4786. [CrossRef]
28. Shigemoto, N.; Shirakami, K.; Hirano, S.; Hayashi, H. Preparation and characterization of zeolites from coal ash. *Nippon Kagaku Kaishi* **1992**, *5*, 484–492. [CrossRef]
29. Jha, V.K.; Matsuda, M.; Miyake, M. Resource recovery from coal fly ash waste: An overview study. *J Ceram. Soc. Jpn.* **2008**, *116*, 167–175.
30. Fotovat, F.; Kazemian, H.; Kazemeini, M. Synthesis of Na-A and faujasitic zeolites from high silicon fly ash. *Mater. Res. Bull.* **2009**, *44*, 913–917. [CrossRef]
31. Jha, V.K.; Matsuda, M.; Miyake, M. Sorption properties of the activated carbon-zeolite composite prepared from coal fly ash for Ni^{2+}, Cu^{2+}, Cd^{2+} and Pb^{2+}. *J. Hazard. Mater.* **2008**, *160*, 148–153. [CrossRef]
32. Klamrassamee, T.; Pavasant, P.; Laosiripojana, N. Synthesis of zeolite from coal fly ash: Its application as water sorbent. *Eng. J.* **2010**, *14*, 37–44. [CrossRef]
33. Shoumkova, A. Zeolites for water and wastewater treatment: An overview. *Res. Bull. Aust. Inst. High Energetic Mater. Spec. Issue Glob. Fresh Water Short.* **2011**, *2*, 10–70.
34. Boycheva, S.; Zgureva, D.; Miteva, S.; Marinov, I.; Behunová, D.M.; Trendafilova, I.; Popova, M.; Václaviková, M. Studies on the potential of nonmodified and metal oxide-modified coal fly ash zeolites for adsorption of heavy metals and catalytic degradation of organics for waste water recovery. *Processes* **2020**, *8*, 778. [CrossRef]
35. Srivastava, S.; Sinha, R.; Roy, D. Toxicological effects of malachite green. *Aquat. Toxicol.* **2004**, *66*, 319–329. [CrossRef] [PubMed]
36. Shoumkova, A.; Stoyanova, V.; Tsacheva, T. Preliminary study on the zeolitization of coal fly ashes from six Bulgarian thermal power plants. SEM-EDX analyses. *Compt. Rend. L'Acad. Bulg. Sci.* **2011**, *64*, 937–944.
37. Pascova, R.; Stoyanova, V.; Shoumkova, A. Room temperature zeolitization of boiler slag from a Bulgarian thermal power plant. *J. Serb. Chem. Soc.* **2017**, *82*, 227–240. [CrossRef]
38. *ISO 9277:2010 Determination of the Specific Surface Area of Solids by Gas Adsorption—BET Method;* International Organization for Standardization: Geneva, Switzerland, 2010.
39. Wang, Z.; Jiang, X.; Pan, M.; Shi, Y. Nano-scale pore structure and its multi-fractal characteristics of tight sandstone by N_2 adsorption/desorption analyses: A case study of shihezi formation from the sulige gas field, Ordos Basin, China. *Minerals* **2020**, *10*, 377. [CrossRef]
40. Majchrzak-Kucęba, I. A simple thermogravimetric method for the evaluation of the degree of fly ash conversion into zeolite material. *J. Porous Mater.* **2013**, *20*, 407–415. [CrossRef]
41. Ho, Y.S.; McKay, G. Pseudo-second order model for sorption processes. *Process Biochem.* **1999**, *34*, 451–465.
42. Raval, N.P.; Shah, P.U.; Shah, N.K. Malachite green "a cationic dye" and its removal from aqueous solution by adsorption. *Appl. Water Sci.* **2017**, *7*, 3407–3445. [CrossRef]
43. Zanoletti, A.; Federici, S.; Borgese, L.; Bergese, P.; Ferroni, M.; Depero, L.E.; Bontempi, E. Embodied energy as key parameter for sustainable materials selection: The case of reusing coal fly ash for removing anionic surfactants. *J. Clean. Prod.* **2017**, *141*, 230–236. [CrossRef]

Publisher's Note: MDPI stays neutral with regard to jurisdictional claims in published maps and institutional affiliations.

Article

Use of a Copper- and Zinc-Modified Natural Zeolite to Improve Ethylene Removal and Postharvest Quality of Tomato Fruit

Johannes de Bruijn [1,*], Ambar Gómez [1], Cristina Loyola [1], Pedro Melín [1], Víctor Solar [2], Norberto Abreu [3], Federico Azzolina-Jury [4] and Héctor Valdés [2,*]

[1] Departamento de Agroindustrias, Facultad de Ingeniería Agrícola, Universidad de Concepción, Avenida Vicente Méndez 595, 3780000 Chillán, Chile; ambargomez@udec.cl (A.G.); cloyola@udec.cl (C.L.); pmelin@udec.cl (P.M.)

[2] Laboratorio de Tecnologías Limpias, Facultad de Ingeniería, Universidad Católica de la Santísima Concepción, Alonso de Ribera 2850, 4030000 Concepción, Chile; vsolar@ucsc.cl

[3] Center of Waste Management and Bioenergy, BIOREN, Departamento de Ingeniería Química, Facultad de Ingeniería y Ciencias, Universidad de la Frontera, Francisco Salazar 01145, 4780000 Temuco, Chile; norberto.abreu@ufrontera.cl

[4] ENSICAEN, UNICAEN, CNRS, Laboratoire Catalyse et Spectrochimie, Normandie Université, 14000 Caen, France; federico.azzolina-jury@ensicaen.fr

* Correspondence: jdebruij@udec.cl (J.d.B.); hvaldes@ucsc.cl (H.V.); Tel.: +56-42-220-8891 (J.d.B.); +56-41-234-5044 (H.V.)

Received: 23 April 2020; Accepted: 1 June 2020; Published: 3 June 2020

Abstract: Ethylene stimulates ripening and senescence by promoting chlorophyll loss, red pigment synthesis, and softening of tomatoes and diminishes their shelf-life. The aim of this work was to study the performance of a novel copper- and zinc-based ethylene scavenger supported by ion-exchange on a naturally occurring zeolite by analyzing its ethylene adsorption capacity and the influence of ethylene scavenging on quality attributes during the postharvest life of tomatoes. The influence of copper- and zinc-modified zeolites on ethylene and carbon dioxide concentrations and postharvest quality of tomatoes was compared with unmodified zeolite. Interactions among ethylene molecules and zeolite surface were studied by diffuse reflectance infrared Fourier transform spectroscopy in *operando* mode. The percentage of ethylene removal after eight days of storage was 57% and 37% for the modified zeolite and pristine zeolite, respectively. The major ethylene increase appeared at 9.5 days for the modified zeolite treatment. Additionally, modified zeolite delayed carbon dioxide formation by six days. Zeolite modified with copper and zinc cations favors ethylene removal and delays tomato fruit ripening. However, the single use of unmodified zeolite should be reconsidered due to its ripening promoting effects in tomatoes at high moisture storage conditions, as water molecules block active sites for ethylene adsorption.

Keywords: adsorption; DRIFTS *operando*; ethylene scavenging; postharvest quality; tomato; zeolite

---

1. Introduction

Tomato fruit (*Solanum lycopersicum* L.) is one of the most consumed vegetables worldwide with a total production of about 182 million tons per year [1]. Tomatoes are climacteric fruit in which ripening is accompanied by a quickly increased respiration and ethylene production. Ethylene (C_2H_4) is a simple, naturally produced plant-growth-regulating substance that has numerous effects during the postharvest life of fruit and vegetables. Ethylene stimulates ripening and senescence that may result in detrimental effects by promoting unwanted softening and a grainy structure in tomatoes [2].

In addition, ethylene accelerates pigment synthesis and chlorophyll loss in tomatoes [2]. Complete ripening within just a few days is not desirable and should be avoided because it limits commercial shelf-life of horticultural products by accelerating their quality loss. Commercial postharvest strategies are often based on avoiding the exposure of climacteric fruit to ethylene, attempting to minimize ethylene production, inhibit its action, and by the removal of C_2H_4 not only from the postharvest facilities but also from the inside of storage packages [3–5].

Adequate ventilation with fresh air is an effective way to remove C_2H_4 from storage rooms. However, this method means an enormous energy and moisture loss and is usually impractical in the case of controlled or modified atmosphere [6]. Another postharvest storage technology is the use of active packaging for ethylene scavenging, including the adsorption and breakdown of ethylene [7]. Sorption is a cheap and relatively simple alternative process. Moreover, the sorbent can be recovered avoiding waste disposal, proving to be a cost-efficient technology in the elimination of a large variety of compounds [8]. Commercial ethylene adsorbents consist of sachets with an active agent impregnated on a porous material. Several materials have been used as ethylene scavengers such as halloysite nanotubes [9], active packaging of low-density polyethylene coated with polylactic acid [10], potassium permanganate supported onto porous inert material with a high surface area such as montmorillonite [11], silica, alumina [12], zeolite, vermiculite, and activated carbon [5], among others, to extend the shelf-life of fresh fruit. Zeolites are aluminosilicates with a negative framework charge that is balanced with alkali or alkaline earth elements and are often used as supporting material. They have a stable structure, a surface area up to 3000 $m^2\ g^{-1}$ and pore radii ranging from 3 Å to 12 Å [5,13]. Moreover, the choice of zeolites for fresh fruit makes them suitable at low or room temperature and high humidity storage conditions [14]. It has been indicated that the removal of C_2H_4 can be promoted by the impregnation of an inert support with potassium permanganate (3.5%–12%) [15], which oxidized ethylene to carbon dioxide (CO_2) and water (H_2O), simultaneously forming manganese dioxide and releasing potassium hydroxide [16]. In addition, several pure metallic elements and metal oxides fixed on supports have been tested as catalysts for the oxidation of ethylene to carbon dioxide and water. According to Terry et al. [17], the ethylene adsorption and removal capacity of palladium (Pd)–impregnated zeolite was far superior to potassium permanganate-based scavengers when used in low amounts at 20 °C and high relative humidity. Moreover, ethylene removal rate by a palladium chloride ($PdCl_2$)-impregnated acidified activated carbon scavenger was further promoted by the addition of copper sulfate [18]. Indeed, copper (Cu) has been shown to be helpful in promoting ethylene removal [13]. Additionally, photocatalytic degradation of C_2H_4 under UV light irradiation using a mixture of titanium dioxide and silica (TiO_2/SiO_2) or a nanofiber film containing TiO_2 nanoparticles was effective in reducing the concentration of ethylene in the atmosphere surrounding of climacteric fruit, thereby delaying their ripening [19,20]. Complete ethylene removal was achieved by a photocatalytic oxidation into carbon dioxide and water [21]. Doping TiO_2 with Cu^{2+} may further improve the photocatalytic activity and minimize deactivation. Besides, the use of zinc oxide (ZnO) as photocatalyst is a suitable alternative to the more expensive TiO_2. However, careful attention should be paid in the application of photocatalysis as a process treatment to remove C_2H_4. Direct artificial light during the storage of fruit promotes ripening [22]. Hence, photocatalytic treatment should be redesigned and cannot be used directly on fruit as a straightforward technology. The combination of zinc oxide with copper oxide for ethylene scavenging in tomatoes has not been reported. So far, most of these studies mainly concern the characterization of the prepared materials and the application of tests to evaluate fruit maturity and the effect of the ethylene scavenger. However, there is a lack of studies that evidence the chemical interactions among ethylene and the material active sites giving a relationship between such processes with fruit ripening. Therefore, the aim of the present work was to develop a novel copper- and zinc-based ethylene scavenger supported on naturally occurring zeolite and to study the surface interactions among adsorbed molecules and the active sites of the scavenger and their implications on postharvest attributes during the shelf-life of tomatoes.

2. Materials and Methods

2.1. Materials

Tomato fruit (Medano cultivar) were purchased from a greenhouse in Pueblo Seco, Ñuble, Chile. Natural zeolite, composed of 53% clinoptilolite, 40% mordenite, and 7% quartz with an average grain size of 5.28 ± 0.28 mm, was provided by the mining company of *Minera Formas* (Parral, Chile). Argon, oxygen, and carbon dioxide of 99.9% purity were supplied by Praxair (Santiago, Chile). Ethylene of 99.99% purity was purchased from Air Liquide S.A. (Houston, TX, USA). Zinc nitrate was obtained from Sigma-Aldrich Corporation (St. Louis, MO, USA), while copper nitrate, ammonium sulfate, copper sulfate, sodium hydroxide, butylated hydroxytoluene, sodium potassium tartrate, ethanol, acetone, and hexane were provided by Merck (Darmstadt, Germany) and were of analytical grade (≥99.0% purity). Sodium benzoate and potassium sorbate (food grade) were supplied by Furet (Chillán, Chile). Deionized water (≥18.0 MΩ cm) was used to prepare solutions using a Thermo Scientific Barnstead Easypure II RF portable ultrapure water system (Waltham, MA, USA).

2.2. Preparation of Zeolites

After the grinding and sieving of natural zeolite, a zeolite sample with an average particle size of 0.36 ± 0.06 mm was rinsed with deionized water, filtrated and then dried at 105 °C for 24 h (Z sample). Prior to the treatment of zeolite with copper and zinc cations, zeolite samples (0.10 w/v) were dispersed in a 0.1 M ammonium sulfate solution and modified by ion-exchange at 90 °C for 2 h at a shaking rate of 27 rpm. The excess of salt was removed by washing with deionized water during 2 h at the same conditions as mentioned before. This rinsing procedure was repeated twice for 1 h. The ion-exchange procedure was repeated once again, including the washing step. Afterwards, zeolite samples were thermally degassed in a U-shaped quartz fixed bed furnace at 350 °C (heating rate of 3 °C min^{-1}) during 2 h under argon flow (100 mL min^{-1}). After this step, pre-treated samples (0.10 w/v) were dispersed in an aqueous solution of both copper and zinc nitrate (66.4 mM) and modified by metal ion-exchange for 24 h at 90 °C under shaking at 27 rpm to prepare the modified zeolite doped with copper and zinc cations. After repeating the rinsing and degassing procedures, samples were stored in a desiccator before use. Prior to the adsorption experiments, samples were heated under oxygen flow (100 mL min^{-1}) at 350 °C (with a heating rate of 1 °C min^{-1}) for 4 h before quenching to room temperature. The physical and chemical surface properties of natural (Z sample) and modified natural zeolite (Z–Cu/Zn sample) used in this study have been recently published by Abreu et al. [23] (Table 1).

Table 1. Physical and chemical characteristics of natural and modified zeolite samples [1] [23].

Sample	S_{BET} [m^2 g^{-1}]	V_{meso} [cm^3 g^{-1}]	V_{micro} [cm^3 g^{-1}]	Si [%]	Al [%]	Ca [%]	Fe [%]	Na [%]	K [%]	Ti [%]	Mg [%]	Mn [%]	Sr [%]	S [%]	Zr [%]	Zn [%]	Cu [%]
Z	281	0.14	0.07	65.4	12.3	11.1	5.3	2.1	1.9	0.9	0.5	0.1	0.1	0.1	0.1	0.0	-
Z- Cu/Zn	337	0.10	0.11	68.3	13.2	4.1	4.7	-	0.9	0.8	0.2	0.1	0.1	0.1	0.1	3.3	4.3

[1] S_{ET}, V_{meso} and V_{micro} were obtained from nitrogen adsorption and desorption at 77 K, while mineral contents were determined by X-ray fluorescence spectroscopy.

2.3. Characterization of Zeolites

2.3.1. DRIFTS *Operando* Study

Interactions among ethylene molecules and zeolite surface, in the presence of air humidity, were studied by diffuse reflectance infrared Fourier transform spectroscopy (DRIFTS) in *operando* mode (see Figure 1). Experiments were performed using a JASCO FT/IR 4700 spectrometer provided by JASCO International Co., Ltd. (Tokyo, Japan) equipped with a mercury cadmium telluride detector (MCTD). Zeolite sample (0.08 g) was loaded in a Praying Mantis diffuse reflectance cell supplied by Harrick Scientific Products Inc. (New York, NY, USA). Then, a pure ethylene stream was continuously sent to the diffuse reflectance cell operating at 20 °C and spectra were registered every 1 min, from 4000 cm^{-1} to 400 cm^{-1} with an optical resolution of 1 cm^{-1}.

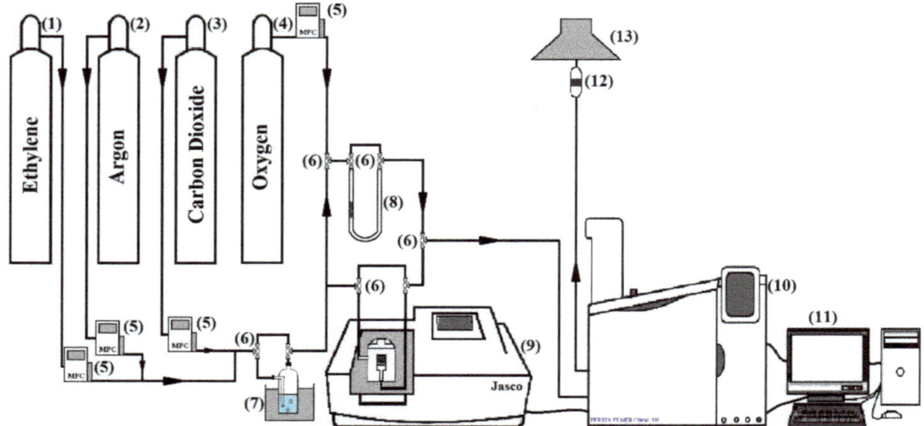

Figure 1. General experimental setup used to study surface interactions among ethylene and zeolite by diffuse reflectance infrared Fourier transform spectroscopy (DRIFTS) in *operando* mode and the retention of ethylene and carbon dioxide using a quartz U-shaped fixed bed adsorber at room temperature: (1) ethylene cylinder, (2) argon cylinder, (3) carbon dioxide cylinder, (4) oxygen cylinder, (5) mass flow controller, (6) three way valve, (7) water saturator, (8) U-type fixed bed adsorber, (9) Praying Mantis diffuse reflectance cell set in a JASCO Fourier transform infrared spectroscopy (FT/IR 4700) spectrometer, (10) Perkin Elmer Clarus 500 gas chromatograph, (11) computer, (12) $KMnO_2$ trap, and (13) extractor hood.

2.3.2. Breakthrough and Regeneration Experiments

To further evaluate the influence of natural and copper- and zinc-modified zeolites toward ethylene and carbon dioxide retention, breakthrough adsorption experiments of ethylene or carbon dioxide were performed in two identical quartz U-shaped fixed bed columns manually packed at atmospheric pressure and 20 °C with 0.30 g of natural zeolite and zeolite doped with copper and zinc cations, respectively (see Figure 1). Before breakthrough adsorption experiments, samples were heated with an oxygen flow (100 mL min^{-1}, heating rate of 1 °C min^{-1}) for 4 h, reaching a final temperature of 350 °C and then cooling to room temperature. Inlet concentration of ethylene was fixed at 120 μg L^{-1} and at 1500 μg L^{-1} for carbon dioxide by diluting each pure gas with a flow of argon. When starting the breakthrough experiment, ethylene was continuously introduced to the zeolite fixed bed at a flow rate of 25 mL min^{-1} at 20 °C, while the outlet gas from the column was analyzed on-line by gas chromatography using a Perkin Elmer Clarus 500 gas chromatograph (Shelton, CT, USA) as a function of time, whereby adimensional concentration ($C_{i,t}/C_{i,0}$) was expressed as a function of treated bed volumes (L gas L^{-1} adsorbent). Similarly, in the case of the breakthrough adsorption experiments of carbon dioxide, it was introduced to the zeolite columns at a flow rate of 300 mL min^{-1} monitoring the outlet concentration on-line by gas chromatography using a Perkin Elmer Clarus 500 chromatograph (Shelton, CT, USA), as indicated previously [24]. Breakthrough experiments were stopped when zeolite samples reached saturation. Adsorbed quantity of component *i* was calculated by integration of the area from the breakthrough curve according to the following equation:

$$q_i = \frac{FC_{i,in}}{m}\int_0^{t_s}\left(1 - \frac{C_{i,t}}{C_{i,in}}\right)dt \quad (1)$$

where q_i is the adsorbed amount of component i (μg i g^{-1} adsorbent), F is the flow rate of feed (L min^{-1}), m is the mass of adsorbent (g), $C_{i,\,in}$ is the inlet concentration of component i (μg L^{-1}), $C_{i,\,t}$ is the outlet concentration of component i (μg L^{-1}) at specific adsorption time t, and t_s is the time to reach the saturation of the adsorbent with component i (min).

After reaching saturation, zeolite samples were submitted to temperature-programmed regeneration experiments in order to gain information about the strength of adsorption of ethylene and carbon dioxide to zeolite active sites. Zeolite samples were heated under argon flow (45 mL min^{-1}) from 30 °C to 550 °C (heating rate of 3 °C min^{-1}) and the concentration of ethylene or carbon dioxide in the outlet gas was analyzed by gas chromatography, following the same procedures as explained previously.

2.4. Postharvest Treatments

Tomato fruit, *Solanum lycopersicum*, were harvested at breaker stage in December 2017 and immediately transported to the postharvest laboratory of *Universidad de Concepción*. Tomatoes were sorted for uniformity according to size and color. Fruit were dipped for 15 min in 0.1 % (w/v) potassium sorbate and 0.1 % (w/v) sodium benzoate solution. After drainage, tomatoes were washed with sterile deionized water and air-dried in a laminar flow hood at room temperature for 1 h, followed by storage in the dark at 10 °C for 12 h.

Fruit were separated into three groups, and each group consisted of three lots. For each determination, two replicates (r) per treatment and sampling period were used. The first group consisted of tomatoes stored without ethylene sorbent material (control), the second group of fruit stored with 10 g natural zeolite (Z) per container, and the third group comprised tomatoes with 10 g zeolite doped with copper and zinc (Z–Cu/Zn) per container. Tomatoes (approximately 0.75 kg per experimental unit) were stored in hermetically sealed glass desiccators (10 L) at 20 °C in the dark and a constant relative humidity (RH) of 88 % provided by a saturated solution of sodium benzoate (C_6H_5OONa). Zeolite samples were deposited on glass fiber within a stainless-steel basket hanging in the center of each desiccator (see Figure 2). Quality fruit parameters (size, weight, skin color, texture characteristics, titratable acidity, soluble solids, moisture content, and reducing sugars and lycopene content) were analyzed at the beginning of the experiment, and after 8 d and 15 d in accordance with tomatoes shelf-life and preliminary experiments.

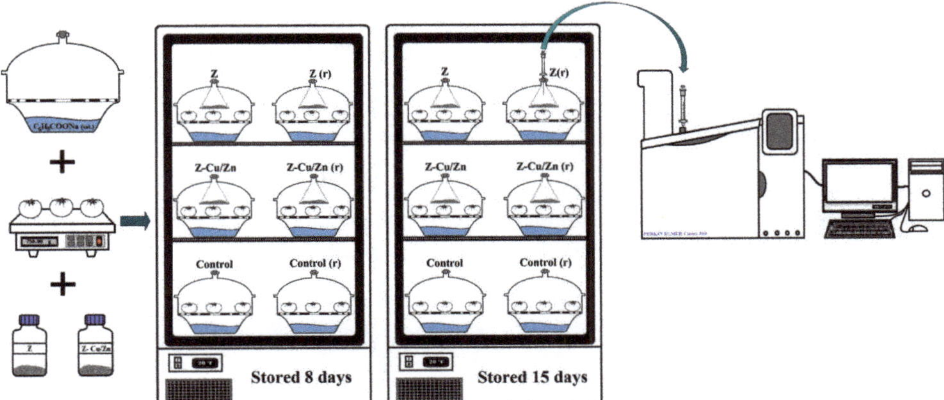

Figure 2. Schematic experimental setup for tomato storage over time for different treatments.

2.5. Gas Composition in the Atmosphere of Tomato Containers

Concentrations of C_2H_4 and CO_2, O_2, and N_2 were determined during the whole storage time of tomato fruit. Samples were daily taken from the headspace atmospheres of each desiccator and analyzed by gas chromatography using a Perkin Elmer Clarus 500 chromatograph (Shelton, CT, USA) equipped with a sample injector operated at 200 °C and featuring two parallel connected columns, a flame ionization detector (FID) operating at 250 °C for C_2H_4 detection and a thermal conductivity detector (TCD) operating at 200 °C for CO_2, O_2, and N_2 determination. A sample of 2.5 mL of the headspace atmosphere from each desiccator was withdrawn using a syringe and injected into the gas chromatograph. A flow of 7 mL min^{-1} of helium was used as a carrier gas for ethylene determination, using a capillary column (VOCOL™, 60 m length × 0.53 mm i.d., with 3.0 μm film thickness) coupled to the FID detector. Separations started at an initial oven temperature of 40 °C; then temperature increased to 170 °C at 2 °C/min^{-1} and held at 170 °C for 5 min. After each separation, elution proceeded for further 2 min before cooling down the oven and to stabilize the column for the next run. Carbon dioxide, oxygen and nitrogen were analyzed by injecting 1 mL of the headspace atmosphere of each desiccator into the gas chromatograph. Separation was conducted using a packed column (4.5 m length × 2.1 mm i.d., containing 60/80 mesh Carboxen-1000 packing) coupled to TCD, applying 30 mL helium min^{-1}. The temperature program was started at 40 °C and increased to 170 °C at 4 °C/min^{-1} with a holding time of 6 min.

2.6. Determination of Physical and Chemical Properties of Tomato Fruit

Equatorial and polar diameters of individual fruit were measured with a digital vernier caliper (150 × 0.02 mm; Stanford professional). Weight of each fruit lot was recorded using an analytical balance (Shimadzu BL-320H, Kyoto, Japan) with 0.001 g precision. Weight loss was expressed as percentage loss of the original weight using the same samples (n = 3) at each sampling time throughout the experiment [25]. Skin color was measured in triplicate at three different points in the equatorial area of fruit using the CIE L^* a^* b^* color space of a HunterLab colorimeter Color Quest II (Hunter Associates Laboratory, Reston, VA, USA). Hue angle (h_{ab}) was quantified by h_{ab} = arc tangent (b^*/a^*), where green = 180°, yellow = 90° and red = 0°. Texture characteristics of tomato samples were measured using the Instron Universal Testing Machine (ID 4467 H 1998, Instron Co., Norwood, MA, USA) by force-deformation measurements, using the compression test for food materials of convex shape [26]. Stylar, equatorial and calyx regions were marked on the fruit. The compression test on each fruit involved six measurements, that is, two determinations in each region using a cylindrical stainless-steel plunger (3.21 mm) and a crosshead speed of 20 mm min^{-1} [27]. Bioyield or critical force (F_{crit}) and apparent modulus of elasticity (E) were calculated according to ASAE S368.2 standards [26], using a Poisson's ratio of 0.45 [28].

Then, each fruit was liquefied in a commercial Oster 4172-051 blender (México City, México), and titratable acidity, soluble solids, and moisture contents were determined. Moisture content of tomato pulp samples was determined by gravimetric method. About 20 g of samples were dried on Petri dishes in a Binder BL-320H oven (Tuttlingen, Germany) at 70 °C until constant weight. Moisture content was expressed as percentage of loss of initial weight (WL) (wet basis). Soluble solids (SS) content was measured with a BOECO 32,195 digital refractometer (Hamburg, Germany), expressed as percentage of sucrose equivalent. Titratable acidity (TA) was determined by titration with 0.1 N NaOH and expressed as percentage of citric acid equivalent [29]. Reducing sugars (RS) were determined by the Layne and Eynon method based on the reduction of Cu^{2+}-complex with tartaric acid in alkaline solutions [30]. The parameter was expressed as percentage of invert sugar weight/fresh sample weight. The maturity index (MI) was expressed as the ratio between reducing sugars and titratable acidity. Lycopene content was determined according to the method proposed by Fish et al. [31] with minor modifications. Approximately 0.6 g of homogenized tomato sample was introduced into a 50 mL amber screw-top vial containing 5 mL acetone with 0.05 % (w/v) butylated hydroxytoluene, 5 mL 95 % (v/v) ethanol and 10 mL hexane. All solvents were refrigerated previously at 4 °C and stored on ice before use. Sample was stirred at 1500 rpm for 10 min using a VM-3000 mini-vortexer (Thorofare, NJ, USA). After shaking, 3 mL of distilled water were added to the vial and sample was shaken for additional 5 min. Then, vial was left at 4 °C in the dark for 30 min to allow for phase separation. Absorbance of the hexane layer was measured at 503 nm versus a blank of hexane using a Merck UV-visible spectrophotometer Pharo 300 Spectroquant®(Darmstadt, Germany) and results were expressed as mg lycopene per kg dry weight of tomato. All measurements were assayed in triplicate.

2.7. Statistical Analysis

Analysis of variance (ANOVA) at P = 0.05 was performed to assess statistically significant differences among treatments for tomato quality attributes by applying Duncan's multiple range test using Statgraphics © Centurion XVI, version 16.1.15 provided by Statpoint Technologies (Warrington, VA, USA).

3. Results and Discussion

3.1. Ethylene Concentration in the Atmosphere of Tomatoes Containers

Ethylene concentration diminished by 37% for natural zeolite and 57% for modified zeolite after 8 d of storage, followed by an increase of ethylene concentrations reaching a plateau value of about 40 μg L^{-1} after 10 d for all treatments, and apparently associated to the saturation of active sorption sites (Figure 3). A comparison of the ethylene removal efficiency using natural and modified natural zeolite doped with copper and zinc to some other ethylene scrubbers reported in the literature is presented in Table 2. As can be seen, the obtained result using modified natural zeolite doped with copper and zinc is within the range of values reported using different scrubbers for ethylene removal for different horticultural products. Control treatment showed just one pronounced increase of ethylene concentration at 2.5 d of tomato storage, while a second increase appeared at 9.5 d for zeolite treatments (Figure 3). A maximum ethylene concentration for control was obtained of about 43.8 ± 8.2 μg L^{-1} per kg of tomato. In particular, the ion-exchange of zeolite with copper and zinc cations increased the capacity to remove ethylene from an environment with 88% RH and 15% (v/v) O_2 during the first week of tomato storage. More efficient ethylene scrubbing promotes ethylene gradient formation and, thus, diffusion across the tomato skin and flesh barriers, diminishing the internal ethylene concentration and retarding the main ethylene burst for about seven days. However, autocatalytic ethylene synthesis may result finally in a sharp increase of the internal C_2H_4 concentration that can exceed 100 ppm [2]. This may be the consequence of unsteady state conditions between an accelerated ethylene production and a relatively low ethylene adsorption rate due to mass transfer limitations.

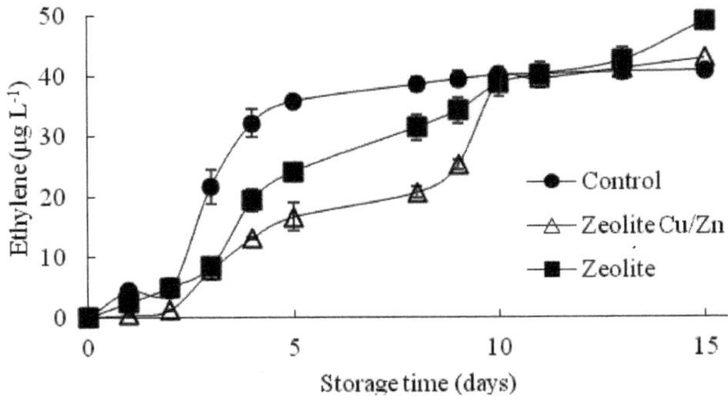

Figure 3. Ethylene concentration for tomato storage over time for different treatments.

Table 2. Comparison of the removal of ethylene using natural (Z) and modified natural zeolite doped with copper and zinc (Z–Cu/Zn) to some other ethylene scrubbers reported in the literature.

Ethylene scrubbers	Removal (%)	Horticultural Products	References
Z	37	Tomato	This work
Z–Cu/Zn	57	Tomato	This work
Potassium permanganate loaded on protonated montmorillonite	79	Blueberry	[11]
Palladium activated carbon	65	Tomato	[13]
Pd-impregnated zeolite	100	Banana, avocado, strawberry	[17]
Acidified activated carbon powder impregnated with $PdCl_2$ and $CuSO_4$	80	Broccoli	[18]
Nanofibers containing TiO_2 nanoparticles	45	Banana	[20]
Titanium dioxide coated glass	66	Apple	[21]
Granular—activated carbon impregnated with palladium	56	Tomato	[32]

Although C_2H_4-scavenging was not very successful at the end of the storage period, probably due to an insufficient adsorption capacity of zeolites at high relative humidity, ethylene removal might be improved by increasing the amount of copper and zinc combined with ultraviolet irradiation, using the modified zeolites as photocatalytic material to promote ethylene degradation and to free the active adsorption sites on the zeolite surface.

3.2. Interaction between Ethylene and Zeolites

Figure 4 displays the evolution of DRIFTS spectra as a function of time for samples of natural zeolite (4a) and modified natural zeolite doped with copper and zinc (4b) continuously exposed to a stream of pure moistened ethylene. According to Figure 4A,B, the intensity in OH vibrating bands such as Si-OH-Al bridge groups (3590 cm^{-1}) and external Si-OH groups (3745 cm^{-1}) decreased as zeolites were exposed to ethylene. In the case of modified zeolite (Figure 4B), the interactions with Si-OH-Al groups (3590 cm^{-1}) are greater than in natural zeolite (Figure 4A). Such findings are related to the formation of new Brønsted active sites during the modification process that increases ethylene adsorption. Our results confirm that the adsorption mechanism takes place by the interaction of ethylene molecules with Brønsted active sites present at the zeolite surface. Moreover, DRIFTS analysis to the modified zeolite sample suggests that ethylene molecules interact not only with Brønsted active sites but also with metal cations incorporated in the zeolite surface. Lower intensity can be clearly observed in the bands located at 3590 cm^{-1} (Si-OH-Al bridge groups) in the modified zeolite sample as the contact time elapsed. Such results are in agreement with those presented in Figure 3, since ethylene concentration remains lower in the gas headspace of stored tomatoes when modified zeolite is used. Ethylene molecules seem to be adsorbed at the new active sites generated in the zeolite after the applied modification treatments. The IR bands registered at 2835 cm^{-1}, 2807 cm^{-1}, and 2820 cm^{-1} related to C–H stretching vibrations in that sample have been associated to the interaction of metals introduced in the zeolite surface with ethylene molecules. These IR bands could be attributed to modifications in the symmetry of ethylene molecules due to the interactions of cations with π-electrons of ethylene molecules. Meanwhile, after some minutes of contact time, other characteristic bands of gas phase ethylene appear in the range around 950 cm^{-1}, 1420 cm^{-1}, 1700–1900 cm^{-1}, and 2970–3125 cm^{-1}, suggesting the saturation of the surface. Additionally, water adsorption on the zeolite surface can be observed by the increase of the band located at 1635 cm^{-1} and a broad band between 3200 cm^{-1} and 3400 cm^{-1} as contact time increased, having higher intensities in the case of the natural zeolite sample.

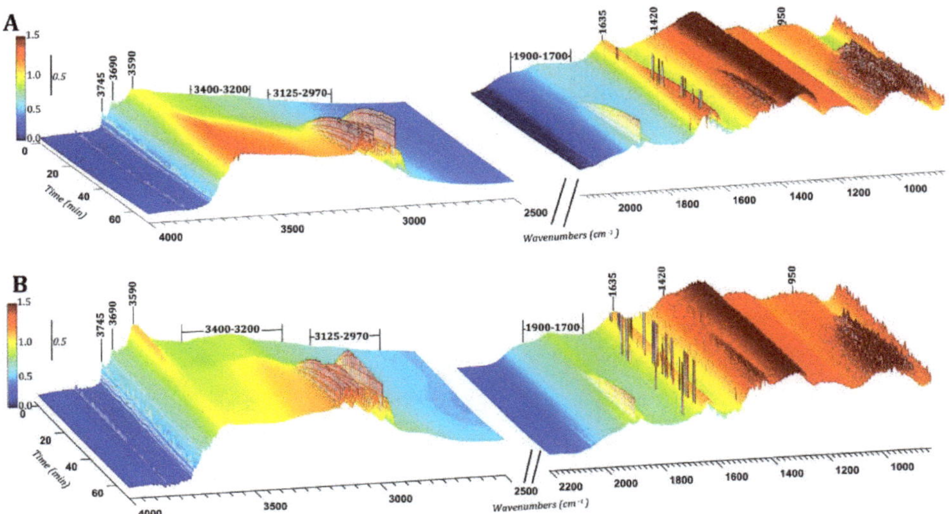

Figure 4. DRIFTS *operando* spectra of ethylene adsorption onto natural and modified natural zeolite samples in the presence of humidity. (**A**) Natural zeolite; (**B**) zeolite modified with copper and zinc.

However, high moisture conditions might also affect negatively the adsorption capacity of the modified zeolite, particularly at the end of tomato storage. It has been indicated that both water and oxygen reduced the adsorption capacity of zeolite, while the greatest removal of C_2H_4 occurs in the presence of nitrogen [14]. In the absence of water and oxygen, ethylene is converted to ethylidyne (CCH_3) species, whereby carbon forms three bonds to surface metal atoms. In contrast, the presence of adsorbed oxygen favored π-bonded ethylene over di-σ bonded ethylene adsorption [14]. The adsorption process in the case of natural zeolite has been indicated to take place mainly by weak hydrogen bonding interactions among CH atoms of ethylene and O atoms of terminal Si-OH and Si-OH-Al bridges of zeolite [23,33]. In the presence of moisture, water and ethylene molecules compete for the same active adsorption sites of zeolites [23] as the CH-O interactions for ethylene are weaker than the OH–O interactions between water and zeolite [34]. As a result, water gets involved in the adsorption mechanism blocking ethylene interactions and diminishing considerably the ethylene adsorption at the final stage of tomato storage.

3.3. Breakthrough and Zeolite Regeneration

According to Figure 5a, ethylene was retained until 75×10^3 bed volume (BV) by modified zeolite with a saturation of sorption material after 140×10^3 BV and a higher adsorbed amount of ethylene on modified zeolite (4683 $\mu g\ g^{-1}$) compared to natural zeolite (4263 $\mu g\ g^{-1}$). Thus, the incorporation of copper and zinc cations in zeolite favored ethylene adsorption. At the same time, the ion-exchange of natural zeolite with copper and zinc cations diminished the retention of carbon dioxide, decreasing the adsorbed amount of CO_2 from 52.5 mg g^{-1} for natural zeolite to 31.7 mg g^{-1} for modified zeolite (Figure 5b). As shown in Table 1, compensating cations of basic nature originally present at natural zeolite surface were removed. These results indicate that the applied sequential modification treatment produced a modified zeolite that is less susceptible to CO_2 adsorption.

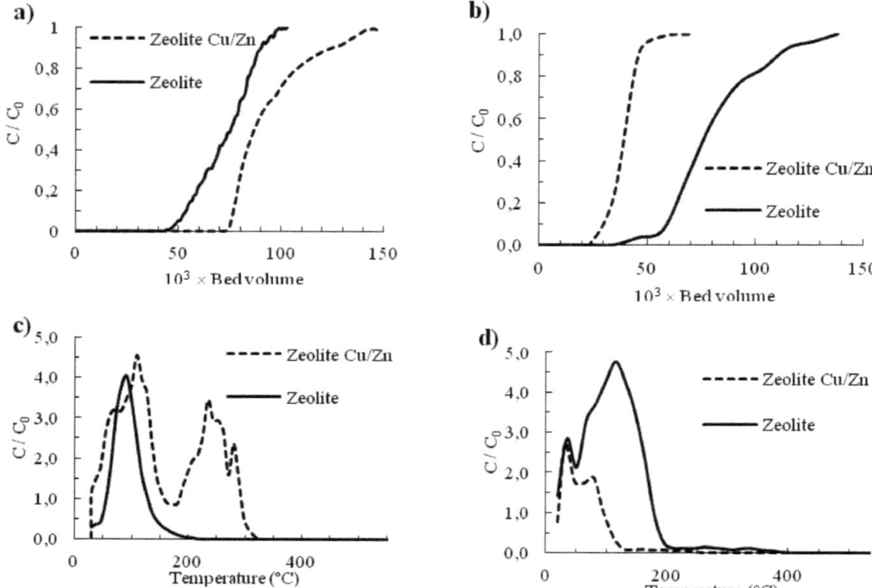

Figure 5. Experimental breakthrough curves for ethylene (**a**) and carbon dioxide (**b**) and desorption curves for ethylene (**c**) and carbon dioxide (**d**) on natural zeolite and modified zeolite with copper and zinc.

On the other hand, desorption of ethylene from natural zeolite took place at relatively low temperature (Figure 5c), which indicates weak interactions between adsorbate and adsorbent, probably due to hydrogen bonds among ethylene molecules and OH groups of the zeolite framework. In the case of modified zeolite (Figure 5c), ethylene evolved as two main desorption peaks that can be related to different adsorption strength among ethylene molecules and active sites of the zeolite surface. The thermal desorption of ethylene detected at low temperature (30–150 °C) means weak interactions between adsorbate and adsorbent, while the desorption of ethylene at high temperature (170–300 °C) suggests stronger chemical interactions that probably involve the copper and zinc compensation cations incorporated in the zeolite surface. According to the desorption profile of carbon dioxide, natural zeolite had a higher adsorption capacity toward CO_2 compared to modified zeolite (Figure 5d). This behavior could be related to strong electrostatic interactions of CO_2 quadrupoles with vacant zeolitic cations of basic character present at the surface of natural zeolite [35]. Moreover, these desorption data suggest a higher amount of basic surface sites in the natural zeolite compared to the modified sample [36]. A great part of these basic sites were initially present in the natural zeolite as compensating cations and were removed during the first step of the applied modification process using ammonium sulfate followed by thermal degasification at 350 °C, having an impact in the increase of microporous surface area and microporous volume [23].

Moreover, the incorporation of Cu^{2+} and Zn^{2+} ions after the ion-exchange treatments into the zeolite structure also contributes to the decrease in the content of compensating cations (Table 1). This generates additional, ion-induced dipole interactions between the ethylene molecules and the new compensating cations located on the zeolite surface. It also improves ethylene binding via the interaction among the π-electrons of ethylene and the orbitals of compensating cations within the zeolite framework [34].

3.4. Carbon Dioxide Concentration in the Atmosphere of Tomatoes Containers

During the experiments, CO_2 levels in the atmosphere of glass desiccators increased continuously during the storage of control tomatoes with a more pronounced increase of CO_2 at the beginning, yielding a maximum CO_2 concentration of about 300 mg L^{-1} at day 10.5 (Figure 6). However, the CO_2 profile is more complex in the presence of zeolites because of the combination of tomato respiration, and CO_2 adsorption and desorption phenomena. First, a strong increase of CO_2 concentrations was observed at day 0.5, which may be due to the accommodation of our system, whereby water vapor in the headspace would promote CO_2 desorption. The zeolites used here having an atomic ratio of silica to alumina of roughly five, are substantially hydrophilic with a water adsorption capacity much higher than the one for CO_2 [37]. In addition, carbon dioxide production has been delayed during the first six days in the presence of modified zeolite. This adsorbent is able to shift the CO_2 increase for six days compared to control treatment due to the presence of copper and zinc (Figure 6). However, carbon dioxide concentrations increased strongly to approximately 585 mg L^{-1} after 14 d of storage at 20 °C, which exceeds the recommended range of CO_2 for tomatoes of 2%–5% (v/v). High levels of CO_2 reduce the sensitivity to ethylene and inhibit ethylene action in tomatoes [6].

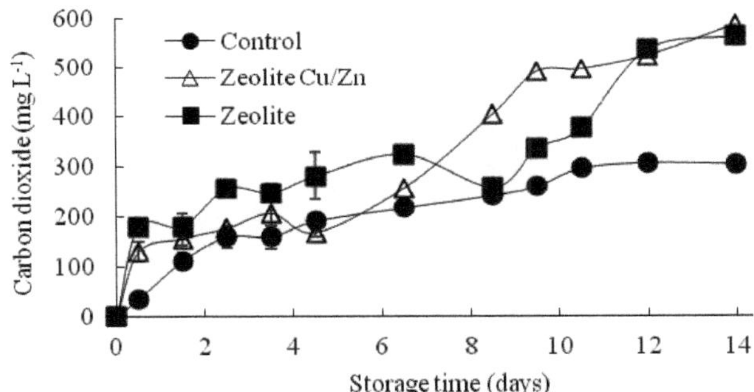

Figure 6. Carbon dioxide concentration as a function of storage time for different treatments.

3.5. Effect of Zeolites on Tomato Quality

Excessive loss of moisture in fresh fruit results in a considerable weight reduction with a negative impact on fruit quality, as it is generally associated with loss of freshness. In this study, no significant effect of sorbents on weight loss of postharvest tomatoes was found although CO_2 production was retarded for 6 d of tomato storage in the presence of modified zeolite (Table 3).

Table 3. Physicochemical and ripening indices of tomatoes stored at 20 °C without (control), in the presence of zeolite (Z) or modified zeolite (Z–Cu/Zn) [1].

	WL [%]		TA [%]		SS [%]		RS [%]		MI		Lycopene [mg kg^{-1}]	
Initial	-		0.47 ± 0.05		3.01 ± 0.42		2.85 ± 0.12		6.27 ± 0.70		9.22 ± 1.88	
Days	8	15	8	15	8	15	8	15	8	15	8	15
C	1.22	2.25	0.35	0.36b	2.78	2.77b	2.81	2.75	7.99	6.69	20.4a	52.9ab
Z	1.40	2.56	0.38	0.34a	2.37	2.02a	2.80	2.80	7.44	7.25	32.1b	62.5b
Z–Cu/Zn	1.40	2.36	0.39	0.34a	2.52	2.08a	2.82	2.79	7.33	7.31	21.3a	49.5a

[1] Values are the mean ± standard error for the replicates. Different letters for the same column indicate significant differences ($P < 0.05$).

Our results indicate that weight loss is mainly a consequence of mass transfer of water vapor from fruit into the headspace of glass desiccators without a significant influence of respiration or zeolite materials. On the other hand, Bailén et al. [32] and Domínguez et al. [38] reported lower weight losses for tomatoes preserved under modified atmosphere packaging using ethylene sorbents in comparison with our results.

Approximately 45% of the total solids in tomato fruit is composed of soluble solids, predominantly reducing sugars (Table 3). Zeolites did not alter the evolution of reducing sugars during tomato storage. A previous study has shown that the main free sugars of commercial tomato varieties are reducing sugars with a negligible amount of sucrose [39]. Moreover, both titratable acidity and soluble solids content are essential quality parameters for tomatoes showing a similar behavior with a significant decrease during the second half of postharvest storage in case of natural and modified zeolites (Table 3). Increased CO_2 and C_2H_4 production, after one week for tomatoes stored in the presence of both zeolites, may trigger the decomposition of organic acids and part of the soluble solids. In particular, citrate and isocitrate concentrations diminished at later stages of fruit ripening [40]. In addition, stimulation of ethylene production by bruising of tomato fruit yielded a significant decrease of 15.3% of titratable acidity in locule tissue compared with non-impacted tissue [41]. According to the initial maturity index of 6.27, the quality of raw material belongs to suitable red fruit genotype class [39] with a slight increase due to the loss of titratable acidity during ripening (Table 3).

Fresh-market tomatoes commonly harvested at mature green or breaker stage become soft as ripening proceeds during storage. For this kind of tomato, texture is one of the most important quality attributes, influenced by both skin strength and flesh firmness. Skin strength depends on the force at the bioyield point (F_{crit}), required to punch through the pericarp of tomatoes. Skin strength drops rapidly during the first stage of storage followed by a slower decrease without significant difference between treatments (Table 4). Flesh firmness or mesocarp strength can be calculated from the apparent modulus of elasticity and the Poisson's ratio [42]. Natural zeolite significantly reduced apparent modulus of elasticity at the end of storage (Table 4), which could be attributed to an increased ethylene concentration of about 20% compared to control tomatoes. Ethylene is able to stimulate the activity of cell wall degrading enzymes by the induction of ethylene responsive genes encoding for pectin methyl esterase, cellulase and polygalacturonase [43]. Softening of fruit flesh is mainly attributed to the degradation of protopectins, which may result in reduced integrity of cell wall components, loss of membrane integrity, and cell turgor. This, in turn, affects shelf-life and consumers' acceptance of tomatoes.

Table 4. Texture and color evolution of tomatoes stored at 20 °C without (control), in the presence of zeolite (Z) or modified zeolite (Z–Cu/Zn) [1].

	F_{crit} [N]		E [MPa]		L*		a*		b*		h_{ab} [°]	
Initial	12.97 ± 2.01		1.31 ± 0.26		54.5 ± 1.8		−5.2 ± 1.5		28.2 ± 1.0		100.4 ± 1.9	
Days	8	15	8	15	8	15	8	15	8	15	8	15
C	7.05	6.98	1.55	1.38b	46.9	42.0	8.3a	19.4a	24.2	21.4a	70.5b	48.1
Z	7.25	7.36	1.19	0.89a	45.4	43.0	21.5b	26.1b	23.2	23.9b	47.3a	42.6
Z–Cu/Zn	7.70	6.75	1.40	1.29b	47.3	44.1	7.8a	20.8a	25.1	24.1b	72.3b	49.1

[1] Values are the mean ± standard error for the replicates. Different letters for the same column indicate significant differences ($P < 0.05$).

Color is the most important external attribute to assess tomato ripeness and shelf-life, being the major factor for consumers' purchase decision. In this study, ripening is accompanied by a loss of green orange color (0 d) associated with the breaker stage and the development of pink (8 d) and light red color characteristics (15 d) for modified zeolite and control tomatoes. Moreover, tomatoes treated with natural zeolite evolve an accelerated color change expressed by the development of light red (8 d) and red color characteristics (15 d) according to USDA color standards [44]. In addition, the greatest effect on delaying color development in tomatoes was found for zeolite that incorporates copper and zinc cations, except for the color parameter a* after 15 d (Table 4). Red color development is the result of chlorophyll degradation, as well as the synthesis of carotenoids, as chloroplasts are converted into chromoplasts [40]. The degree of greenness or redness (a*) is changing from negative (green) to positive (red) values during tomato storage. In particular, red color evolution is promoted by natural zeolite with strong increase of a* values during the first eight days in contrast to modified zeolite and control treatments (Table 4). The degree of yellowness (b*) decreased just a little during fruit ripening with significantly higher values for zeolite samples at the final ripening stage (Table 4). This suggests better retention of yellow pigments due to the presence of zeolites. Lightness (L*) gradually declined during postharvest storage, where zeolites did not influence this quality parameter (Table 4). Moreover, hue angle (h_{ab}) correlates well with consumers' perception of fruit color. A hue angle greater than 90° in the second quadrant of the L* a* b* color space represents the initial condition of raw material and indicates a green orange color of tomato surface (Table 4). During ripening, hue decreased rapidly up to 47° for zeolite treated tomatoes corresponding to light red color, while hue of the other tomato samples was in the 70–72° range after 8 d of storage, which corresponds to yellowish-pink color (Table 4). Natural zeolite promotes red color evolution of tomatoes, which coincides with an increase of 100% in the amount of CO_2 produced after 8 d of storage in comparison to control. The inclusion of copper and zinc cations within the zeolite framework is favorable to avoid excessive red color development of tomatoes treated by natural zeolite.

Health benefits were associated to the consumption of a diet rich in lycopene, a fat-soluble carotenoid with excellent antioxidant properties [45]. The accumulation of lycopene occurs due to the conversion of chloroplasts into chromoplasts, coupled to the synthesis of this red pigment from 2 geranyl-geranyl pyrophosphate via phytoene, phytofluene, ξ-carotene, and prolycopene intermediaries [40]. According to Sipos et al. [45], lycopene content and antioxidant capacity of tomatoes correlated positively to both surface and puree colors. This suggests that lycopene content depends on tomato ripeness. Our results indicate that lycopene content of tomatoes increased progressively over time ranging between 9.2 mg kg^{-1} and 62.5 mg kg^{-1} (Table 4). Moreover, lycopene content had a strong inverse correlation with hue having a Pearson's coefficient of –0.887 ($P < 0.0001$), while lycopene synthesis was also affected by ethylene and CO_2 levels inside the gas atmosphere of the glass desiccators with Pearson's correlation coefficients of 0.787 ($P = 0.063$) and 0.782 ($P = 0.066$), respectively. In addition, natural zeolite yields a significantly higher increase of lycopene synthesis compared to tomatoes stored in the presence of modified zeolite (Table 4).

Finally, improved ripening of tomatoes in the presence of natural zeolite is unfavorable if our goal is to increase postharvest shelf-life, while the incorporation of copper and zinc cations to the support lattice slows down fruit ripening being therefore an emergent postharvest technology that may create new commercial opportunities.

4. Conclusions

Natural zeolite composed of 53% clinoptilolite, 40% mordenite, and 7% quartz promoted tomato ripening according to the quality parameters titratable acidity, soluble solids, apparent modulus of elasticity and greenness or redness degree after 15 d of tomato storage. The introduction by ion-exchange of copper and zinc cations in zeolite followed by calcination treatments resulted in a modified zeolite that was able to remove 57% of ethylene after 8 d of storage and to shift the CO_2 production for 6 d compared to control. Chromatic characteristics, such as hue and greenness or redness degree, and lycopene content improved after 8 d and apparent modulus of elasticity and lycopene content after 15 d of storage. DRIFTS *operando* studies showed that the adsorption mechanism takes mainly place by the interaction of ethylene molecules with Brønsted active sites. However, DRIFTS results also indicated that water molecules compete with ethylene for the active sorption sites, decreasing the adsorption capacity towards ethylene. Doping the zeolite with copper and zinc cations seems to favor ethylene adsorption mechanism by cation-π interactions, improving postharvest quality of tomatoes. Natural zeolite modified with copper and zinc cations may be used as a low cost and non-toxic alternative ethylene scavenger to commercial ones available in the market. Finally, the use of ultraviolet light could be explored to degrade bound ethylene and make free the active sorption sites of modified zeolite.

Author Contributions: Conceptualization, H.V. and J.d.B.; methodology, V.S., C.L., and P.M.; validation, H.V., J.d.B., and P.M.; formal analysis, F.A.-J. and N.A.; investigation, A.G., C.L., N.A., and F.A.-J.; data curation, V.S. and F.A.-J.; writing—original draft preparation, J.d.B.; writing—review and editing, H.V. and P.M.; supervision, H.V.; project administration, H.V.; funding acquisition, H.V. All authors have read and agreed to the published version of the manuscript.

Funding: This research was funded by *Comisión Nacional de Investigación Científica y Tecnológica* of Chile CONICYT, FONDECYT Regular grant number 1170694. The mobility of F.A. was funded by *Fondo Especial de Actividades Académicas* UCSC 2019-II.

Acknowledgments: The authors would like to acknowledge Jaime Fuentealba for providing the tomatoes.

Conflicts of Interest: The authors declare no conflict of interest. The funders had no role in the design of the study; in the collection, analyses, or interpretation of data; in the writing of the manuscript, or in the decision to publish the results.

References

1. FAO. Food and Agricultural Data. Available online: http://www.fao.org/faostat/en/ (accessed on 2 November 2019).
2. Saltveit, M.E. Effect of ethylene on quality of fresh fruits and vegetables. *Postharvest Biol. Technol.* **1999**, *15*, 279–299. [CrossRef]
3. Pathak, N.; Caleb, O.J.; Geyer, M.; Herppich, W.B.; Rauh, C.; Mahajan, P.V. Photocatalytic and photochemical oxidation of ethylene: Potential for storage of fresh produce—A review. *Food Bioprocess Technol.* **2017**, *10*, 982–1001. [CrossRef]
4. Zhang, J.; Cheng, D.; Wang, B.; Khan, I.; Ni, Y. Ethylene control technologies in extending postharvest shelf life of climateric fruit. *J. Agric. Food Chem.* **2017**, *65*, 7308–7319. [CrossRef]
5. Álvarez-Hernández, M.H.; Artés-Hernández, F.; Ávalos-Belmontes, F.; Castillo-Campohermoso, M.A.; Contreras-Esquivel, J.C.; Ventura-Sobrevilla, J.M.; Martínez-Hernández, G.B. Current scenario of adsorbent materials used in ethylene scavenging systems to extend fruit and vegetable postharvest life. *Food Bioprocess Technol.* **2018**, *11*, 511–525. [CrossRef]
6. Saltveit, M.E. Is it possible to find an optimal controlled atmosphere? *Postharvest Biol. Technol.* **2003**, *27*, 3–13. [CrossRef]
7. Vermeiren, L.; Devlieghere, F.; Van Beest, M.; De Kruijf, N.; Debevere, J. Developments in the active packaging of foods. *Trends Food Sci. Technol.* **1999**, *10*, 77–86. [CrossRef]
8. Attar, K.; Demey, H.; Bouazza, D.; Sastre, A.M. Sorption and desorption studies of Pb(II) and Ni(II) from aqueous solutions by a new composite based on alginate and magadiite materials. *Polymers* **2019**, *11*, 340. [CrossRef] [PubMed]
9. Tas, C.E.; Hendessi, S.; Baysal, M.; Unal, S.; Cebeci, F.C.; Menceloglu, Y.Z.; Unal, H. Halloysite nanotubes/polyethylene nanocomposites for active food packaging materials with ethylene scavenging and gas barrier properties. *Food Bioprocess Technol.* **2017**, *10*, 789–798. [CrossRef]
10. García-García, I.; Taboada-Rodríguez, A.; López-Gomez, A.; Marín-Iniesta, F. Active packaging of cardboard to extend the shelf life of tomatoes. *Food Bioprocess Technol.* **2013**, *6*, 754–761. [CrossRef]
11. Álvarez-Hernández, M.H.; Martínez-Hernández, G.B.; Ávalos-Belmontes, F.; Rodríguez-Hernández, A.M.; Castillo-Campohermoso, M.A.; Artés-Hernández, F. An innovative ethylene scrubber made of potassium permanganate loaded on a protonated montmorillonite: A case study on blueberries. *Food Bioprocess Technol.* **2019**, *12*, 524–538. [CrossRef]
12. Spricigo, P.C.; Foschini, M.M.; Ribiero, C.; Corrêa, D.S.; Ferreira, M.D. Nanoscaled platforms based on SiO_2 and Al_2O_3 impregnated with potassium permanganate use color changes to indicate ethylene removal. *Food Bioprocess Technol.* **2017**, *10*, 1622–1630. [CrossRef]
13. Martínez-Romero, D.; Bailén, G.; Serrano, M.; Guillén, F.; Valverde, J.M.; Zapata, P.; Casillo, S.; Valero, D. Tools to maintain postharvest fruit and vegetable quality through the inhibition of ethylene action: A review. *Crit. Rev. Food Sci. Nutr.* **2007**, *47*, 543–560. [CrossRef] [PubMed]
14. Smith, A.W.J.; Poulston, S.; Rowsell, L.; Terry, L.A.; Anderson, J.A. A new-palladium-based ethylene scavenger to control ethylene-induced ripening of climacteric fruit. *Platin. Met. Rev.* **2009**, *53*, 112–122. [CrossRef]

15. Álvarez-Hernández, M.H.; Martínez-Hernández, G.B.; Avalos-Belmontes, F.; Castillo-Campohermoso, M.A.; Contreras-Esquivel, J.C.; Artés-Hernández, F. Potassium permanganate-based ethylene scavengers for fresh horticultural produce as an active packaging. *Food Eng. Rev.* **2019**, *11*, 159–183. [CrossRef]
16. Keller, N.; Ducamp, M.N.; Robert, D.; Keller, V. Ethylene removal and fresh product storage: A challenge at the frontiers of chemistry. Toward an approach by photocatalytic oxidation. *Chem. Rev.* **2013**, *113*, 5029–5070. [CrossRef]
17. Terry, L.A.; Ilkenhans, T.; Poulston, S.; Rowsell, L.; Smith, A.W.J. Development of new palladium-promoted ethylene scavenger. *Postharvest Biol. Technol.* **2007**, *45*, 214–220. [CrossRef]
18. Cao, J.; Li, X.; Wu, K.; Jiang, W.; Qu, G. Preparation of a novel PdCl2-CuSO4-based ethylene scavenger supported by acidified activated carbon powder and its effects on quality and ethylene metabolism of broccoli during shelf-life. *Postharvest Biol. Technol.* **2015**, *99*, 50–57. [CrossRef]
19. De Chiara, M.L.V.; Pal, S.; Licciulli, A.; Amodio, M.L.; Colelli, G. Photocatalytic degradation of ethylene on mesoporous TiO_2/SiO_2 nanocomposites: Effects on the ripening of mature green tomatoes. *Biosyst. Eng.* **2015**, *132*, 61–70. [CrossRef]
20. Zhu, Z.; Zhang, Y.; Shang, Y.; Wen, Y. Electrospun nanofibers containing TiO_2 for the photocatalytic degradation of ethylene and delaying postharvest ripening of bananas. *Food Bioprocess Technol.* **2019**, *12*, 281–287. [CrossRef]
21. Pathak, N.; Caleb, O.J.; Rauh, C.; Mahajan, P.V. Efficacy of photocatalysis and photolysis systems for the removal of ethylene under different storage conditions. *Postharvest Biol. Technol.* **2019**, *147*, 68–77. [CrossRef]
22. Nájera, C.; Guil-Guerrero, J.L.; Enríquez, L.J.; Álvaro, J.E.; Urrestarazu, M. LED-enhanced dietary and organoleptic qualities in postharvest tomato fruit. *Postharvest Biol. Technol.* **2018**, *145*, 151–156. [CrossRef]
23. Abreu, N.J.; Valdés, H.; Zaror, C.A.; Azzolina-Jury, F.; Meléndrez, M.F. Ethylene adsorption onto natural and transition metal modified Chilean zeolite: An operando DRIFTS approach. *Microporous Mesoporous Mater.* **2019**, *274*, 138–148. [CrossRef]
24. Valdés, H.; Solar, V.A.; Cabrera, E.H.; Veloso, A.F.; Zaror, C.A. Control of released volatile organic compounds from industrial facilities using natural and acid-treated mordenites: The role of acidic surface sites on the adsorption mechanism. *Chem. Eng. J.* **2014**, *244*, 117–127. [CrossRef]
25. Zhu, Z.; Chen, Y.; Shi, G.; Zhang, X. Selenium delays tomato fruit ripening by inhibiting ethylene biosynthesis and enhancing the antioxidant defense system. *Food Chem.* **2017**, *219*, 179–184. [CrossRef]
26. ASAE. ASAE S368.2: Compression test of food materials of convex shape. In *ASAE Standards 1991: Standards, Engineering Practices, and Data*; American Society of Agricultural Engineers: Joseph, MI, USA, 1991; pp. 402–405.
27. Adegoroye, A.S.; Jolliffe, P.A.; Tung, M.A. Textural characteristics of tomato fruit (Lycopersicon esculentum) affected by sunscald. *J. Sci. Food Agric.* **1989**, *49*, 95–102. [CrossRef]
28. Li, Z.; Li, P.; Yang, H.; Liu, J. Internal mechanical damage prediction in tomato compression using multiscale finite element models. *J. Food Eng.* **2013**, *116*, 639–647. [CrossRef]
29. AOAC. *Official Methods of Analysis*; Association of Official Analytical Chemists: Washington, DC, USA, 1990; p. 1141.
30. Schneider, F. *Sugar Analysis: Official and Tentative Methods Recommended by the International Commission for Uniform Methods of Sugar Analysis (ICUMSA)*; The International Commission for Uniform Methods of Sugar Analysis: Peterborough, UK, 1979; pp. 45–55.
31. Fish, W.W.; Perkins-Veazie, P.; Collins, J.K. A quantitative assay for lycopene that utilizes reduced volumes of organic solvents. *J. Food Compos. Anal.* **2002**, *15*, 309–317. [CrossRef]
32. Bailén, G.; Guillén, F.; Castillo, S.; Serrano, M.; Valero, D.; Martínez-Romero, D. Use of activated carbon inside modified atmosphere packages to maintain tomato fruit quality during cold storage. *J. Agric. Food Chem.* **2006**, *54*, 2229–2235. [CrossRef]
33. Patdhanagul, N.; Seithanratana, T.; Rangsriwatananon, K. Ethylene adsorption on cationic surfactant modified zeolite NaY. *Microporous Mesoporous Mater.* **2010**, *131*, 97–102. [CrossRef]
34. Sue-aok, N.; Srithanratana, T.; Rangsriwatananon, K.; Hengrasmee, S. Study of ethylene adsorption on zeolite NaY modified with group I metal ions. *Appl. Surf. Sci.* **2010**, *256*, 3997–4002. [CrossRef]
35. Salehi, S.; Anbia, M. Characterization of CPs/Ca-exchanged FAU-and LTA-type zeolite nanocomposites and their selectivity for CO_2 and N_2 adsorption. *J. Phys. Chem. Solids* **2017**, *110*, 116–128. [CrossRef]

36. Lavalley, J.C. Infrared spectrometric studies of the surface basicity of metal oxides and zeolites using adsorbed probe molecules. *Catal. Today* **1996**, *27*, 377–401. [CrossRef]
37. Hefti, M.; Marx, D.; Joss, L.; Mazzotti, M. Model-based process design of adsorption processes for CO_2 capture in the presence of moisture. *Energy Procedia* **2014**, *63*, 2152–2159. [CrossRef]
38. Domínguez, I.; Lafuente, M.T.; Hernández-Muñoz, P.; Gavara, R. Influence of modified atmosphere and ethylene levels on quality attributes of fresh tomatoes (Lycopersicon esculentum Mill.). *Food Chem.* **2016**, *209*, 211–219. [CrossRef] [PubMed]
39. Raiola, A.; Pizzolongo, F.; Manzo, N.; Montefusco, I.; Spigno, P.; Romano, R.; Barone, A. A comparative study of the physico-chemical properties affecting the organoleptic quality of fresh and thermally treated yellow tomato ecotype fruit. *Int. J. Food Sci. Technol.* **2018**, *53*, 1219–1226. [CrossRef]
40. Aivalakis, G.; Katinakis, P. Biochemistry and molecular physiology of tomato and pepper fruit ripening. In *The Fruiting Species of the Solanaceae. The European Journal of Plant Science and Biotechnology*; Passam, H., Ed.; Global Science Books: Isleworth, UK, 2008; Volume 2, pp. 145–155.
41. Moretti, C.L.; Sargent, S.A.; Huber, D.J.; Calbo, A.G.; Puschmann, R. Chemical composition and physical properties of pericarp, locule, and placental tissues of tomatoes with internal bruising. *J. Am. Soc. Hortic. Sci.* **1998**, *123*, 656–660. [CrossRef]
42. Figura, L.O.; Teixeira, A.A. *Food Physics. Physical Properties–Measurement and Applications*; Springer: Berlin, Germany, 2007; pp. 121–203.
43. Bu, J.; Yu, Y.; Aisikaer, G.; Ying, T. Postharvest UV-C irradiation inhibits the production of ethylene and the activity of cell wall-degrading enzymes during softening of tomato (Lycopersicon esculentum L.) fruit. *Postharvest Biol. Technol.* **2013**, *86*, 337–345. [CrossRef]
44. Batu, A. Determination of acceptable firmness and colour values of tomatoes. *J. Food Eng.* **2004**, *61*, 471–475. [CrossRef]
45. Sipos, L.; Orbán, C.; Bálint, I.; Csambalik, L.; Divéky-Ertsey, A.; Gere, A. Colour parameters as indicators of lycopene and antioxidant activity traits of cherry tomatoes (Solanum lycopersicum L.). *Eur. Food Res. Technol.* **2017**, *243*, 1533–1543. [CrossRef]

Article

Assessing the Sustainability and Acceptance Rate of Cost-Effective Household Water Treatment Systems in Rural Communities of Makwane Village, South Africa

Resoketswe Charlotte Moropeng * and Maggy Ndombo Benteke Momba

Department of Environmental, Water and Earth Sciences, Tshwane University of Technology, Arcadia Campus, P/B X 680, Pretoria 0001, South Africa; mombamnb@tut.ac.za

* Correspondence: moropengrc@tut.ac.za; Tel.: +27-12-382-6365

Received: 14 July 2020; Accepted: 23 September 2020; Published: 26 September 2020 

Abstract: The current study investigated the acceptance rate and long-term effectiveness of cost-effective household water treatment systems deployed in Makwane Village. A structured questionnaire was used prior to implementation to collect information such as level of education, level of employment, and knowledge about point-of-use water treatment systems in the target area. The long-term effectiveness was determined by factors such as the *Escherichia coli* removal efficiency, turbidity reduction, silver leached, and flow rate of the household water treatment devices. The results of the survey prior to deployment revealed that only 4.3% of the community had a tertiary qualification. Moreover, 54.3% of the community were unemployed. The results further revealed that 65.9% of the community were knowledgeable about other point-of-use water treatment methods. The acceptance rate, which was found to be initially higher (100%), reduced after three months of implantation (biosand filter with zeolite-silver clay granular—82.9%; silver-impregnated porous pot filters—97.1%). Moreover, the long-term effectiveness was determined, taking into consideration the adoption rate, and it was found that silver-impregnated porous pot filters have a long life compared to biosand filter with zeolite-silver clay granular. Although household water treatment systems can effectively reduce the burden of waterborne diseases in impoverished communities, the success of adoption is dependent on the targeted group. This study highlights the significance of involving community members when making the decision to scale up household water treatment devices in rural areas for successful adoption.

Keywords: adoption; acceptance; household water treatment systems; long-term effectiveness

---

1. Introduction

Access to piped water supply through house connections is the ideal solution to counteract water-related illnesses. Nonetheless, with the financial and political challenges faced by most developing countries, coupled with the high capital and maintenance costs of piped supply systems, centralized safe piped water is likely decades away for most developing regions [1]. According to the WHO [2], an estimated 502,000 people die each year due to diarrhea as a result of drinking unclean water. Early childhood death, especially in the poorest rural areas, is ascribed to an inadequate supply of safe drinking water [3–9]. The WHO [10] has highlighted that properly managed water, sanitation, and hygiene (WASH) services are an indispensable part of preventing disease and protecting human health, especially during infectious disease outbreaks. It is of paramount significant for the government to invest in water and sanitation systems in preparation for disastrous situations. The most important aspect in improving public health is to provide communities with safe and clean water. Point-of-use water treatment systems are, therefore, a solution to addressing water-related diseases which result from the pollution of water sources.

Lack of access to piped water supply systems has forced most underserved rural dwellers of developing countries to utilize common practices, such as water collection from any available source (rivers, springs, community standpipes, and boreholes) and the storage of water in their homes. In most cases, these communities store drinking water in jerry cans, buckets, drums, basins, or local pots to maintain the supply in their homes [11–13]. Even if the drinking water is supplied through piped systems in homes, it is not always available on a regular basis, and therefore the storage of water is still a necessity. However, reports have highlighted that the contamination of safe drinking water collected from a reliable source may happen during transport, handling, and storage, and this has resulted in poor health outcomes [11,13–15].

In South Africa, despite the effort made by the government in terms of the provision of clean water for all and the stipulations of "access to safe drinking water for all" in the South African Constitution [16], access to a sustainable potable water supply is still lacking, especially in rural areas [17]. In spite of reports highlighting that the country achieved the Millennium Development Goal (MDG) 7, which aimed to halve the number of people without access to safe drinking water [18], the survey conducted by Statistics South Africa in 2016 showed that almost 2.6 million of the 16.8 million households surveyed did not have acceptable access to safe drinking water [19]. The South African communities without an adequate water supply are left with no choice but to collect water from any available sources, which may pose a health risk to their lives. The aim of the United Nations Sustainable Development Goal (SDG) 6 is to ensure the availability and sustainable management of water and sanitation for all [20]; however, water that meets the international standards for quality might not be achieved due to financial, infrastructure, and human capital constraints.

The need to control waterborne diseases is of paramount importance to ensure the protection of public health in rural areas of the developing world. Consequently, the scientific community has developed a large number of household water treatment systems. These point-of-use (PoU) water treatment technologies coupled with safe storage have long been proposed as a short-term solution for the provision of safe drinking water and a reduction in the waterborne disease burden in rural communities without access to improved water sources [21–24]. However, achieving the potential of household water treatment systems (HWTS) depends not only on them being made available to the target population but also on them being used correctly and consistently on a sustained basis. Like most health interventions, HWTS must actually reach the target population with safe, effective, appropriate, and affordable solutions. Nevertheless, this can be a formidable challenge, even for an intervention such as a vaccine. Unlike vaccines, HWTS require people to use it on a daily basis to provide maximum protection from waterborne diseases. Even occasionally, the drinking of untreated water may cancel out the potential health benefits of HWTS [24,25]. Allowing the target groups to understand the key characteristics, such as the perception of water quality and usefulness of HWTS as well as added factors such as household income and/or parental education, are essential to enhance the successful adoption of HWTS in developing countries. Thus, the challenge of implementing HWTS lies in providing sustainable treatment methods that can be implemented in a wide range of locations and that are accepted by the end-users. For that reason, this study was carried out between April 2015 and March 2016 to investigate the long-term effectiveness of, attitude to, and acceptance rate of two cost-effective household water treatment systems (biosand filter with zeolite-silver clay granular (BSZ-SICG) and silver impregnated porous pot (SIPP) filters) by the Makwane community in the Limpopo Province of South Africa.

2. Materials and Methodology

2.1. Ethical

This study was conducted taking into consideration the requirement of the ethics clearance approved by the Faculty of Science Research Ethics Committee (FCRE) at the Tshwane University of Technology (TUT), where the study was registered. Access to Makwane Village was obtained through the local pastor and community leaders. Furthermore, authorization to conduct the study was also obtained from the municipal manager, the municipal councilor, and the local municipal committee. All of the households that were selected for participation were given informed consent forms to sign at the beginning of the project. The project expectations and respective obligations by both the participants and investigators were explained and any questions were answered. The participants were not subjected to risks of any kind as a result of the project.

2.2. Household Water Treatment Systems Description

The HWTS [35x BSZ-SICG and 55x SIPP filters (CSIR and Tshwane University of Technology, Pretoria, South Africa)] in this study were modified and implemented in the Makwane community, as previously described [26]. However, for the purpose of this study 70 HWTS (35 SIPP and 35 BSZ-SICG) were assessed for their adoption/acceptance and effectiveness in the Makwane community. These two sets of HWTS assessed in this study were formerly tested for their ability to remove waterborne pathogens [26] prior to being implemented in the Makwane community. In brief, the BSZ-SICG filters consisted of layers of gravel, coarse sand, natural zeolite (Clinoptilolite) (Pratley minerals (PTY) LTD, Johannesburg, Krugersdorp, South Africa), silver impregnated clay granular, fine sand, and two diffusion disks (Figure 1). The natural zeolite particle size used in this study ranged between 1 and 3 mm, with a chemical composition of $(Na,K,Ca)_{2\text{-}3}Al_3(Al,Si)_2Si_{13}O_{36}\cdot 12H_2O$, and was used without any modification. The SIPP filters consisted of a clay pot incorporating silver nitrite and inserted inside a 5-liter plastic bucket, which was mounted on top of a 10-liter plastic bucket that was used as a receiving container (Figure 1). Figure 2 shows the types of HWTS implemented in Makwane Village. Both water filters were constructed by the Tshwane University of Technology with the help of the CERMA Lab (CSIR, Pretoria, South Africa). The two sets of HWTS were found to produce water of good quality prior to their implementation in Makwane Village. A total of 70 households committed to participate in the study, and each of them was given one type of water treatment device (free of charge) randomly. The follow-up was conducted on a weekly basis from the time of implementation for a period of 12 months. In addition, one member of each household was trained on how to maintain the HWTS depending on the type of treatment device given. Briefly, the BSZ-SICG filters were maintained by removing all the layers and washing them individually, and the layers were allowed to dry before being packaged back into the device (Figure 3). In contrast, the SIPP filters were washed by rinsing off the ceramic pot to avoid clogging. In addition, households were given 2 × 25 L (one improved storage container with a spigot installed 5 cm from the bottom and an unimproved storage container without a spigot) for the storage of treated water. The cleaning of the filters was performed by the householders when necessary, and this also depended on the volume of water filtered.

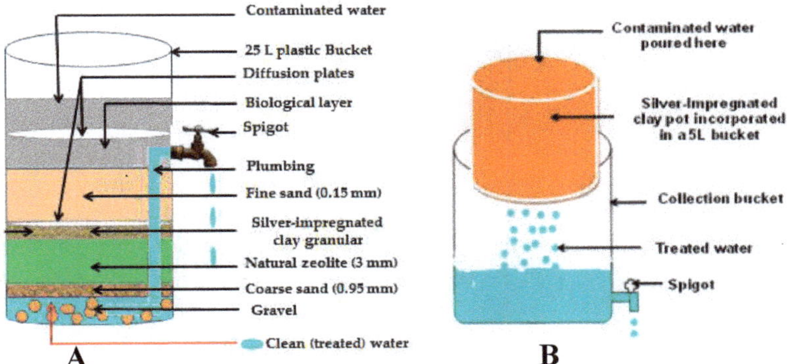

Figure 1. Schematic representation of the Biosand-zeolite silver impregnated clay granular (BSZ-SICG) (**A**) and Silver-impregnated porous pot (SIPP) (**B**) filters showing all layers within the device. Adapted from [26].

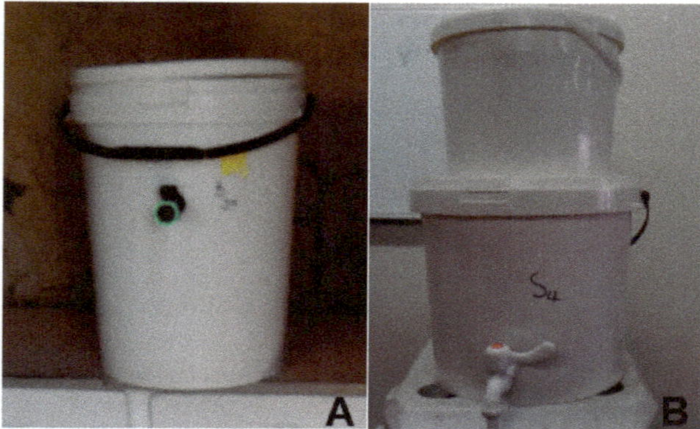

Figure 2. The Household water treatment systems (HWTS) implemented in Makwane Village; (**A**) BSZ-SICG filter and (**B**) SIPP filter.

Figure 3. Maintenance of the BSZ-SICG filter implemented in Makwane Village. (**A**) Training of one of the householders on how to wash the layers of the BSZ-SICG filter; (**B**) all five layers after being washed (gravel, coarse sand, fine sand, natural zeolite, and silver-impregnated clay granules); and (**C**) filtered water after the washing of the BSZ-SICG filters.

2.3. Data Collection

The cohort study was conducted between 2015 April and 2016 March (12 months) subsequent to the deployment of the HWTS devices in Makwane Village. Prior to the implementation, a questionnaire was used to collect information on the community, such as level of education, level of employment, and knowledge about PoU water treatment systems. To determine the long-term effectiveness and acceptance rate of HWTS in this village, a survey questionnaire in combination with observations was used to collect information such as (1) "How often do they treat their water with implemented HWTS?", (2) "How do they store treated water?", (3) "How often do they wash the storage containers?", and (4) "Are they willing to buy one of the HWTS?". A scale of 1 to 10 was used (with 10 being the highest and 1 the lowest score) to determine the knowledge of water treatment methods, whereby good knowledge was assigned a score of 7 to 10, fair knowledge a score of 4 to 6, and poor knowledge a score of 3 or less. For the determination of the long-term effectiveness, the HWTS devices were assessed in terms of their performance in flow rate to supply the required water volume of 25 L/person/day, in removing pathogenic bacteria (*Escherichia coli (E. coli)*) and turbidity from untreated water sources, and the level of silver leaching into the treated water over a period of 12 months. Observations and questionnaires were used for determining the number of filters still in use during the study period and the reasons for not being in use (for those that were not in use). All the surveys were conducted in Sepedi, which is the local language of the target community. The respondents included in this study were aged between 17 and ≥37 years and were unemployed during the period of the study, and they were therefore always available to answer the questions. The survey was conducted in all 70 households which showed interest in using the HWTS devices deployed by the TUT Water Research Group.

2.4. Water Quality Assessment

In each household, the flow rates, turbidity, microbial quality of water (*E. coli*), and leaching of silver into the final drinking water were assessed during weekly visits. In brief, the turbidity level of the water samples before and after filtration was determined using a portable turbidity meter (2100P Hach). The turbidity reduction percentages achieved by both HWTS were calculated using Equation (1). The flow rates of both HWTS were measured by recording the volume of water collected from all devices after a period of 1 hour, and calculated using Equation (2). Moreover, the concentration of silver in water treated by both HWTS was monitored on a weekly basis throughout the study period. The SPECTRO ARCOS ICP spectrometer (SPECTRO Analytical Instruments (Pty) Ltd., Kempton Park, Johannesburg, South Africa) was used to detect and determine the concentration of silver in the treated water samples.

$$\%\ turbidity\ reduction = \frac{turbidity\ unfiltered - turbidity\ filtered}{turbidity\ unfildered}\ X\ 100 \tag{1}$$

$$Flow\ rate\ of\ HWTS = \frac{Volume\ of\ water\ filtered}{Time\ (1\ hour)} \tag{2}$$

The enumeration of presumptive *E. coli* before and after treatment was conducted using standard methods (APHA, 2001). Briefly, the spread plate technique was used in this study, whereby a 250 µL aliquot of each water sample (untreated and treated) was spread on MacConkey agar plates (Merck, Johannesburg, South Africa). The plates were then incubated overnight at 37 °C, and thereafter presumptive *E. coli* colonies were counted and recorded as colony-forming units per milliliter (CFU/mL). The arithmetic mean Log bacterial reductions were calculated using Equation (3) and were converted to the *E. coli* percentage removed (Equation (4)), as previously described by [27]:

$$\begin{aligned} Log\ reduction\ &= (Log10\ bacterial\ counts\ before\ filtration \\ &\quad -Log10\ bacterial\ counts\ after\ filtration) \end{aligned} \tag{3}$$

$$\%\ E.\ coli\ removal = 100 - \frac{survival\ counts}{initial\ counts} \tag{4}$$

3. Results

3.1. Level of Education in Makwane Community during the Study Period

Table 1 below provides a summary of the level of education in the Makwane community during the study period. The results revealed that most of the Makwane community were uneducated, with only 4.3% of the households surveyed either having a tertiary qualification or still being at university/college. Of all the surveyed households, 52.9% had dropped out of secondary/high school, while 31.4% dropped out of primary school. Moreover, 11.4% of the surveyed households did not attend school.

Table 1. Level of education in the Makwane community during the study period.

N = 70		
Category	**Frequency**	**Percentage**
Primary school	30	31.4
Secondary/high school	47	52.9
Tertiary institution	3	4.3
None	8	11.4

3.2. Level of Employment in Makwane Community during the Study Period

The results of the survey revealed that most of the Makwane community were unemployed and they depended on social welfare grants (54.3%) and other sources of income (5.7%). Only 4.3% of the Makwane community had professional jobs during the study period, while 35.7% had non-professional jobs. All the results are illustrated in Table 2 below.

Table 2. Level of employment in the Makwane community during the study period.

N = 70			
Category		**Frequency**	**Percentage**
Employed	Professional jobs	3	4.3
	Non-professional jobs	25	35.7
Unemployed	Social welfare grant	38	54.3
	Other	4	5.7

3.3. Knowledge of Water Treatment Methods and Practice in Makwane Village Prior to the Implementation of HWTS

The results of the survey revealed that the majority (65.9%) of the households knew about household water treatment methods (boiling and use of liquid bleach) and 34% of the community members had no knowledge of any household water treatment (HWT) methods. Furthermore, the results revealed that almost the entire community did not treat their drinking water prior to use (81%). Of all the surveyed households of Makwane community, only 11% were found to treat their drinking water with liquid bleach, while 8% used the boiling method. Figure 4 depicts the results of the water treatment methods used by the Makwane community.

3.4. Relationship between the Knowledge of Water Treatment Methods and Selected Demographic Profiles Prior to Implementation

Table 3 shows the results of the relationship between the knowledge of water treatment methods practiced by the Makwane community and the selected demographic profile. The results showed a relationship between the age of the participant/household and the knowledge of the water treatment methods. It was found that 57% of the group aged 32–36 years had a good knowledge of the household water treatment methods used in Makwane Village, as compared to the age group of 17–21 years

(22.1%). Moreover, the level of education was also found to be associated with knowledge of water treatment systems. Participants/households with a secondary/high school qualification (77.1%) were found to be more knowledgeable about the treatment methods as opposed to participants/households with a primary school education (8.6%). Almost none of the participants/households with a primary school education (81.8%) knew about the water treatment methods used in the Makwane community.

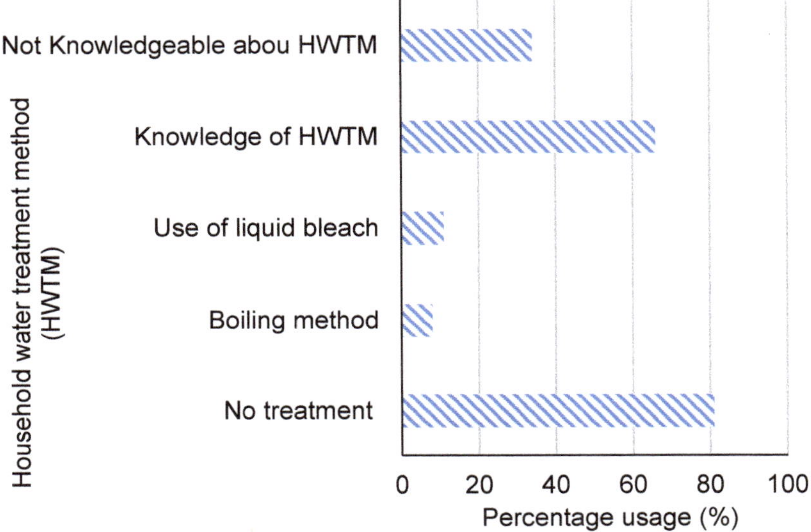

Figure 4. Water treatment methods practised in the households of Makwane Village during the study period.

Table 3. Relationship between the knowledge of practised household water treatment methods and selected demographic profiles in Makwane Village prior to the implementation of HWTS.

Variables	Knowledge of Boiling and Liquid Bleach Methods N = 70		
Age Group	Good	Fair	Poor
	Frequency (%)	Frequency (%)	Frequency (%)
17–21	2 (22.1)	1 (11.1)	**6 (66.7)**
22–26	8 (44.4)	3 (16.7)	**7 (38.9)**
27–31	7 (50)	2 (14.3)	5 (35.7)
32–36	12 (57)	5 (24)	4 (19)
≥37	3 (37.5)	1 (12.5)	4 (50)
Level of education			
Primary	3 (12.5)	3 (12.5)	18(75)
Secondary/High school	27 (71.1)	8 (21.0)	3 (7.9)
Tertiary level	3 (100)	0 (0)	0 (0)
none	2 (40)	2 (40)	1 (20)

3.4.1. Number of Household Water Treatment Systems in Use during the Study Period and Reasons for Not Being in Use for the Determination of Acceptance Rate

Table 4 provides a summary of the results obtained during visits to the Makwane community. The number of HWTS devices in use was obtained through observations during sampling, while the reasons for not being in use were obtained through a questionnaire. The results revealed a greater loss of interest in using the BSZ-SICG filters in the village, from 35 systems (100%) during the first three months of the study down to five systems (20.0%) during the last three months of the study. Moreover, the reasons for the BSZ-SICG filters not being in use were almost the same, with the majority of the households indicating that the water had a bad smell. In contrast, the results indicated that the decrease in the use of the number of SIPP filters (from 35 (100%) to 19 (54.3%) filters) was due to the filters being damaged.

Table 4. Number of household water treatment systems in use throughout the study period and the reasons for not being in use.

Assessment Period	Bsz-Sicg Filters N = 35	Sipp Filters N = 35	Reason for Not in Use
April–June 2015	35 (100%)	35 (100%)	N/A
July–September 2015	29 (82.9%)	35 (97.1%)	**BSZ-SICG**: Water had bad smell. **SIPP**: Filters got damaged.
October–December 2015	16 (45.7%)	22 (66.9%)	**BSZ-SICG**: Bad smell, no time for maintenance (time consuming), broken spigot. **SIPP**: Damaged.
January–March 2016	07 (20.0%)	19 (54.3%)	**BSZ-SICG**: Bad smell, time consuming during maintenance, no time for treating water. **SIPP**: Damaged, flow rate too slow (time consuming).

3.4.2. Survey Subsequent to the Implementation of the HWTS Devices in Makwane Village for the Determination of Acceptance and Adoption Rates

A survey was conducted using a questionnaire subsequent to the implementation of the HWTS devices in order to determine the acceptance and adoption rate. The results (Table 5) revealed that the majority of the Makwane community members used the implemented HWTS only when they needed to use water (77.1%). The results further showed that even though they treated the water, 72.9% stored treated water in other storage containers (any available containers other than the improved or unimproved containers they received) rather than in the improved storage containers (11.4%). Moreover, the results highlighted that the majority of the community washed their storage containers only when dirt was visible (80%). It was also shown that, of the two HWTS devices (SIPP and BSZ-SICG filters) implemented in Makwane Village, the majority of the community members preferred the SIPP filter (80%) to the BSZ-SICG filter (20%). Nonetheless, the majority of the community members showed no willingness to purchase either of the two HWT devices (84.3%).

Table 5. Responds obtained during the survey subsequent to the implementation of HWTS in Makwane community.

Survey Questions		Participant's Responds N = 70
How often do they treat their water with implemented HWTS?	When needed	54 (77.1%)
	On a daily basis	4 (5.7%)
	Never	12 (17.1%)
How do they store treated water?	Improved storage container	8 (11.4%)
	Unimproved storage container	11 (15.7%)
	Other	51 (72.9%)
How often do they wash the storage containers?	When dirty (visible dirt)	56 (80%)
	Once a week	8 (11.4%)
	Never	6 (8.6%)
Are they willing to buy one of the HWTS?	Yes	4 (5.7%)
	No	59 (84.3%)
	Not sure	7 (10%)

3.5. Long-Term Effectiveness of the HWTS Based on Their Performance

3.5.1. Long-Term Effectiveness Based on the Flow Rate and Turbidity Removal

Figure 5 below depicts the long-term effectiveness of the flow rate and the turbidity removal of the BSZ-SICG and SIPP filters for the period of the study (12 months). The results revealed that the flow rates of both water treatment systems were fluctuating. This was because of the maintenance (washing) of the systems (Figure 3). The flow rates were shown to have decreased from 38.7 and 27.5 L/h to 17.6 and 18.4 L/h for the BSZ-SICG and SIPP filters, respectively. The results for turbidity removal showed a gradual decrease from 99.5% to 95.2% for SIPP filters, while for the BSZ-SICG filters there was a rapid decrease (98.6% to 63.5%).

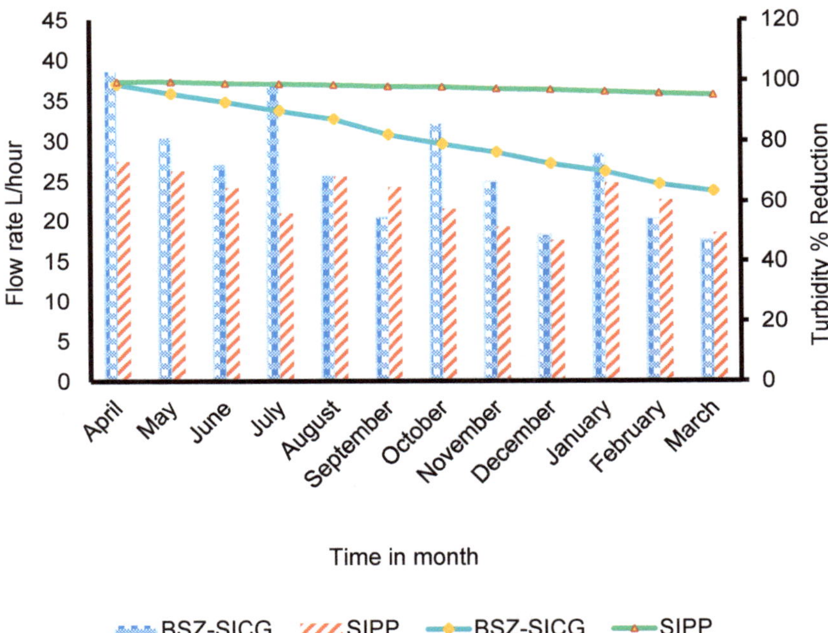

Figure 5. The arithmetic mean turbidity reduction (%) and flow rate in L/hour of the BSZ-SICG and SIPP filters versus time in months.

In addition, Table 6 showes the arithmetic mean results of the efluent and infflluent turbidity. The turbidity of the influent was found to be high form April (86.7 NTU) to September (26.82 NTU) and started to decrease from October (12.24 NTU).The turbidity for both water treatment devices was found to be within the recommended limit of less than 5 NTU for drinking water.

Table 6. The arithmetic mean results of the influent and effluent turbidity during the study period.

Time in Month	Influent Turbidity (NTU)	Effluent Turbidity (NTU)	
		SIPP Filters	BSZ-SIGC Filters
April	86.7	0.43	1.21
May	66.9	0.33	2.94
June	65.0	0.65	4.68
July	33	0.40	3.3
August	28.67	0.43	3.67

Table 6. *Cont.*

Time in Month	Influent Turbidity (NTU)	Effluent Turbidity (NTU)	
		SIPP Filters	BSZ-SIGC Filters
September	26.82	0.54	4.8
October	**12.24**	**0.23**	**2.57**
November	16.67	0.47	3.95
December	15.75	0.49	4.33
January	12.33	0.47	3.7
February	13.37	0.57	4.68
March	13.42	0.64	4.9

3.5.2. Long-Term Effectiveness Based on the *E. coli* Removal Efficiency and Leaching of Silver in Treated Water

Figure 6 illustrates the continuous removal of *E. coli* and the level of silver leaching into treated water. During the deployment of the HWTS devices, the BSZ-SICG and SIPP filters were shown to have an *E. coli* removal efficiency of 99.99% and 100%, respectively, and the leaching of silver for both the filters was within the standard limits set by the WHO (0.1 mg/L). However, the removal of *E. coli* by BSZ-SICG was characterized by fluctuations, while a progressive decrease in silver concentrations was observed between April and December; thereafter, no silver residual was detected for the rest of the study period. Overall, it was noted that the performance of the BSZ-SICG increased after it was washed at three-month intervals. The efficiency of SIPP filters at removing *E. coli* remained almost constant from April to December, even though the silver concentration decreased progressively. On the 12th month, the SIPP filters showed to be more effective in removing *E. coli* (96.6%) compared to the BSZ-SICG filter, which showed a decrease of up to 50.2%. Furthermore, over the period of 12 months, the concentration of silver leaching from the filters was observed to be very low, less than 0.002 mg/L for both the BSZ-SICG and SIPP filters.

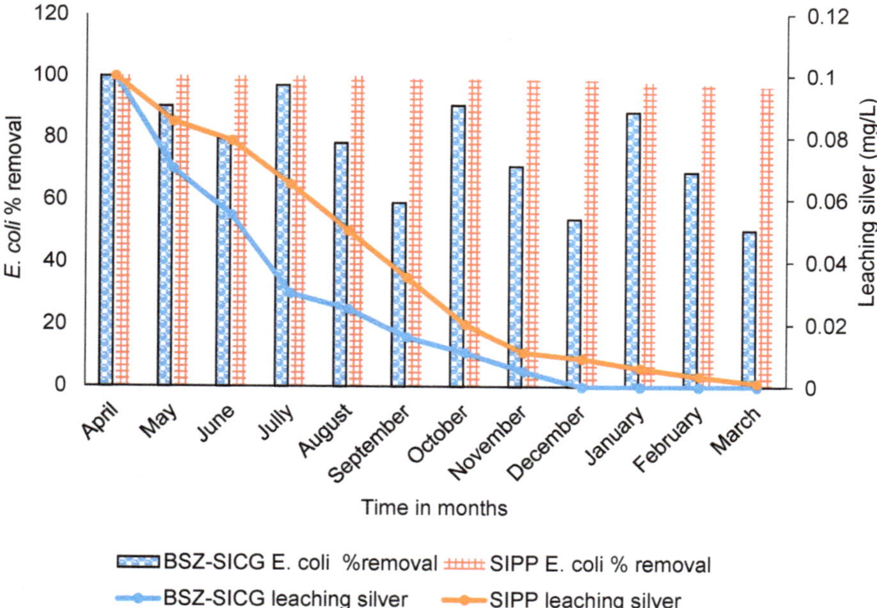

Figure 6. Arithmetic mean of *E. coli* % removal and the concentration of silver leached from the SIPP and BSZ-SICG into treated water versus the time in month.

4. Discussion

Household water treatment coupled with safe storage has long been proposed as an interim solution for the provision of safer drinking water and a reduction in the burden of water-related disease [28]. However, the adoption of HWTS is dependent on the user's preferences. This study assessed the adoption/acceptance rate and the long-term effectiveness of the cost-effective HWTS implemented in the Makwane community. The first approach of this study was to determine the educational and employment level and to assess the knowledge and adoption of water treatment methods practised by the Makwane community. The results of the survey highlighted that the majority of the community members had dropped out of secondary school (52.9%), with only 4.3% of the members having completed tertiary education (Table 1). The high drop-out percentage observed in this study might have contributed to the low adoption rate of known water treatment methods by Makwane community members due to lack of knowledge. In addition, it was also found that more than half (54.3%) of the households in Makwane community depended on social welfare grants (Table 2). The government's intervention is required to subsidize underserved communities with HWT devices in order to improve the health status of rural communities which have no access to a piped water supply.

It was further highlighted in this study that most of the people in the Makwane community were knowledgeable (65.9%) about other methods of treating water at PoU (i.e., boiling and the use of liquid bleach). However, the results showed that only 19% of the Makwane community were applying those methods at PoU (Figure 4). In most cases, the reasons for adoption were dependent on the user's interests and preferences. Furthermore, in order to understand the reasons of why the Makwane community members were not using the known methods to treat their water before drinking, the participants were asked a number of questions. Some of the reasons given by participants for not using the PoU methods are as follows: (i) Affordability—most of them are unemployed; the limited income from the government grant does not allow them to purchase the liquid bleach that is used for the treatment of drinking water in the dwellings. Although the liquid bleach might be seen to be inexpensive to others, it might be expensive to rural dwellers who solely depend on social grants. (ii) The lack of energy supply in dwellings (it takes a lot of energy to boil water and it is exhausting to fetch firewood from the forest). (iii) The majority of the community members have a perception that their water is clean; therefore, it is not important to treat it before use. In spite of this overall perception, some respondents were aware of the fact that their water is of poor quality, but they could not treat it because they could not afford to purchase the liquid bleach. These findings show that there is still a dire need to educate rural dwellers about the importance of treating water before use.

Furthermore, to ascertain the determining factor that might have contributed to the low adoption of household water treatment methods (boiling and the use of liquid bleach) by the Makwane community, selected demographic features were compared with the knowledge of household water treatment methods. The participants were grouped according to their age, and it was found that the participants of the age group 32 to 36 years had a good knowledge (57%) of HWT methods used by the Makwane community as compared to the other age groups (Table 3). This could be related to the fact that most of the participants in this age group had matric and some had tertiary education, which implies that education is a powerful tool in the community and can save lives in vulnerable populations. In addition, the relationship between the level of education and knowledge of the HWT methods was also determined. The results showed that people with a tertiary education were highly knowledgeable (100%) about the methods of treating water at PoU, follwed by the group with a secondary education (71.1%). These results, therefore, prove that education plays a vital role in the community. The level of education can therefore be a contributive factor to accelerate the adoption rate for HWTS in ruralcommunities.

The reasons for the poor acceptance of water treatment technologies are quite complex and understudied [29–31]. In this study, various determinant factors including questionnaire responses and selected demographic characteristics were used to determine the adoption and acceptance rate of the household water treatment systems. The results of the survey revealed that all the Makwane community accepted the installation (100%) of both HWTS devices (Table 4). However, after three months of implementation, it was observed that some of the households had stopped using the devices. One of the reasons given by the participants who had withdrawn was that the spigots were broken and they could not replace these because of their lack of jobs. Other reasons given were that the BSZ-SICG produced water with a bad smell and that it takes time to treat water. According to Walch (1992), biofilm formation in drinking water systems can influence the taste and odor of drinking water. Therefore, the bad smell of drinking water produced by the BSZ-SICG filter in this study can be attributed to the formation of biofilm on the inner surface of the filter due to the fact that the householders were not using these systems on a daily basis. Nonetheless, the importance of the taste and smell of drinking water has been previously highlighted [29,32,33]. In a study by Wright and co-authors (2012), some respondents reported that the water was tasteless, while others described the water as bitter. This feedback clearly indicates the participants' preference for drinking water to have a taste similar to that of untreated water [33]. It is therefore of utmost importance to address this sensitive aspect for a successful acceptance of water treatment technology interventions.

Although the community accepted the HWTS, they were not utilising them fully, as it was observed that only 5.7% treated their water on a daily basis, while 77.1% treated their water only when needed (Table 5). Previous studies have shown that only those households that regularly treat their water will experience the maximum health benefits of household water treatment methods [34]. However, it has been reported that households often do not treat water regularly and even abandon household water treatment methods over time [25,35,36]. The results of this study therefore corroborate the findings of those earlier studies. Moreover, more than three quarters of those community members who were treating their water were found to store it in other containers (72.9%) rather than in the improved containers (11.4%) provided to them. As a result, this inappropriate practice could cancel the health benefits of the HWTS devices. It was highlighted in the literature that the microbiological quality of water deteriorates in homes during storage due to unhygienic practices of storing water in homes [37–39], leaving the water unsafe for human consumption [37,38,40]. Moreover, the results of the survey in this study showed that 80% of the participants washed their storage containers only when dirt was visible. This could contribute to the formation of the biofilm on the inner surface of storage vessels, which has been reported to offer a suitable medium for the growth of microorganisms and consequently to contribute to the deterioration of drinking water quality in homes [41,42]. The results thus indicate that there is still a need for educating rural dwellers on good hygiene practices in the home for better health.

In order to determine the adoption rate of the HWTS, the systems must first be accepted. Moreover, the acceptance of the systems by the users does not mean that they will automatically adopt them. In this study, the adoption rate was determined by the number of HWTS devices that were still in use during the last month (12th) of the survey. It was found that the rate of adoption for the BSZ-SICG (20.8%) and SIPP (54.3%) filters differed (Table 4). The variation in adoption rates for the two HWTS devices could be attributed to the users' preferences based on the appearance and portability of the systems. The SIPP filters were much lighter compared to the BSZ-SICG filters, therefore it was very easy for the users to carry them and to maintain these filters (maintenance in terms of cleaning the system), unlike the BSZ-SICG filters, which are heavy. This aspect added to the preferences of the users. Moreover, in addition to the preferences, willingness to purchase the HWTS was also found to contribute to the adoption rate. When the participants were asked which HWTS device they preferred, most of the respondents (80%) indicated that they preferred the SIPP filter, while 20% preferred the BSZ-SICG filter. This further proves that the appearance and portability of HWTS are of crucial importance during implementation. However, when the participants were asked whether they would purchase any of these HWTS devices in future, 84.3% said no, while only 5.7% said yes (Table 5). This could have been influenced by the lack of sufficient income due to the high unemployment rate in the village (60%), which contributed to the community's inability to afford the devices. These findings show that there is a dire need for the governments in African countries to subsidise household water treatment systems or collaborate with non-governmental organisations and the private sector in manufacturing and implementing HWTS devices in underserved communities.

The key factors influencing the sustainable use of HWTS devices are the acceptance and adoption rate by the end-users. In this study, however, the long-term effectiveness was determined by the flow rate produced and turbidity reduction achieved by these HWTS devices during the study period of 12 months. The flow rate of the HWTS can play a role in acceptance and adoption of the systems by the users simply because of the large quantity of water produced and the fact that it saves time. Upon implementation, the flow rate of the BSZ-SICG filter was 38.9 L/h and that of the SIPP filter was 27.5 L/h; the amount of drinking water produced was thus within the recommended limits of 25 L/person/day set by the WHO (2006). The flow rate of both the systems was observed to decrease over time, and' during the third month of the survey, the flow rate of the BSZ-SICG and SIPP filters decreased to 27.2 and 24.2 L/h, respectively (Figure 5). The decrease in flow rates for the BSC-SICG was caused by the fine sand which accumulated at the bottom of the water treatment systems, while the decrease in flow rates for the SIPP filters was found to be caused by clogging caused by small/fine particles found in the source water. These findings were in line with the findings of previous investigators [43], who found that the flow rates of the BSF-S and BSF-Z declined after filtering 40 L of a water sample. The BSZ-SICG filters were washed during the third month of being in use when the flow rate was at 27.2 L/h, and a drastic increase in the flow rate to 37.4 L/h was observed. Moreover, the SIPP filters also showed a decrease in the flow rate; however, the decrease was slower than that of the BSZ-SICG. The SIPP filters, unlike the BSZ-SICG filters, were washed (Figure 5) during the fourth month of being in use and they showed an increase in flow rate to 25.6 L/h. In addition, the results in this study showed that decreases in flow rates has a negative influence on the turbidity of infflunt (Table 6). Futhermore, it was noted that the turbidity of the influent was very hihg between April and September, and start to decrease from October. In most cases, the influent turbididy variation is attributed to change in seasons whereby, it is reported to be mostly high in wet seasons [44]. However, in this study, the decrease in influent turbidity from October was due to sedimentation, as the householders where adviced to allow their water to settle before filterring for the effective results. Moreover, the turbidity of the effluent was found to decreases subsequent to the washing of the water treatment devices. This, however, was noted in the BSZ-SICG filters, whereby the decrease in turbidity was attributed to the fine sand that penetrates all the layers to the bottom of the device. These findings imply that the BSZ-SICG and SIPP filters must be washed every two to three months of being in use, depending on the amount of water filtered and the quality in terms of turbidity for the production of safe clean drinking water.

Moreover, the long-term effectiveness of the systems in this study was also measured by the ability of HWTS to remove *E. coli* from source water and the level of silver leached in treated water (Figure 6). Although cleaning/maintenance of the BSZ-SICG filters was shown to have a positive effect on increasing the removal efficiency of *E. coli* from source water, the percentage removal of *E. coli* was shown to gradually decrease with a decrease in the silver concentration. These results simply imply that the silver concentrations in HWTS have a significant effect in removing *E. coli* from source water, provided that the water treatment devices are maintained. Furthermore, the decrease in the flow rate of the HWTS was found to have a negative impact on the *E. coli* removal efficiency. Thus, it was observed in this study that when the flow rate decreased, the *E. coli* removal efficiency of the HWTS also decreased. During implementation, the *E. coli* removal efficiency of the BSZ-SICG filter was 99.99%, but it decreased to 79.6% after three months of use, and then again increased to 96.9% during the fourth month after being washed. The presence of *E. coli* in the treated water was attributed to the accumulation of fine sand at the bottom of the systems, which was removed by washing the systems. Furthermore, the *E. coli* removal efficiency of the BSZ-SICG filter gradually decreased as the silver concentration was depleted in the systems (Figure 6). These findings further prove that, with proper maintenance (the washing of the HWTS), the BSZ-SICG filters can produce safe drinking water and continue to save lives in vulnerable communities. In contrast, the washing of the SIPP filters did not have any effect on the *E. coli* removal efficiency, as the results showed continuous and gradual decrease in the *E. coli* % removal even after washing the filters.

5. Conclusions

In this study, the acceptance/adoption rate together with the long-term effectiveness of the HWTS implemented in Makwane Village was determined. It was found that the user's preference is the key factor in enhancing the acceptance and adoption rate of HWTS devices in communities. It is thus vital for consumera' preferences, choices, and aspirations to be taken into consideration for a successful scale-up of HWTS in underserved communities. Above all, finance was found to play a major role in the adoption rate, as it was noted in this study that the majority of the community showed no willingness to purchase HWT devices in the future due to the high unemployment rate in the community. The intervention of the government is therefore needed for subsidizing HWTS in impoverished rural communities. In addition, the life span of the HWTS depends on the maintanance of the devices; thus, the microbiological quality of water deteriorates over time when the BSZ-SICG filters are in use and improves when the filters are washed. However, the washing of the SIPP filters did not have an impact on the quality of water. Maintenance, therefore, plays a key role in the long-term effectiveness of the HWTS. Consequently, it is important to train the users on how to maintain the HWTS devices for sustainable use.

Author Contributions: R.C.M. and M.N.B.M. conceived and designed the experiments; R.C.M. performed the experiments; R.C.M. and M.N.B.M. analyzed the data and wrote the paper. All authors have read and agreed to the published version of the manuscript.

Funding: This research was funded by the Department of Science and Technology—National Research Fund (Grand number 87310 and 111104).

Acknowledgments: The authors would like to extend their gratitude to the National Research Foundation and the SARChI (South African Research Chair Initiative) Chair for Water Quality and Wastewater Management and the Department of Science and Technology—National Research Fund for funding this project and supporting the postgraduate student for the scholarship. Lastly, we would like thank the household caregivers of the communities of Makwane for their participation in this study.

Conflicts of Interest: The authors declare that there are no conflicts of interest.

References

1. Ojomo, E.; Elliott, M.; Goodyear, L.; Forson, M.; Bartram, J. Effectiveness and scale-up of household water treatment and safe storage practices: Enablers and barriers to effective implementation. *Int. J. Hyg. Environ. Health* **2015**, *8*, 704–713. [CrossRef] [PubMed]
2. World Health Organization (WHO). *Fact Sheet, Drinking Water*; World Health Organization: Geneva, Switzeland, 2018.
3. Craun, M.F.; Craun, G.F.; Calderon, R.L.; Beach, M.J. Waterborne outbreaks reported in the United States. *J. Water Health* **2006**, *4*, 19–30. [CrossRef] [PubMed]
4. Bessong, P.O.; Odiyo, J.O.; Musekene, J.N.; Tessema, A. Spatial distribution of diarrhoea and microbial quality of domestic water during an outbreak of diarrhoea in the Tshikuwi community in Venda, South Africa. *J. Health. Popul. Nutr.* **2009**, *27*, 652. [CrossRef] [PubMed]
5. Allam, R.R.; Uthappa, C.K.; Nalini, C.; Udaragudi, P.R.; Tadi, G.P.; Murhekar, M. V An outbreak of cholera due to contaminated water, Medak District, Andhra Pradesh, India, 2013. *Indian J. Community Med. Off. Publ. Indian Assoc. Prev. Soc. Med.* **2015**, *40*, 283.
6. Momtaz, H.; Dehkordi, F.S.; Rahimi, E.; Asgarifar, A. Detection of Escherichia coli, Salmonella species, and Vibrio cholerae in tap water and bottled drinking water in Isfahan, Iran. *BMC Public Health* **2013**, *13*, 556. [CrossRef]
7. Diouf, K.; Tabatabai, P.; Rudolph, J.; Marx, M. Diarrhoea prevalence in children under five years of age in rural Burundi: An assessment of social and behavioural factors at the household level. *Glob. Health Action* **2014**, *7*, 24895. [CrossRef]
8. Ezeh, O.; Agho, K.; Dibley, M.; Hall, J.; Page, A. The impact of water and sanitation on childhood mortality in Nigeria: Evidence from demographic and health surveys, 2003–2013. *Int. J. Environ. Res. Public Health* **2014**, *11*, 9256–9272. [CrossRef]

9. Ding, Z.; Zhai, Y.; Wu, C.; Wu, H.; Lu, Q.; Lin, J.; He, F. Infectious diarrheal disease caused by contaminated well water in Chinese schools: A systematic review and meta-analysis. *J. Epidemiol.* **2017**, *27*, 274–281. [CrossRef]
10. World Health Organization (WHO) WASH (Water, Sanitation & Hygiene) and COVID-19; Water, Sanitation, Hygiene, and Waste Management for SARS-CoV-2, the Virus that Causes COVID-19. 2020. Ref. no. WHO/2019-nCoV/IPC_WASH/2020.4. Available online: https://www.who.int/publications/i/item/WHO-2019-nCoV-IPC-WASH-2020.4 (accessed on 26 September 2020).
11. Steele, A.; Clarke, B.; Watkins, O. Impact of jerry can disinfection in a camp environment–experiences in an IDP camp in Northern Uganda. *J. Water Health* **2008**, *6*, 559–564. [CrossRef]
12. Rufener, S.; Mäusezahl, D.; Mosler, H.-J.; Weingartner, R. Quality of drinking-water at source and point-of-consumption—Drinking cup as a high potential recontamination risk: A field study in Bolivia. *J. Health. Popul. Nutr.* **2010**, *28*, 34. [CrossRef]
13. García-Betancourt, T.; Higuera-Mendieta, D.R.; González-Uribe, C.; Cortés, S.; Quintero, J. Understanding water storage practices of urban residents of an endemic dengue area in Colombia: Perceptions, rationale and socio-demographic characteristics. *PLoS ONE* **2015**, *10*, e0129054. [CrossRef] [PubMed]
14. Healy-Profitós, J.; Lee, S.; Mouhaman, A.; Garabed, R.; Moritz, M.; Piperata, B.; Lee, J. Neighborhood diversity of potentially pathogenic bacteria in drinking water from the city of Maroua, Cameroon. *J. Water Health* **2016**, *14*, 559–570. [CrossRef] [PubMed]
15. Too, J.K.; Kipkemboi Sang, W.; Ng'ang'a, Z.; Ngayo, M.O. Fecal contamination of drinking water in Kericho District, Western Kenya: Role of source and household water handling and hygiene practices. *J. Water Health* **2016**, *14*, 662–671. [CrossRef] [PubMed]
16. Assembly, C. Constitution of the Republic of South Africa, 1996 (Act 108 of 1996). *Pretoria Gov. Print.* **1996**. Available online: http://www.gov.za/aboutsa/people.htm (accessed on 26 September 2020).
17. Isa, M. Heading for Murky Waters. FinWeek; Fin24, South Africa. 2016. Available online: https://www.news24.com/fin24/finweek/featured/heading-for-murky-waters-20161124 (accessed on 26 September 2020).
18. WHO/UNICEF. *Progress on Sanitation and Drinking Water—2015 Update and MDG Assessment*; WHO: Geneva, Switzerland, 2015.
19. StatSA. The State of Basic Service Delivery in South Africa: In-Depth Analysis of the Community Survey 2016 Data, Report 03-01-22/; Pretoria Gov. Print, South Africa. 2016. Available online: http://www.statssa.gov.za/publications/Report%2003-01-22/Report%2003-01-222016.pdf (accessed on 26 September 2020).
20. UNDP. Sustainable Development Goals (SDGs). Goal 6: Ensure Access to Water and Sanitation for All. New York, NY, USA, 2015. Available online: https://www.un.org/sustainabledevelopment/water-and-sanitation/ (accessed on 26 September 2020).
21. Momba, M.N.B.; Offringa, G.; Nameni, G.; Brouckaert, B. *Development of a Prototype Nanotechnology-Based Clay Filter Pot to Purify Water for Drinking and Cooking in Rural Homes*; WRC Report No. KV 244/10; Water Research Commission: Pretoria, South Africa, 2010; pp. 27–32.
22. Mwabi, J.K.; Adeyemo, F.E.; Mahlangu, T.O.; Mamba, B.B.; Brouckaert, B.M.; Swartz, C.D.; Offringa, G.; Mpenyana-Monyatsi, L.; Momba, M.N.B. Household water treatment systems: A solution to the production of safe drinking water by the low-income communities of Southern Africa. *Phys. Chem. Earth Parts A/B/C* **2011**, *36*, 1120–1128. [CrossRef]
23. Mahlangu, T.O.; Mamba, B.B.; Momba, M.N.B. A comparative assessment of chemical contaminant removal by three household water treatment filters. *Water SA* **2012**, *38*, 39–48. [CrossRef]
24. Brown, J.; Clasen, T. High adherence is necessary to realize health gains from water quality interventions. *PLoS ONE* **2012**, *7*, e36735. [CrossRef]
25. Hunter, P.R. Household water treatment in developing countries: Comparing different intervention types using meta-regression. *Environ. Sci. Technol.* **2009**, *43*, 8991–8997. [CrossRef]
26. Moropeng, R.; Budeli, P.; Mpenyana-Monyatsi, L.; Momba, M. Dramatic Reduction in Diarrhoeal Diseases through Implementation of Cost-Effective Household Drinking Water Treatment Systems in Makwane Village, Limpopo Province, South Africa. *Int. J. Environ. Res. Public Health* **2018**, *15*, 410. [CrossRef]
27. Brözel, V.S.; Cloete, T.E. Effect of Storage Time and Temperature on the Aerobic Plate Count and on the Community Structure of Two Water Samples; Institutional Repository of the University of Pretoria, South Africa. 1991. Available online: http://hdl.handle.net/2263/4198 (accessed on 26 September 2020).

28. Wolf, J.; Prüss-Ustün, A.; Cumming, O.; Bartram, J.; Bonjour, S.; Cairncross, S.; Clasen, T.; Colford, J.M., Jr.; Curtis, V.; De France, J. Systematic review: Assessing the impact of drinking water and sanitation on diarrhoeal disease in low-and middle-income settings: Systematic review and meta-regression. *Trop. Med. Int. Health* **2014**, *19*, 928–942. [CrossRef]
29. Makutsa, P.; Nzaku, K.; Ogutu, P.; Barasa, P.; Ombeki, S.; Mwaki, A.; Quick, R.E. Challenges in implementing a point-of-use water quality intervention in rural Kenya. *Am. J. Public Health* **2001**, *91*, 1571–1573. [CrossRef] [PubMed]
30. Luby, S.P.; Mendoza, C.; Keswick, B.H.; Chiller, T.M.; Hoekstra, R.M. Difficulties in bringing point-of-use water treatment to scale in rural Guatemala. *Am. J. Trop. Med. Hyg.* **2008**, *78*, 382–387. [CrossRef] [PubMed]
31. Arnold, B.; Arana, B.; Mäusezahl, D.; Hubbard, A.; Colford, J.M., Jr. Evaluation of a pre-existing, 3-year household water treatment and handwashing intervention in rural Guatemala. *Int. J. Epidemiol.* **2009**, *38*, 1651–1661. [CrossRef] [PubMed]
32. Firth, J.; Balraj, V.; Muliyil, J.; Roy, S.; Rani, L.M.; Chandresekhar, R.; Kang, G. Point-of-use interventions to decrease contamination of drinking water: A randomized, controlled pilot study on efficacy, effectiveness, and acceptability of closed containers, Moringa oleifera, and in-home chlorination in rural South India. *Am. J. Trop. Med. Hyg.* **2010**, *82*, 759–765. [CrossRef]
33. Wright, J.A.; Cronin, A.; Okotto-Okotto, J.; Yang, H.; Pedley, S.; Gundry, S.W. A spatial analysis of pit latrine density and groundwater source contamination. *Environ. Monit. Assess.* **2013**, *185*, 4261–4272. [CrossRef]
34. Clasen, T.; Schmidt, W.-P.; Rabie, T.; Roberts, I.; Cairncross, S. Interventions to improve water quality for preventing diarrhoea: Systematic review and meta-analysis. *BMJ* **2007**, *334*, 782. [CrossRef]
35. Waddington, H.; Snilstveit, B. Effectiveness and sustainability of water, sanitation, and hygiene interventions in combating diarrhoea. *J. Dev. Eff.* **2009**, *1*, 295–335. [CrossRef]
36. Cairncross, S.; Hunt, C.; Boisson, S.; Bostoen, K.; Curtis, V.; Fung, I.C.H.; Schmidt, W.-P. Water, sanitation and hygiene for the prevention of diarrhoea. *Int. J. Epidemiol.* **2010**, *39*, i193–i205. [CrossRef]
37. Trevett, A.F.; Carter, R.C.; Tyrrel, S.F. The importance of domestic water quality management in the context of faecal–oral disease transmission. *J. Water Health* **2005**, *3*, 259–270. [CrossRef]
38. Gundry, S.W.; Wright, J.A.; Conroy, R.; Du Preez, M.; Genthe, B.; Moyo, S.; Mutisi, C.; Ndamba, J.; Potgieter, N. Contamination of drinking water between source and point-of-use in rural households of South Africa and Zimbabwe: Implications for monitoring the Millennium Development Goal for water. *Water Pract. Technol.* **2006**, *1*. [CrossRef]
39. Onabolu, B.; Jimoh, O.D.; Igboro, S.B.; Sridhar, M.K.C.; Onyilo, G.; Gege, A.; Ilya, R. Source to point of use drinking water changes and knowledge, attitude and practices in Katsina State, Northern Nigeria. *Phys. Chem. Earth Parts A/B/C* **2011**, *36*, 1189–1196. [CrossRef]
40. Momba, M.N.B.; Tyafa, Z.; Makala, N. Rural water treatment plants fail to provide potable water to their consumers: The Alice water treatment plant in the Eastern Cape province of South Africa. *S. Afr. J. Sci.* **2004**, *100*, 307–310.
41. Momba, M.N.B.; Notshe, T.L. The microbiological quality of groundwater-derived drinking water after long storage in household containers in a rural community of South Africa. *J. Water Supply Res. Technol.* **2003**, *52*, 67–77. [CrossRef]
42. Budeli, P.; Moropeng, R.C.; Mpenyana-Monyatsi, L.; Momba, M.N.B. Inhibition of biofilm formation on the surface of water storage containers using biosand zeolite silver-impregnated clay granular and silver impregnated porous pot filtration systems. *PLoS ONE* **2018**, *13*, e0194715. [CrossRef]
43. Mwabi, J.K.; Mamba, B.B.; Momba, M.M.B. Removal of waterborne bacteria from surface water and groundwater by cost-effective household water treatment systems (HWTS): A sustainable solution for improving water quality in rural communities of Africa. *Water SA* **2013**, *39*, 445–456. [CrossRef]
44. Edokpayi, J.N.; Odiyo, J.O.; Msagati, T.; Potgieter, N. Temporal variations in physico-chemical and microbiological characteristics of Mvudi River, South Africa. *Int. J. Environ. Res. Public Health.* **2015**, *12*, 4128–4140. [CrossRef]

MDPI
St. Alban-Anlage 66
4052 Basel
Switzerland
Tel. +41 61 683 77 34
Fax +41 61 302 89 18
www.mdpi.com

Crystals Editorial Office
E-mail: crystals@mdpi.com
www.mdpi.com/journal/crystals

www.ingramcontent.com/pod-product-compliance
Lightning Source LLC
LaVergne TN
LVHW071004170726
843515LV00006B/1067

* 9 7 8 3 0 3 6 5 3 4 1 2 1 *